Marine Auxiliary
Machinery
– 6th edition

Marine Engineering Series

Marine Auxiliary Machinery — 6th edition
David W. Smith, CEng, MIMarE

Pounder's Marine Diesel Engines — 6th edition
C. T. Wilbur, CEng, MIMarE and
D. A. Wight, BSc, CEng, MIMechE, FIMarE

Marine Electrical Practice — 5th edition
G. O. Watson, FIEE, FAIEE, FIMarE

Marine and Offshore Pumping and Piping Systems
J. Crawford, CEng, FMarE

Marine Steam Boilers — 4th edition
J. H. Milton, CEng, FIMarE and
Roy M. Leach, CEng, MIMechE, FIMarE

Marine Steam Engines and Turbines — 4th edition
S. C. McBirnie, CEng, FIMechE

Merchant Ship Stability
Alan Lester, Extra Master, BA (Hons), MRINA, MNI

Marine Auxiliary Machinery
– 6th edition

David W. Smith, C.Eng., M.I.Mar.E.
Principal Surveyor, Bureau Veritas

Specialist Contributors
J. Crawford, C.Eng., F.I.Mar.E.
P. S. Moore
A. D. Third, B.Sc., Ph.D., A.R.C.S.T.
P. T. C. Wilkinson
M. K. Forbes, M.A., C.Eng., M.I.Mech.E.
J. H. Gilbertson, M.B.E., C.Eng., F.I.Mar.E., M.I.Mech.E.

BUTTERWORTHS
London – Boston – Durban – Singapore – Sydney – Toronto – Wellington

First published in 1952 by George Newnes Ltd.
Second edition 1955
Third edition 1963
Fourth edition published in 1968 by Newnes-Butterworths,
    reprinted 1971, 1973
Fifth edition 1975
    reprinted 1976, 1979
Sixth edition 1983
    reprinted 1987

© Butterworth & Co (Publishers) Ltd, 1983

British Library Cataloguing in Publication Data

Marine auxiliary machinery. — 6th ed.
  1. Ships — Equipment and supplies
  2. Marine engines
  I. Smith, David W.
  623.8'2   VM781

  ISBN 0–408–01123–8

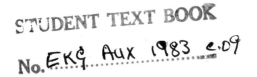
Typeset by Scribe Design, Gillingham, Kent
Printed in England by Butler & Tanner Ltd, Frome and London

Preface

The correct operation and maintenance of marine auxiliary machinery is essential if the efficiency of the whole installation is to be assured. Marine engineers will find this book valuable both as a work of reference and as a textbook when preparing for Certificates of Competency to the level of the First Class Certificate.

In this new edition due attention has been given to the changes brought about by the Ratification of the 1974 IMO Convention of the Safety of Life at Sea and some areas of the text have been expanded. Additional items not covered in previous editions include pressure regulating valves, mechanically operated hatch covers and a hydraulic engine cranking system. The chapter on Control Engineering has been completely rewritten and it is hoped that this will be of greater practical help to those sailing with automated machinery.

I am indebted to Ernest Souchotte for the spadework he did in the fifth edition, to those who contributed sections of the fifth edition and to all who have helped on this occasion with new or revised material.

David W. Smith

Contents

Acknowledgements

The editors and publishers would like to acknowledge the cooperation of the following who have assisted in the preparation of the book by supplying information and illustrations.

Alfa-Laval Ltd.
APE-Allen Ltd.
ASEA.
Auto-Klean Strainers Ltd.
Bell & Howell Cons. Electro dynamics.
Blakeborough & Sons Ltd.
Blohm & Voss A.G.
Brown Bros & Co. Ltd.
B.S.R.A.
Bureau Veritas.
Caird & Rayner Ltd.
Caterpillar Traction Co.
Chubb Fire Security Ltd.
Clarke, Chapman Ltd.
W. Crockatt & Sons Ltd.
Deep Sea Seals Ltd.
The Distillers Co. Ltd (CO_2 Div.).
Donkin & Co. Ltd.
Fire Fighting Enterprises Ltd.
G. & M. Firkins Ltd.
Foxboro-Yoxall Ltd.
G.E.C.-Elliott Control Valves Ltd.
Germannischer Lloyd.
Glacier Metal Ltd.
Hall Thermotank Ltd.
The Henri Kummerman Foundation
Howden Godfrey Ltd.
Hamworthy Engineering Ltd.
Harland & Wolff Ltd.
John Hastie & Co. Ltd.
Hattersley Newman Hender Ltd.
Hawthorn Leslie (Engineers) Ltd.
F. A. Hughes & Co. Ltd.
W. C. Holmes & Co. Ltd.
Howaldtswerke-Deutche Werft A.G.
Hydraulics & Pneumatics Ltd.
IMI-Bailey Valves Ltd.
IMO Industri.

International Maritime Organisation
KaMeWa.
Richard Klinger Ltd.
Kockums (Sweden).
K.D.G. Instruments Ltd.
Lister Blackstone Mirrlees Marine Ltd.
Lloyds Register of Shipping.
Mather & Platt Ltd.
Metering Pumps Ltd.
Michel Bearings Ltd.
Nash Engineering (G.B.) Ltd.
Norwinch
Peabody Ltd.
Pennwalt Ltd.
Peter Brotherhood Ltd.
Petters Ltd.
Phillips Electrical Ltd.
Thos. Reid & Sons (Paisley) Ltd.
Ross-Turnbull Ltd.
Royles Ltd.
Ruston Paxman Diesels Ltd.
Simplex-Turbulo Marine Ltd.
Serck Heat Exchangers Ltd.
Spirax-Sarco Ltd.
Sperry Marine Systems Ltd.
Stella-Meta Filters Ltd.
Stone Manganese Marine Ltd.
Svanhhoj, Denmark
United Filters & Engineering Ltd.
Vickers Ltd.
Vokes Ltd.
Vosper Ltd.
The Walter Kidde Co. Ltd.
Weir Pumps Ltd.
Welin Davit & Engineering Ltd.
Wilson-Elsan Ltd.
Worthington-Simpson Ltd.

1 Engine and boiler room arrangements

Auxiliary machinery covers everything on board except the main engines and boilers, and includes almost all the pipes and fittings as well as many items of equipment providing the following functions.

1. Supply the requirements of the main engines and boilers — circulating water, forced lubrication, feed and feed heating, coolers, condensers, air compressors, oil fuel reception, transfer and treatment.
2. Keep the ship dry and trimmed — bilge and ballast systems.
3. Supply the domestic requirements — fresh, salt, sanitary and sewage systems, refrigeration, heating and ventilating.
4. Apply the main power for propulsion and manoeuvring — shafting, propellers, steering gear, stabilisers.
5. Supply the ship with electric power and lighting — steam and diesel generating engines.
6. Moor the ship and handle cargo — windlass, capstan, winches, cargo oil pumps.
7. Provide for safety — fire detection and fighting, lifeboat engines and launching gear, watertight doors.
8. Provide for data logging, remote control and automatic action — pneumatic electro-pneumatic, and electronic equipment, self-regulating apparatus, etc.

The chapters of the book fall therefore into a natural order, five of a general nature (subject to certain exceptions), covering the items under 1, 2, and 3 above, then one dealing with the application of the main power. The remaining eight chapters are devoted to equipment having specific purposes, i.e. propellers, steering gears, deck machinery, etc. The final chapter, under the heading 'Control Engineering' covers the items listed in 8 above.

The seaman making his first voyage may not at first realise that the essentials of certain arrangements do not vary greatly from ship to ship. After some experience he will find that he can familiarise himself in a strange ship quite rapidly with, for example, bilge, ballast, oil fuel transfer, fresh water, sanitary water and other important systems. It is a good idea to be painstaking and to trace the piping carefully, even laboriously, until the arrangements are fully understood. The knowledge so gained may be very useful on one wild night in the darkness.

Remote control

In the last ten years, developments have been numerous and rapid; the powers of single engines have increased dramatically, remote and automated controls (with provision for local manual operation) operated from air-conditioned, sound-proof compartments have come into common use. As these devices are now reliable, engine rooms are often unmanned for long periods. These changes have naturally been accompanied by some simplification of main systems, but have brought about a great increase in auxiliary equipment. Low temperature (and therefore very low pressure) cascade evaporators are usual, often utilising waste heat. Relatively elaborate treatment of oil fuel, lubricating oil, boiler feed water and drinking water have become normal practice.

A large amount of control equipment demands clean, dry, oil-free air and the compressors to provide it. The oil-lubricated stern-tube is general, as is underwater discharge of sewage, often treated and the refrigeration load for air-conditioning commonly exceeds that for cargo. Tables 1.1 and 1.2 list the more important auxiliaries in four modern vessels; ships A and B are single-screw motor-ships, ship C single-screw and ship D twin-screw turbine steamers. The proliferation of tanks, small pumps and heat exchangers is apparent. Note that there is a tendency to provide pumps in duplicate for essential services and to keep systems simple and separate, though it is not unusual to instal three pumps, one dual-connected, instead of four, to give effective duplication.

It is usual to have only two sea inlet openings, one port, one starboard, with a pipe connection from a valve at a higher level, for use in shallow water. Each opening on which all sea suction valves are mounted, is fitted with a strainer and weed-clearing valve (also sometimes a heating connection). The freshwater engine-cooling systems in motorships work under positive head, i.e. a header tank placed at a greater height than the highest engine cooling water outlet is connected in the system; all air vents are led to this tank and to it all make-up water is added.

Table 1.1 Auxiliaries found in single-screw motor ships

	Ship A	Ship B
Main L.O. pumps	2	2
Main S.W. pumps	2	2
Auxiliary pumps	2	2
Main jacket cooling pumps	2	2
Main piston cooling pumps	2	2
O.F. transfer (heavy) pumps	1	2
O.F. transfer (diesel) pumps	1	—
Heavy O.F. separators	2	2
L.O. separators	1	1
Sludge pumps	2	3
Boiler feed pumps	2	2
Fire wash-deck pumps	1	2
Gen. service pumps	1	1
Ballast pumps	2	1
E.R. Bilge pumps	1	1
Refrigerating circulating pumps	2	1
Fresh water pumps	2	3
Sanitary pumps	1	1
Starting air compressors	2	2
Starting air reservoirs	2	2
Main L.O. coolers	1	1
Main jacket water coolers	1	1
Main piston water coolers	1	1
Generating engines	4	4
Storage, drain, sludge, etc. tanks	22	40
Heat exchangers	9	20
Small pumps	10	23

Table 1.2 Auxiliaries found in single- and twin-screw turbine steamers

	Ship C	Ship D
Main circulating pumps	2	4
Main feed pumps	2	4
Auxiliary pumps	1	2
Main condensate pumps	2	4
Main L.O. pumps	2	4
Oil fuel transfer pumps	2	2
Diesel pumps	—	2
Auxiliary bilge pumps	1	1
Bilge and ballast pumps	2	2
Fire and washdeck pumps	2	2
Auxiliary F.W. pumps	—	2
Auxiliary S.W. pumps	—	2
Sanitary pumps	1	1
Fresh water pumps	2	2
Turbo-alternator/generator	1	1
Diesel alternator/generator	3	3
Main L.O. coolers	1	2
Storage, drain, sludge, etc. tanks	35	45
Heat exchangers	23	25
Small pumps	17	20

CIRCULATING WATER SYSTEMS

In motorships there are two main circuits, one salt, one fresh. The salt circuit is: sea inlet box-pump(s) — oil cooler(s) — f.w. cooler(s) in series — turbo-blower aftercooler(s) in parallel — overboard. Branches may be taken to blower oil cooler(s), fuel valve cooling oil cooler(s) and from the outlet side, to evaporator sea inlets or domestic warm water systems, baths, etc. There may be a blanked connection to the fresh water circuit.

The fresh water circuit being under positive head, is closed, i.e. pump — fresh water coolers — cylinder jackets — cylinder heads — exhaust valves (if any) — turbo-blower(s) (if any), by branches — pump. There may be a closed circuit branch to evaporator primary-stage heating inlet(s). If the engine pistons are water-cooled, the circuit may be in parallel with the jacket circuit; it is more likely to be separate, the circuit being: pump — cooler — inlet manifold — telescopic tubes — outlet manifold — pump. The f.w. temperature should be kept as high as practicable by the use of the salt water bypass valves on oil and f.w. coolers. These may be butterfly valves controlled by thermo-pneumatic devices. It is usual to provide for warming the fresh circulating water before the main engines are started, either by steam or more usually, by bleeding from the auxiliary fresh cooling circuit.

The auxiliary salt cooling circuit is: sea inlet — pump — oil cooler(s) — fresh water cooler(s) — overboard. The air compressor inter- and after-coolers are likely to be supplied in parallel; alternatively, their fresh water coolers. It is unusual for supercharger blowers to have aftercoolers; if they have, they will be circulated similarly. The auxiliary fresh water system is similar to the main and may use the same header tank if the resulting head is not too great; if it is, a separate header tank will be provided at a lower level.

If the pumps for each service are not duplicated, they will be dual-connected. If one pump fails, the survivor becomes the fresh water pump and a clean service e.g. ballast pump circulates the sea water.

In steamers the main circulating system is: sea inlet-pump(s) — main condenser — overboard, with branches to the main oil coolers and possibly, turbo-generator condensers. Pumps are usually centrifugal, engine, turbine or motor driven. In some fast, high-powered ships, axial-flow pumps are used in conjunction with scoops. At speed, the scoop gives an adequate flow of water, the pump impeller idling; the pump is used in other circumstances. If the largest clean service pump (e.g. ballast pump) has adequate capacity, it will act as stand-by circulator, if not, two main pumps are usual.

The use of auxiliary condensers is limited to small services and port use and they will have separate circulating pumps, which may serve also the oil coolers; wherever possible feed water is used for cooling the auxiliary exhausts. Many steamers have diesel-engined generators, the circulating systems being as for motorships.

FORCED LUBRICATION

In motorships the oil, after passing through main and motion bearings and oil-cooled pistons, etc., is either retained in the bedplate or flows by gravity to a drain tank. In either case, the oil pumps, which may be self-priming, submerged centrifugal or rotary positive displacement, draw the oil through magnetic strainers and discharge to: filters – oil coolers – engine lubricating and/or cooling manifolds.

There are branch pipes to the separators and occasionally to the settling tanks (blanked). Other branch pipes lead to the deck filling lines (also blanked) so that the oil can be discharged ashore for reconditioning. Low-pressure alarms (bell siren and light) are installed; these may be arranged to operate upon the engine governor or to cut off the fuel.

Turbo-blowers, compressors and generators invariably have their own dependent lubricating oil systems. In steamers, the pumps and circuit are similar to motorships. The circuit is: drain tank (or gear case) – magnetic strainer – pump(s) – filters – oil coolers – main engine bearings and gearcase sprayers – drain tank. There are branch pipes to separate thrust blocks and to separators and there will be similar low-pressure. Also, a header tank at an upper level is connected to the discharge system to ensure that the engines are supplied with oil while coming to rest after the governor has acted. Turbo-generators have their own lubricating oil pumps. Two main lubricating oil pumps are usual and are obligatory in passenger ships.

FEED AND FEED HEATING

To ensure the efficient operation of water-tube boilers and steam turbines there is a compelling demand for clean, neutral boiler feed, free from gases in solution, (e.g. air, oxygen) at the highest temperature attainable economically. In practical terms, this means minimal contact between water and air at every stage, the dissociation and removal of gases in solution and maximum recovery of the latent heat in steam. Regenerative condensers, de-aerators and feed systems of some complexity have therefore evolved.

6

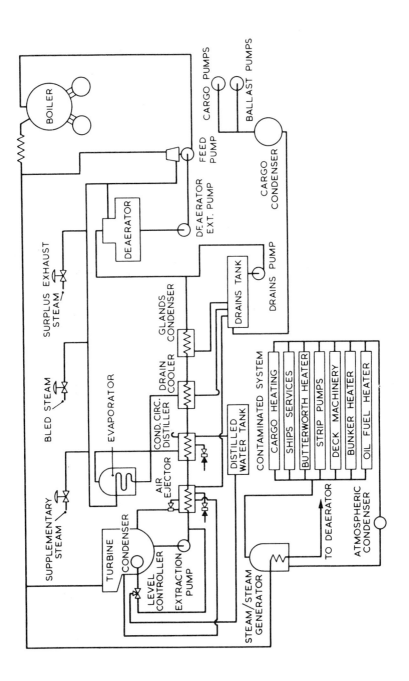

Figure 1.1 Feed system diagram

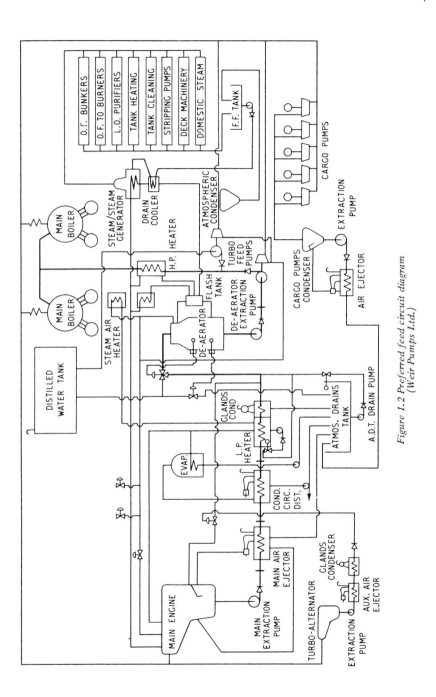

Figure 1.2 Preferred feed circuit diagram
(Weir Pumps Ltd.)

Figures 1.1 and 1.2 show that the non-condensible gases are removed from the condensers by ejectors (or air pumps), cooled by the main condensate and released in the ejector condenser. The condensed ejector steam passes with other clean drains, (gland steam condenser, l.p. feed heater, evaporator, etc.) to a drains tank from which a pump draws, to discharge, with the mains condensate, to the de-aerator. It is common practice to reflux these drains, i.e. to return them to the main condenser in the form of a spray at a high level where, meeting the turbine exhaust, they are de-aerated before mixing with the main body of the condensate and being removed by the condensate extraction pump. It will be seen too, that the heating steam for the steam-generator (in which l.p. steam is produced for services whose condensate may be contaminated), the de-aerator and the l.p. feed heater is bled from the main turbines at appropriate stages, so that all the latent heat is recovered, the feed pump exhaust being treated similarly.

Extraction pumps

A centrifugal extraction pump either motor- or turbine-driven, draws from the condenser under level control and delivers to a de-aerating heater through the heat exchangers already mentioned (see also Chapter 5). Another extraction pump passes the de-aerated feed to a multi-stage centrifugal pump which can be motor or more often turbine driven. The pump delivers the feed to the boilers at a temperature approaching that of saturation through a high pressure feed heater, supplied with steam bled from the h.p. turbine. Make-up feed is produced by evaporation and distillation (sometimes double) at sub-atmospheric pressure, stored in a tank and introduced to the boilers from the main condenser with the refluxed drains or through the de-aerator. The feed pumps, feed piping and fittings must be in duplicate, i.e. there must be a main and auxiliary feed system.

'Dirty', i.e. possibly contaminated exhausts and drains pass as water direct and as steam through an auxiliary condenser, to a drain tank, thence by pump to the steam generator to be re-evaporated.

COMPRESSED AIR SYSTEMS

Air at high pressure, i.e. at 25 bar or more, is required for starting main and auxiliary engines in motorships and in steamers, often for the latter purpose. In both, air at low pressure, say at 6 bar or less is required for much remote or temperature or pressure controlled equipment, for which purpose it must be clean and oil-free.

A typical simple h.p. air system is shown in Figure 1.3; the arrangement of a larger system can be readily envisaged from this diagram. It is a requirement that there shall be at least two reservoirs whose combined capacity allows (all) the main engines to be started twelve times if they are reversible and six times if they are not, while the compressors are idle.

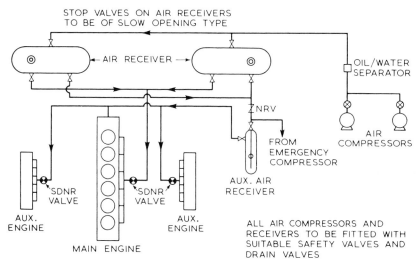

Figure 1.3 Compressed air diagram

Safety valves must be fitted; these may be common to the compressors but if they can be isolated from the reservoirs, the latter must have fusible plugs fitted to prevent rupture in the event of fire. Reservoirs are designed, built and tested similarly to boiler drums. Compressors are discussed in Chapter 4. Two or more, independently driven, are usually fitted, each capable of charging one reservoir from zero to working pressures in a reasonable time.

Explosions can and do occur in starting air pipes caused by the ignition of oil or vapour gaining access from compressors or defective engine starting valves. The need for careful maintenance, cleaning and draining is important. Bursting discs are fitted (usually in the form of a thin copper closed-ended cylinder) to minimise and localise the effects.

Starting air for auxiliary engines may be taken directly from the main reservoirs or through an auxiliary reservoir (which can be kept at full pressure at all times). Low pressure air is usually taken through reducing valves or pressure regulators to a low pressure reservoir supplying a ring main through driers. Alternatively, the l.p. reservoir may be charged by an oil-free compressor, rotary or reciprocating.

In steamers, the starting air arrangements for diesel auxiliary engines are similar to those in motorships; the l.p. systems for instrumentation and remote control devices are likely to be the alternative described above. It should be noted that air manifolds or ring mains must be proportioned generously and must approach closely the point where the air is to be used, so that the connecting pipes are short. This will ensure satisfactory operation.

OIL FUEL TRANSFER

Oil fuel transfer arrangements provide for receiving stations at an upper deck level, port and starboard, furnished with valves, elbows, pressure gauges, filters and relief valves. From these receiving stations, the oil flows to double-bottom, peak or deep bunker tanks and can be transferred from forward to aft and from port to starboard (and vice versa) in passenger ships and in many others, to settling tanks and, in motorships, from settling to service tanks via filters, separators and clarifiers, thence to boilers or engines.

Fire is an ever-present hazard and often has its origin in apparently unremarkable happenings. There are stringent rules governing flash points, permissible temperatures, storage and handling of oil fuel. These rules should be carefully studied and observed.

The flash point (closed test) should be above 65°C; oil should not be heated to more than 51°C for settling or purificaiton; if necessary, this may be increased to a figure 20°C beneath its known flash point. Settling tanks must have thermometers and the sounding arrangements must be proof against accidental egress of oil. Drain cocks must be self-closing and the outlet valves should be capable of being closed from safe positions outside the engine or boiler room. In passenger ships, this applies also to suction and levelling valves on deep tanks. Overflow pipes and relief valves not in closed circuit must discharge to an overflow tank having an alarm device, the discharge being visible. Tank air pipes must have 25% more area than their filling pipes and should have their outlets situated clear of fire risks. They should also be fitted with detachable wire gauze diaphragms. Provision should be made for stopping oil fuel pumps from outside the machinery spaces.

From the filling stations, pipes descend to the oil fuel main(s). These will probably be two pipes, one for heavy oil and one for diesel fuel. They extend forward and aft in the engine and boiler

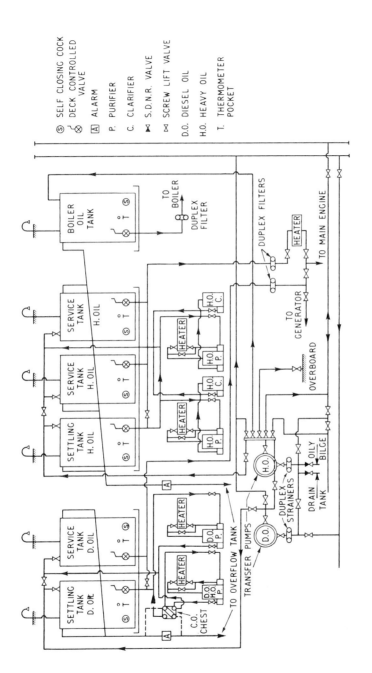

Figure 1.4 Oil fuel service diagram

rooms, possibly extending along the shaft tunnel, and, in some ships, a forward pipe tunnel. These connect to the fuel transfer pump(s) and to distribution valve (or cock) chests, from which pipes run to the fuel tanks. Water ballast is also sometimes carried in these tanks and the chests (see Chapter 2) are arranged so that they cannot be connected to oil and ballast mains at the same time. Midship tanks have centre and wing suctions; the forward and aft peak, and deep tanks have centre suction only.

Transfer pumps draw from the oil main(s), from overflow and drain tanks and from the oily bilges — parts of the engine and boiler room bilges separated from the remainder by coamings — to which oil spillage is led. The pumps discharge to settling tanks, the oily water separator and the oil main(s). In passenger ships, it must be possible to transfer oil from any tank to any other tank without use of the ballast main, so that two oil mains are required. In cargo ships, the ballast pump may act as standby transfer pump, in which case they must be interconnected; alternatively, a diesel oil transfer pump may be the standby. Detail and arrangement will vary with the size, type and trade of the ship.

In steamers, the fuel is heated in the settling tanks by steam coils, to assist water separation, and is then taken to the burners through heaters and filters by the oil fuel pressure pumps. In motorships, after settling in a similar manner the heavy fuel passes through heaters to two separators in series, the first removing the water and most of the solids in suspension; the second, called a clarifier, removes the finer solids remaining, see Chapter 4. The separators, usually having their own pumps, deliver the clean oil to one of two service tanks in turn, from which the oil passes to the engine service pumps and so to the injection pumps, through further heaters. Diesel fuel is treated similarly but more simply, with a single stage of separation and no heating. Sludge from separators passes to a tank, from which it is removed by a pump capable of handling high-viscosity matter. (O.F. service diagram Figure 1.4).

It may be mentioned here, because it is not always understood, that fuel is heated for combustion, not to raise its temperature but to bring it to a viscosity acceptable to the injectors or burners.

BILGE AND BALLAST

The essentials of the bilge and ballast system are simple. They consist of a bilge main in the engine room which is connected to one or more pumps and, also to the hold, tunnel and machinery space bilges by suitably placed valves; the pumps discharge overboard. A ballast main is similarly connected to a pump, the sea and to ballast tanks; the pump discharging overboard or to deep and peak tanks through the ballast main. In practice the use of oil fuel with the attendant need to retain the oil when discharging overboard, the need to provide adequate services in the event of breakdown or casualty without unreasonable duplication and to avoid accidental flooding, have all given rise to some elaboration of the system. Figure 1.5 is a diagrammatic arrangement showing bilge, ballast and oil-fuel filling.

A fire and bilge pump has suctions from sea, bilge main and engine room bilge, with discharges to fire main, oily water separator and overboard. A ballast pump has suctions from sea, ballast main, engine room, bilge direct and bilge main with discharges to overboard. the ballast main, the oily water separator and possibly, the main salt water circulating system. A general service pump has suctions from sea, ballast main, bilge main and engine room bilge with discharges to the fire main, the ballast main, the oily water separator and overboard. In this way, three pumps provide effective alternatives for all essential services in the event of breakdown of one or even two. Many ships will have more generous provision and all passenger ships will have a submersible fire and bilge pump, supplied with power from an emergency dynamo.

There are many differences in arrangement; some ships will be fitted for oil or ballast in all double bottom tanks (except one or two, port and starboard for fresh water) some in only two or three. Other vessels will have one (or more) lower hold(s) fitted as deep tanks and most will have peak ballast tanks forward and aft. Some ships will have a tunnel from the engine room to No. 1 hold aft bulkhead, for bilge, ballast and oil pipes and fittings and others will have a duct keel to carry the pipes forward. In most other ships the bilge suction pipes will pass through the wings of the holds and the ballast pipes through the double bottom. In the two latter cases, the valve chests will be on the engine room (or boiler room) forward bulkhead or in a forward cofferdam. In all cases, the bilge suction valves will be screw-down, non-return, the oil and ballast valves, screw-lift. Ring and blank flanges will be fitted in deep tank suctions, so that ballast cannot be discharged inadvertently by a bilge pump nor the hold be flooded when used for cargo. If liquid cargoes are carried, both will be blanked. Note that double bottom tanks should *never* be pumped up.

The minimum number and capacity of bilge pumps and fire pumps and their dispersement within the ship is governed by:

1 Classification Society Rules
2 National requirements

14

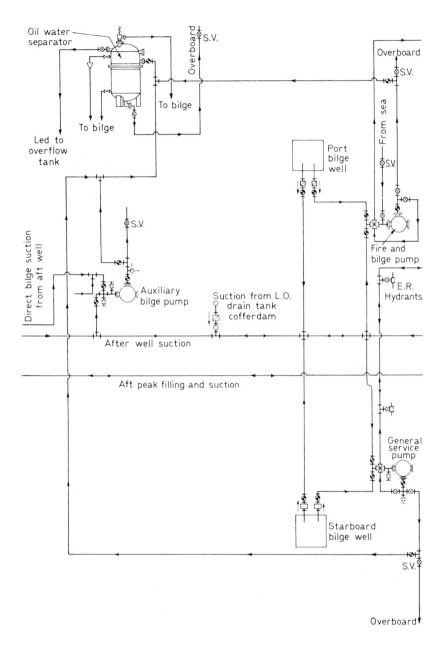

Figure 1.5 Diagram of bilge, ballast and oil fuel filling.
(Hawthorn Leslie (Engineers) Ltd)

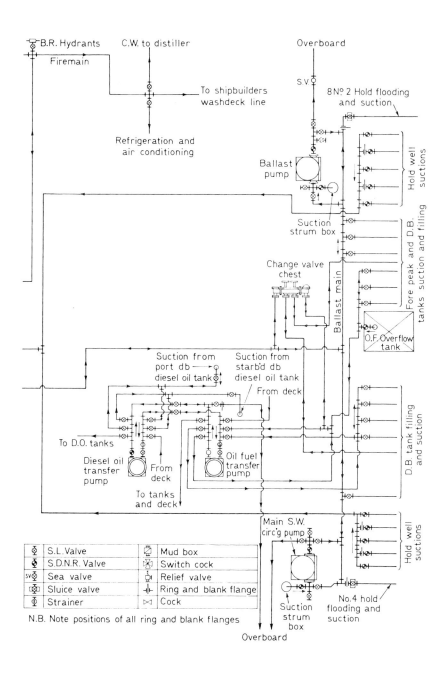

Figure 1.5 (continued)

3 The IMCO International Convention for the Safety of Life at Sea, 1974 (SOLAS 74).

The basic philosophy is similar in all three cases but SOLAS 74 only defines bilge pump capacity for passenger ships and the Convention only applies to vessels trading internationally; more-over it excludes cargo ships of less than 500 gross tons. The Classification Societies generally prescribe the bore of the main bilge line and branch bilge lines and relate the bilge pump capacity of each pump to that required to maintain a minimum water speed in the line; the fire pump capacity is related to the capacity of the bilge pump thus defined e.g.

Bilge main dia. $d_1 = 1.68\sqrt{L(B + D)} + 25$ mm

Branch dia. $d_2 = 2.16\sqrt{C(B + D)} + 25$ mm

d_2 not to be less than 50 mm and need not exceed 100 mm.

d_1 must never be less than d_2

where

L = length of ship in m;

B = Breadth of ship in m;

D = Moulded depth at bulkhead deck in m;

C = Length of compartment in m.

Each pump should have sufficient capacity to give a water speed of 122 m/min through the Rule size mains of this bore. Furthermore each bilge pump should have a capacity of not less than

$$\frac{0.565}{10^3}\, d_1{}^2 \;\; \text{m}^3/\text{h}$$

The fire pumps, excluding any emergency fire pump fitted, must be capable of delivering a total quantity of water at a defined head, not less than two-thirds of the total bilge pumping capacity. The defined head ranges from 3.2 bar in the case of passenger ships of 4000 tons gross or more to 2.4 bar for cargo ships of less than 1000 tons gross.

The following paragraphs are extracted from the International Convention for the Safety of Life at Sea 1974 Chapter 11–1 Regulation 18 which relates to passenger ships:

'The arrangement of the bilge and ballast pumping system shall be such as to prevent the possibility of water passing from the sea and from water ballast spaces into the cargo and machinery spaces, or from one compartment to another. Special provision shall be made to prevent any deep tank having bilge and ballast connexions being inadvertently run up from the sea when containing cargo, or pumped out through a bilge pipe when containing water ballast.'

'Provision shall be made to prevent the compartment served by any bilge suction pipe being flooded in the event of the pipe being severed, or otherwise damaged by collision or grounding in any other compartment. For this purpose, where the pipe is at any part situated nearer the side of the ship than one-fifth the breadth of the ship (measured at right angles to the centre line at the level of the deepest subdivision load line), or in a duct keel, a non-return valve shall be fitted to the pipe in the compartment containing the open end.'

'All the distribution boxes, cocks and valves in connexion with the bilge pumping arrangements shall be in positions which are accessible at all times under ordinary circumstances. They shall be so arranged that, in the event of flooding, one of the bilge pumps may be operative on any compartment; in addition, damage to a pump or its pipe connecting to the bilge main outboard of a line drawn at one-fifth of the breadth of the ship shall not put the bilge system out of action. If there is only one system of pipes common to all the pumps, the necessary cocks or valves for controlling the bilge suctions must be capable of being operated from above the bulkhead deck. Where in addition to the main bilge pumping system an emergency bilge pumping system is provided, it shall be independent of the main system and so arranged that a pump is capable of operating on any compartment under flooding condition; in that case only the cocks and valves necessary for the operation of the emergency system need be capable of being operated from above the bulkhead deck.'

'All cocks and valves mentioned in the above paragraph of this Regulation which can be operated from above the bulkhead deck shall have their controls at their place of operation clearly marked and provided with means to indicate whether they are open or closed.'

DOMESTIC FRESH AND SALT WATER SYSTEMS

Open systems using gravity tanks, etc., are rarely met nowadays and Figure 1.6 shows a characteristic arrangement of pressure tanks, calorifiers and pumps for a large ship. The fresh water pumps are self-priming, drawing from storage tanks and discharging through filters to a rising main, branched to give a cold supply and a hot supply in closed circuit through a calorifier. A tank kept under pressure by an external supply of compressed air rests on the system and provides the head necessary for constant supply throughout the ship.

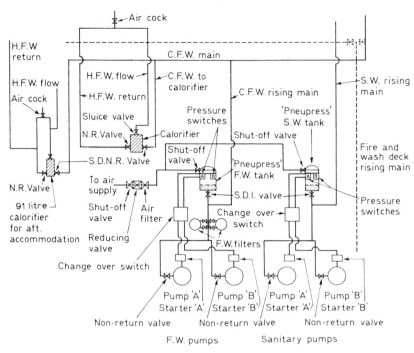

Figure 1.6 Diagrammatic layout of a typical installation
(Worthington-Simpson Ltd.)

The pump starters are controlled by pressure switches which operate when the pressure in the tank varies within predetermined limits as water is drawn off. Distilled water is now largely used on board; this may be corrosive (due to the presence of dissolved CO_2) and flat to the taste. It is not unusual to provide for the introduction of certain chemicals, e.g. Calcium Chloride ($CaCl_2$) and Sodium Bicarbonate ($2NaHCO_3$) which, in solution are converted to Calcium Bicarbonate ($Ca(HCO_3)_2$) and Sodium Chloride

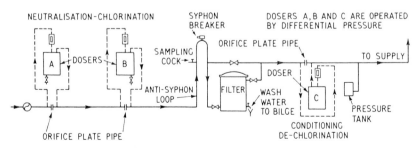

Figure 1.7 Typical marine domestic water treatment system
(United Filters & Engineering Ltd.)

(2NaCl) to render the water neutral and palatable. It may then be chlorinated to ensure bacteriological purity and subsequently de-chlorinated. Figure 1.7 shows such a system diagrammatically.

The salt water pump(s) need not be self-priming, nor are heating nor treatment required, otherwise the system is identical.

2 Pipes and fittings

Pipes, valves etc., carry and control the flow of a number of fluids at various, often varying, pressures and temperatures; the fluids may be corrosive, erosive, flammable or benign. Their functions, requirements arising from ship construction, the nature and arrangement of the machinery and regulations of certifying authorities, create situations in which systems basically simple, become complex and bring into use a variety of materials and fittings. The term 'fittings' covers valves, cocks, branch and bulkhead pieces, reducers, strainers, filters, separators and expansion pieces, in short, everything in a system which is not a pipe. Couplings and unions are used only in small bore pipes: tank heating coils are normally joined by sweated sleeves.

Materials

Cast iron has poor corrosion resistance to sea water, especially in the presence of bronzes; its weakness in tension and under shock loading limits its use to low pressure steam, air, oil and low speed water fittings. Its brittle nature excludes its use for side shell fittings. Its performance is improved by the addition of, for example, nickel to the melt and by cyclic heat-treatment in which the free graphite assumes spherical forms; it is then known as SG Iron and may be used for high pressure services and for steam below 461°C. Lined with rubber, plastic or epoxy compounds it is used for sea water fittings. Mild steel welded fabrications, lined similarly, are used for larger fittings of this sort. Wrought iron has an inherent resistance to corrosion but is not to be had freely; its place has been taken by the cuprous alloys and by mild steel, electric resistance welded (ERW) or hot rolled galvanised by hot dipping.

Seamless mild steel is used for steam (also copper for moderate pressure and temperature) h.p. air, feed discharges and all oil fuel pressure piping: its strength begins to fall at about 460°C above which steels having small additions of molybdenum, chromium or both, come into use. Seamless copper may be used for lubricating oil (*not for oil fuel*) and certain water services, fresh and salt, but it stands up badly to high water speeds, entrained air and polluted river water, so that the cuprous alloys are used increasingly for water and for oil tank heating coils. Flanges are secured to steel pipes by fusion welding or by screwing and expanding, to copper and its alloys, by brazing or sweating.

Cast iron and gunmetal fittings are used freely in small sizes at moderate pressures; large fittings, those for high pressure and temperature and for pressure oil fuel are cast or fusion welded (fabricated) mild steel or SG Iron: for temperatures above 460°C they are usually 0.5% molybdenum steel, (see Tables 2.1 and 2.2).

EXPANSION JOINTS AND PIPELINES

Provision must be made in pipe systems for changes in length due to change of temperature, both in the pipe and the machinery items which they join. For many years this was effected satisfactorily by incorporating in straight length pipe an expansion joint i.e. an anchored sleeve having a stuffing box and gland in which an extension of the joining pipe slid freely within imposed limits (Figure 2.1) or by incorporating pipes of bow form or 'a long and two short legs',

Table 2.1 Jointing materials

JOINTING MATERIAL	MAX. TEMP. RECOMMENDED °C	STEAM	FREON 12/22	CALCIUM CHLORIDE SOLUTION	DIESEL, GAS OIL, GASOLINE, KEROSINE, HEAVY OIL, LIGHT LUB. OIL, AIR, WATER	BENZENE	OCTANE	WATER, L.P. AIR
SPIRAL WOUND GASKET	815	•	•		•	•	•	
COMPRESSED ASBESTOS	510	•	•	•	•	•	•	
P.T.F.E. ENVELOPE	260		•	•	•	•	•	
CARTRIDGE PAPER	200							•
CARTRIDGE PAPER IMPREGNATED					•			
SYNTHETIC RUBBER BONDED CORK	150		•	•	•			
RUBBER-COTTON-REINFORCED	150							•

Table 2.2 Materials used for pipes and fittings

Material

Pipes	*Duty*
Seamless mild steel	Saturated steam, boiler feed, lub. oil > 175 mm, oil fuel pressure
	Superheated steam below 461°C, H.P. air, bilge and ballast (galvanised)
Seamless steel 1%Cr 0.5%Mo	Superheated steam above 460°C
E.R.W. mild steel	Aux. feed, lub. oil <175 mm, oil fuel suction and L.P. discharge
	Circulating and domestic sea and fresh water, bilge and ballast (galvanised)
Electric fusion welded mild steel	Large fresh and sea water (lined or galvanised), diesel exhaust
Copper	L.P. saturated steam <220°C, Aux. feed, lub. oil, small H.P. air, L.P. air
Aluminium brass	Circulating fresh and sea water, firemain, bilge and ballast, condensate, heating coils, domestic fresh and sea water, etc.
Copper-nickel-iron alloy	Circulating fresh and sea water, bilge and ballast

Fittings	
Cast iron	L.P. saturated steam, lub. oil, oil fuel transfer, circulating Sea (lined) and fresh water, bilge and ballast, domestic water
S.G. iron	Steam <461°C, boiler feed, diesel exhaust
Cast steel	Steam <461°C, boiler feed, oil fuel pressure, H.P. air
Cast steel 0.5% Mo	Superheated steam >460°C
Gunmetal	Saturated steam, aux. feed, firemain, H.P. air, small fittings
Electric fusion welded mild steel	Large fittings for fresh and sea water (lined) circulating, oil

Test pressures may be taken as 2WP generally, 2.5WP for boiler feed (or 2 × actual feed pressure) and 2-3WP for alloy steel for temperatures above 460°C.

with 'cold pull up' carefully calculated and applied. Examples are shown in Figure 2.2.

Increasing size of unit, pressures and temperatures with associated greater pipe diameter and thickness have diminished the effectiveness of these arrangements.

Bellows expansion joints

Stainless steel bellows expansion joints are commonly used since they will absorb movement or vibration in several planes, eliminate maintenance and reduce friction and heat losses (Figure 2.3).

Maximum and minimum working temperatures must be considered when choosing the joint, which must be so installed that it is neither

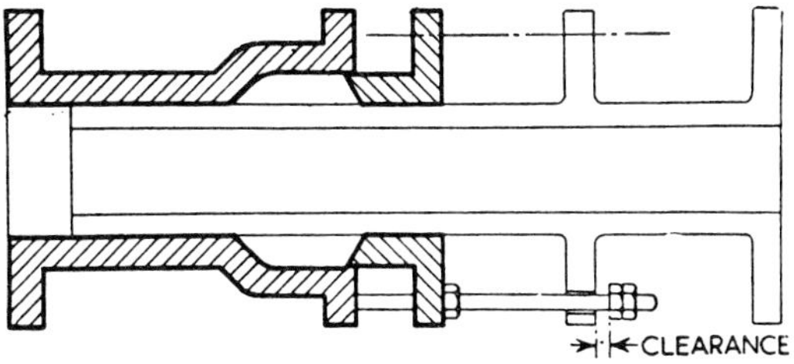

Figure 2.1
Tie-rod expansion joint

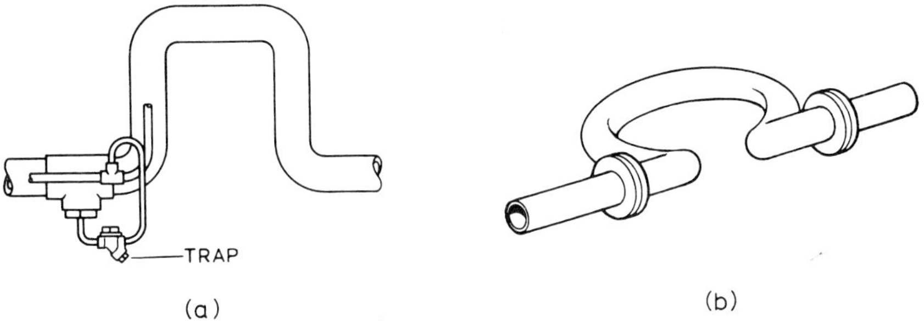

Below:
Figure 2.2 Steam-line drainage arrangements
(a) Expansion loop up-
wards. Large bore drain
pocket fitted before loop
(b) Expansion loop hori-
zontal, no drainage required

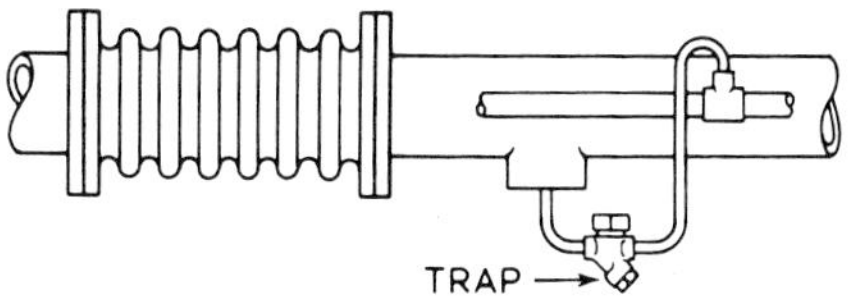

←CLEARANCE

———TRAP

(a)

(b)

Figure 2.3 Bellows expansion
fitting with independent full
bore pocket
(Spirax-Sarco Ltd.)

TRAP →

over-compressed nor over-extended; the installation length must be appropriate to the temperature at installation. Stainless steel is the usual material for temperatures up to 500°C; beyond that and for severe corrosive conditions other materials are required.

Articulated units may be called for in exceptional cases, tied, hinged, gimbal or pressure balanced. These are outside the scope of this book.

By definition, each joint will have one or more bellows, each bellows having a number of convolutions determined by considerations of flexibility and stress; normally each will have an internal sleeve, to give smooth flow, to act as a heat shield and to prevent erosion. If exposed to possible damage, it should have an external cover. In usual marine applications, bellows joints are designed and fitted to accommodate straight-line axial movement only and the associated piping requires adequate anchors and guides to prevent misalignment. It will be apparent that, in certain cases, the end connections will act adequately as anchors and that well-designed hangers will be effective guides.

An axial bellows expansion joint can accommodate compression and extension, usually stated as plus or minus X mm, i.e. it will compress or extend X mm from the free length, at which it is supplied. Thus, a quoted movement of ± 50 mm means a total movement of 100 mm; to obtain this, the bellows must be extended 50 mm when installed. This is 'cold pull-up' or 'cold draw'. It is most important that the unit be installed at its correct length as extension or compression outside its specified limits will cause premature fatigue.

Disastrous explosions have been caused by accumulations of oil or oil vapour in air lines not drained. Similarly, severe damage has been caused by 'water hammer' when steam has been admitted to pipes containing water, especially when the run of the pipe, e.g., being inclined only slightly from the horizontal, allowed the water to have a large free surface area; *this is the important point.*

On steam being admitted, condensation occurs on the cool water surface, a partial vacuum develops and the water moves along the pipe at great speed: if arrested by a bend or a closed valve, very large hydraulic forces are generated and fracture follows. It should be apparent that little or nothing is gained by opening a drain at the same time as or after admitting steam. It is thus imperative that steam pipes be left draining when not in use and that master valves be first eased off their seats or 'cracked' when a line is being brought into use until the pipe is thoroughly warmed. Only then should the valve be opened fully.

It is usual to identify the fluids in pipes by bands of paint at intervals, and the engineer should familiarise himself with the code in use.

Protective fittings and linings

Pipes are carried through bulkheads by fittings such as shown in Figure 2.4. Joints between flanges should be of materials impervious to the fluids carried, e.g. compressed asbestos for steam; rubber, with or without cotton insertion for water; cork, fibre, cardboard or (best of all) liquored leather for oil. They may be sheathed with copper or stainless steel, sometimes grooved finely and lightly in the area adjacent to the pipe bore. Jointing and gasket materials and their applications are shown in Table 2.1. Most deteriorate with time and temperature; graphite compounds assist flexibility. Mating flanges should be parallel, bolts reasonably well fitting with good threads. When brass, gunmetal or copper-nickel-iron flanges are brazed or sweated to copper or copper alloy pipes, the flange must

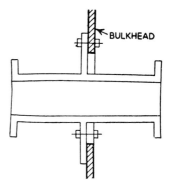

*Figure 2.4 Bulkhead piece for use when a pipe
passes through a watertight bulkhead*

fit the pipe well, forming a capillary space between them: this space
should be filled completely in the brazing or sweating operation.

Corrosion

Corrosion and erosion are unremitting enemies; often they are
encountered simultaneously and the problems they pose demand
consideration when deciding the choice of material. Accelerated
corrosion or corrosion-erosion may arise from any of the following
causes:

Galvanic action when dissimilar metals are associated.
Local deposits or transfer of small metal fragments e.g. weld spatter
in steel pipes, by sulphur or sulphides in weld spatter in steel
pipes, by sulphur or sulphides in polluted waters, gassing in hot
tubes of coolers, fouling and broken cathodic films in pipe bores.
High water speeds, entrained air, cavitation and turbulence caused
by e.g. protuberances, tight bends, abrupt change of direction or
pipe sectional areas. (This form of erosion is usually encountered
as pitting.)
Dezincification or corrosion-stress cracking of brass and some so-
called 'bronzes'. Corrosion stress cracking is a phenonomen
associated chiefly with brass tube stressed by expanding or by
working in the unsoftened condition; in corrosive fluids, e.g. sea
water, splitting occurs suddenly, even violently.
Inadequate protection of steel, pinholes or discontinuities in
protective linings: these should always be carried over the flange
faces.

Engineers should be familiar with the galvanic series. In this, the
more noble metals are placed after the less noble thus: zinc, aluminium,

carbon steels, cast iron, lead-tin alloys, lead, brass, copper, bronze, gunmetal, copper-nickel iron, monel. A metal in contact with one occurring later in the series, e.g. steel in contact with copper, may corrode rapidly in sea water or even in condensate. The action is galvanic and sacrificial anodes give protection when attached in a manner giving good electrical bonding. Examples are soft iron plates in condenser water-boxes and zinc plates around the propeller aperture, adjacent to high-tensile brass propellers. Basically, brass is an alloy of copper and zinc and bronze an alloy of copper and tin. In both cases there may be additions of other metals and there is some confusion of nomenclature; some high-tensile brasses are called 'bronze' and the practice has prevailed for so long as to be accepted.

COCKS AND VALVES

Cocks and valves control or interrupt flow. Basically, they have bodies furnished with flanged or screwed ends (or ends prepared for welding) for connection to the joining pipes and internal passages arranged so that they can be restricted or closed: this is done, in cocks, by rotating the plug, and in valves, by lowering, raising or rotating a disc in relation to a seating surface or by controlling the movement of a ball.

A cock may be straight-through, right-angled or open-bottomed as required by its situation in a pipe system, the plug may be tapered or parallel; tightness may be achieved by grinding the plug to the body or by resilient packing material packed tightly into suitably placed grooves in the body (for low or high pressures respectively). If under pressure, they may be double-glanded, so that the plug

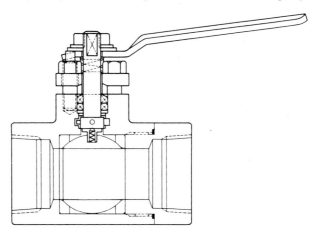

Figure 2.5a
Example of a ball
valve (Weir-Pacific
Valves Ltd.)

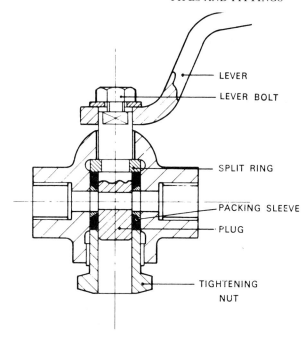

LEVER

LEVER BOLT

SPLIT RING

PACKING SLEEVE

PLUG

TIGHTENING
NUT

STAINLESS
STEEL
EYELET

RIDGE

PACKING SLEEVE

Figure 2.5b
Example of a sleeve-
packed cock
(Richard Klinger Ltd.)

cannot blow out; in some cases e.g. boiler blowdown cocks on ship's
shell, construction is such that the handle can be removed only when
the cock is closed.

The most commonly used valves fall into the following categories:
globe valves, gate and slide valves, butterfly valves, check and non-
return valves and control valves.

Globe valves

The globe valve has a bulbous body, housing a valve seat and screw
down plug or disc arranged at right angles to the axis of the pipe. An
example is shown in Figure 2.6.

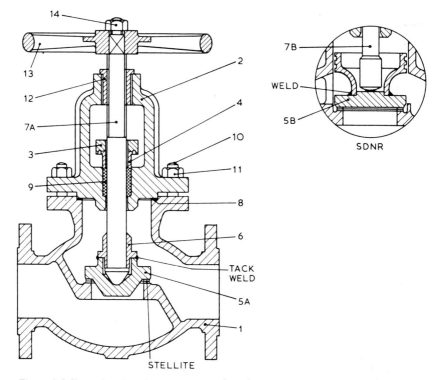

Figure 2.6 Example of a globe valve with (inset) detail of the valve arranged as a screw-down non-return (SDNR) unit

1.	*Body*	8.	*Bonnet gasket*
2.	*Bonnet*	9.	*Gland packing*
3.	*Gland flange*	10.	*Bonnet studs*
4.	*Gland*	11.	*Bonnet stud nuts*
5A	*Disc — stop type*	12.	*Yoke bush*
5B	*Disc — piston SDNR*	13.	*Handwheel*
6.	*Disc stem nut*	14.	*Handwheel nut*
7A	*Stem — stop type*		
7B	*Stem — piston SDNR*		*(Hattersley Newman Hender Ltd.)*

Here both seat and disc faces are stellited and almost indestructible; alternatively, the seat may be renewable and screwed into the valve chest or given a light interference fit and secured by a grub screw. The seatings may be flat or more commonly mitred. The spindle or stem may have a V or square thread, below or above the stuffing box; if the latter it will work in a removable or an integral bridge (bonnet); either may be bushed.

The spindle may be fitted to the valve disc (or 'lid') as shown in Figure 2.6 by a locknut and button; others will be found in which the button locates in a simple horseshoe. Leakage along the valve spindle is prevented by a stuffing box, packed with a suitable

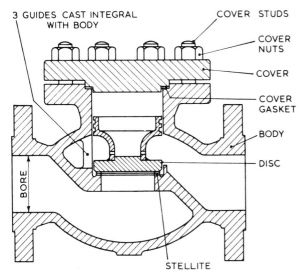

3 GUIDES CAST INTEGRAL
WITH BODY

COVER STUDS

COVER
NUTS

COVER

COVER
GASKET

BODY

DISC

BORE

STELLITE

Figure 2.7
Example of a non-return
valve
(Hattersley Newman
Hender Ltd.)

material and a gland. When the body is so formed as to change direction, e.g. in a bilge suction or on the side of a pressure vessel the valve is referred to as an angle valve. The flow is from below the valve seat, so that the gland is always on the low pressure side.

The type of valve having the disc unattached to the spindle is known as screw-down non-return (SDNR). The disc must be guided by wings or a stem on the underside for location or by a piston, as shown. Such valves are used in bilge suction lines to prevent flooding back in the event of a compartment being flooded or a valve being carelessly left open, also as feed check and boiler stop valves. The greatest lift required is one-quarter of the bore. Figure 2.7 shows a free-lifting non-return valve.

The change in direction of flow may give rise to large pressure drops (see page 55).

Gate valves

Unlike the globe valve, gate (or sluice) valves give full bore flow without change of direction. They consist of chests divided at mid-length by a double membrane having central circular openings, furnished with seats, tapered or parallel on their inner, i.e. facing, sides. A gate, shaped appropriately, can be moved in a direction at right angles to the flow by a screwed spindle working in a nut (Figure 2.8a).

Such a valve is not suitable to partially open operation since wire-drawing of the seat will occur. The bonnets of these valves are

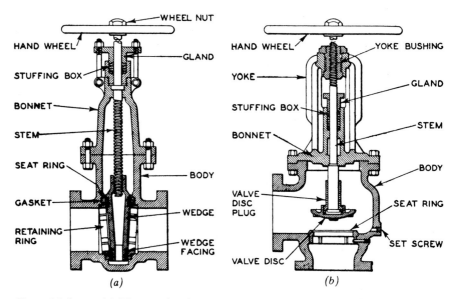

Figure 2.8 Gate and full bore angle valves

frequently of cast iron and care should be taken when overhauling. To ensure tightness, some parallel gates are fitted with twin discs, dimensioned similarly to the chest seats, separated by a spring. Where change of direction is required, a full bore angle valve (Figure 2.8*b*) may be used.

Butterfly valves

A butterfly valve consists basically of a disc pivoted across the bore of a ring body having the same radial dimensions as the pipe in which it is fitted. The full-bore straight-through flow arrangement of this type of valve, especially if combined with a carefully streamlined disc profile, gives this type of valve excellent flow characteristics and low pressure drop. The valve is quick-acting if required, since only a quarter of a turn of the spindle is required to move the valve from the fully open to the fully closed position. Sizes range from 6 mm to over 1000 mm bore.

An isometric view of a typical butterfly valve is shown in Figure 2.9.

Because of the good flow characteristics of this type of valve and the simplicity with which it can be opened or closed by remotely controlled actuators it is widely used in cargo and ballast lines and cooling water systems.

For fine control of cooling water temperature a special type of ganged butterfly valve is used to by-pass occasionally coolers. Known as a diverter valve it consists of a Y or T casting with butterfly valves in two of the legs (Figure 2.10). A pneumatic actuator working from a signal provided by a temperature sensor opens one valve while closing the other. This gives precise control of the flow rate in the

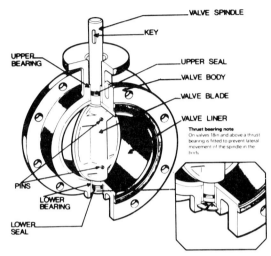

VALVE SPINDLE

KEY

UPPER BEARING

UPPER SEAL

VALVE BODY

VALVE BLADE

VALVE LINER

Thrust bearing note
On valves 18in and above a thrust bearing is fitted to prevent lateral movement of the spindle in the body

PINS

LOWER BEARING

LOWER SEAL

Figure 2.9 Type BA butterfly valve (Hindle Cockburns Ltd)

Figure 2.10 A diverter valve (Cockburn-Rockwell Ltd.)

main and branch lines. In the event of a temperature controller failure, a built-in return spring opens or closes the main and branch lines (as appropriate to the system of operation) to provide maximum cooling flow; manual control is available for emergency use.

Check valves

A check valve is another name for a non-return valve. It allows flow in one direction only. Screw-down and free-lifting non-return valves have already been described (pages 27–29).

Other types, also based on the globe valve have no stem and thus cannot be shut off by hand. A more common type has a hinged flap (Figure 2.11) and is often called a flap valve. Such a valve is commonly used in scupper lines.

Change-over valve chests

Sometimes tanks have a dual purpose. The most frequent examples of these are tanks used for either oil or water ballast and tanks used

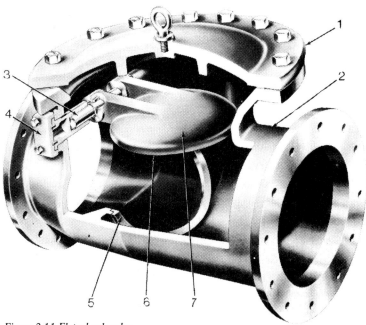

Figure 2.11 Flap check valve
1. *Cover*   5. *Seat ring*
2. *Body*   6. *Face ring*
3. *Hinge shaft*   7. *Door*
4. *Shaft bearing*
(Blakeborough and Sons Ltd.)

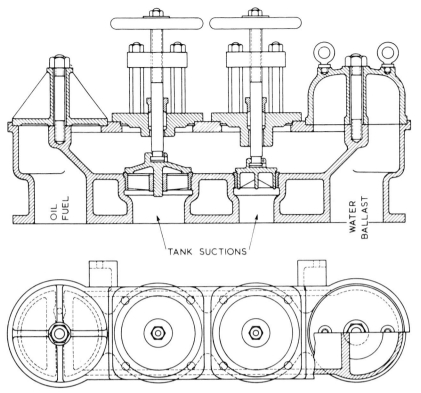

OIL FUEL

WATER BALLAST

TANK SUCTIONS

Figure 2.12 2-valve change-over chest for oil and ballast suctions as arranged when filling or discharging ballast

for either dry cargo or water ballast. In both cases the suction branches will have to be led to more than one pumping system. In the case of the former it may be the oil main and ballast main; in the case of the latter it will be the ballast main and the bilge main.

To prevent accidents occurring, special valve chests are used in such cases in which interchangeable blanks and connecting passages are incorporated. An example is shown in Figure 2.12 where the two suction valves are flanked by a blank on one side and a dome on the other. In the position shown, the two suction valves have access, via the dome, to the water ballast main but not to the oil fuel main. By simply changing over the blank and the dome the situation is reversed.

Valve actuators

A variety of valve actuators to control the opening and closing of globe, gate and butterfly valves may be found. In some types, an

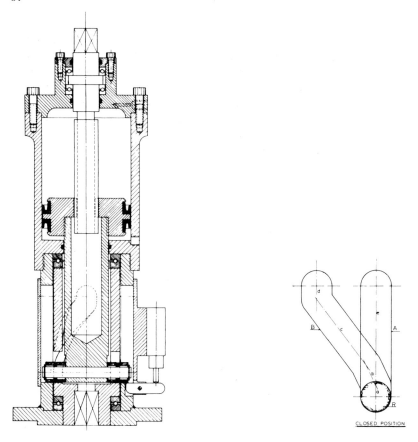

Figure 2.13 Pneumatic butterfly valve actuator showing scroll cam arrangement

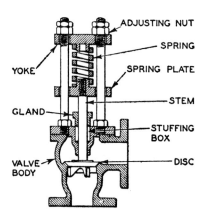

Figure 2.14 Relief valve

electric motor, fitted with limit switches is used to turn a threaded stem through a yoke, purely substituting the action of a handwheel. In most instances, however, pneumatic or hydraulic actuators are used. These give linear motion to a piston which in the case of a globe or gate valve merely pushes or pulls the valve stem although in the case of the former the valve disc is given a slight turn on landing so that the seat is kept cleaned. In the case of a butterfly valve the 90° motion through which the disc needs to be turned can be had from a suitable scroll cam (Figure 2.13).

Control valves

Temperature control may be obtained by diverter valves, automatically operated globe valves or by automated cocks. Pressure of steam, air, or various liquids is regulated by a variety of valve types depending on the fluid and the duty.

Excess pressure is eased by a relief valve (Figure 2.14). This consists of a disc held closed by a spring-loaded stem. The compression on the spring can be adjusted so that the valve opens at the desired pressure. The special case of boiler relief or safety valves is dealt with in *Marine Steam Boilers* by J.H. Milton and R.M. Leach (published by Butterworths).

Selection of a valve of the correct size and loading is important since they have a narrow pressure range.

Under normal conditions a relief valve should operate consistently within reasonable limits of its set pressure. Leakage, failure to close properly, too long blowdown or *simmering* on opening indicate that the valve is improperly adjusted or the valve-seat damaged. If a valve chatters, that is, opens and closes rapidly, this is due to back pressure. Relief valves on water pumps should be eased off their seats manually from time to time since deposits accumulate and seize the valve. They should be opened up and examined whenever the pump is overhauled.

Pressure reducing valves

It is sometimes necessary to provide steam or air at a pressure less than that of the boiler or air reservoir. To maintain the downstream pressure within defined limits over a range of flow and despite any changes in supply pressure, a reducing valve is fitted.

In its simplest form the reducing valve is opened or closed by a pre-tensioned spring which is balanced by the downstream pressure acting on a diaphragm or, in some cases, bellows. A regulating screw is provided for setting the tension on the spring (see Figure 2.15).

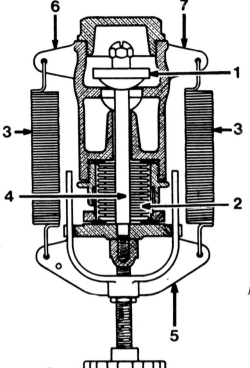

Figure 2.15 Pressure reducing valve
1. *Main valve*
2. *Flexible bellows*
3. *Springs*
4. *Valve spindle*
5. *Yoke*
6, 7. *Anchoring lugs*
8. *Adjusting handwheel*

In the reducing valve shown in Figure 2.15 the inlet pressure acts in an upward direction on the main valve (1) and in a downward direction on the controlling flexible bellows (2) which forms the equivalent of a piston of equal area to that of the valve (1). These two parts are therefore in a state of balance so that variations in pressure on the inlet (i.e. upstream) side do not affect the reduced pressure. The pressure on the outlet side acts in a downward direction on the valve (1) and will rise until just balanced by the pull of the springs (3). It is important that this type of valve is installed in the vertical position and the yoke 5 should be checked regularly for freedom of movement in its guides.

A more sophisticated reducing valve is shown in Figure 2.16. This incorporates a pilot valve which controls the fluid above the valve operating piston in response to downstream pressure. This type of valve may be found with the pilot valve connected to an external sensing point located some distance from the valve. Some are provided to give tight-closing characteristics under conditions of no flow.

While reducing valves such as those described above will be found

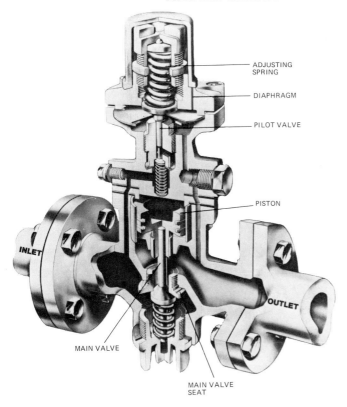

Figure 2.16 Bailey Class G4 pressure-reducing valve (IMI Bailey Valves Ltd)

on non-automated ships the function of the self-regulating valve has largerly been supplanted by the automatic process control valve for fluid pressure control in unmanned machinery spaces. This permits remote control of the set-point of the valve and by careful selection of valve trim (the control industry's term for the internal parts of the valve which come in contact with the controlled fluid and form the actual control portion) a variety of flow characteristics can be achieved. The subject is dealt with further in Chapter 14.

Quick closing valves

Fuel oil gravity tanks must be fitted with valves that can be closed rapidly and remotely in the event of an emergency such as a fire. Figure 2.17 shows wire operated Howden Instanter valves which fulfil this purpose while Figure 2.18 shows a hydraulically-operated quick-closing valve used for the same purpose.

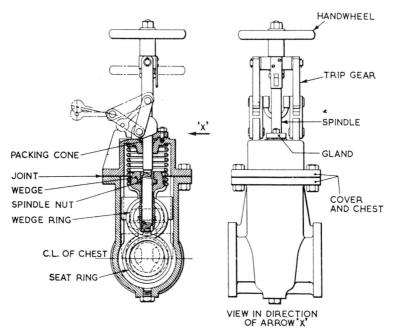

Figure 2.17 Howden Instanter quick closing valve (James Howden & Co. Ltd.)

STEAM TRAPS

A steam trap is a special type of valve which prevents the passage
of steam but allows condensate to pass. It works automatically and is
put into drain lines so that these drain off condensate automatically
without passing any steam. There are three main types, namely
mechanical, thermostatic and thermodynamic. There is also the
vacuum trap, or automatic pump.

Mechanical traps

Mechanically operated traps utilise float-controlled needle valves to
trap steam or air. Some use open floats (Figure 2.20), others ball
type floats (Figure 2.19).

Thermostatic traps

Thermostatic traps use the expansion of an oil-filled element, a
flexible bellows or a bi-metallic strip to actuate a valve. An oil-filled
type in common use is shown in Figure 2.21a. As the condensate

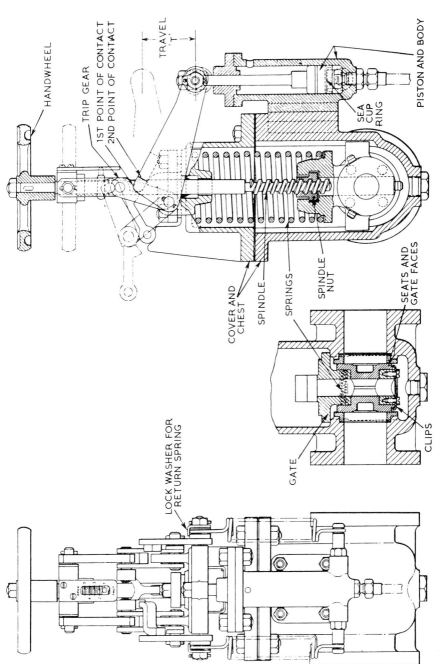

HANDWHEEL

TRIP GEAR

1ST POINT OF CONTACT

2ND POINT OF CONTACT

TRAVEL T

PISTON AND BODY

SEA CUP RING

COVER AND CHEST

SPINDLE

SPRINGS

SPINDLE NUT

SEATS AND GATE FACES

CLIPS

GATE

LOCK WASHER FOR RETURN SPRING

Figure 2.18 Arrangement of Instanter quick closing parallel slide valve with oil operating cylinder

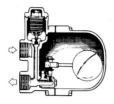

Figure 2.19 Ball float type
mechanical trap

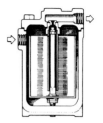

Figure 2.20 Open float
mechanical trap

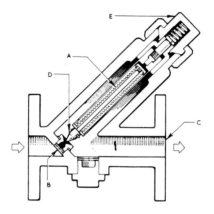

Figure 2.21(a) Oil filled thermostatic steam trap

A. Element   C. Outlet   E. Adjuster
B. Orifice   D. Valve head

(Spirax-Sarco Ltd.)

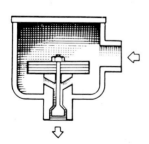

Figure 2.21(b) Bi-metallic steam trap.

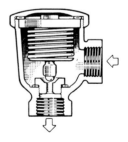

Figure 2.21(c) Bellows type
steam trap
(Spirax-Sarco Ltd.)

temperature rises element A expands to close the valve D. An adjust-ment screw E permits the valve to be set up for condensate release at a specific temperature. Clearly in an application where the pressure varies there could be a broad band of operation in which the trap would be either waterlogged or passing steam.

In the flexible bellows type (Figure 2.21c) the bellows is filled with a mixture which boils at a lower temperature than does steam. The trap self-compensates for operating pressure. It will be damaged if water hammer occurs and will burst if subjected to superheated steam.

Figure 2.21b shows a trap operated by a bi-metallic strip. Deflection of the bi-metallic strip with increasing temperature closes the valve. It will work over a range of pressures without the need for re-adjustment and will work satisfactorily under superheated conditions. It is not particularly prone to damage from either water hammer or vibration.

Thermodynamic traps

This type of trap uses the pressure energy of the steam to close the valve which consists of a simple metal disc. The sequence of

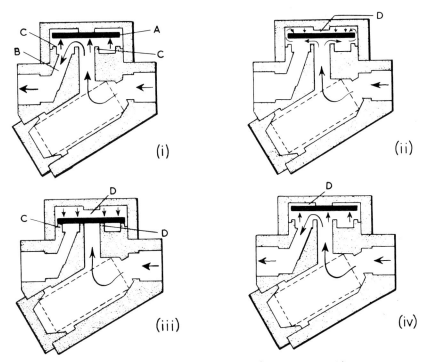

Figure 2.22 Operation of thermodynamic steam trap (Spirax-Sarco Ltd.)

operation is shown in Figure 2.22. In (i), disc A is raised from the seat rings C by incoming pressure allowing discharge of air and condensate through outlet B. As the condensate approaches steam temperature it flashes to steam at the trap orifice. This means that the rate of fluid flow radially outwards under the disc is greatly increased. There is thus an increase in dynamic pressure and a reduction in static pressure. The disc is therefore drawn towards the seat. Due to this alone the disc will never seat. However, steam can flow round the edge of the disc resulting in a pressure build up in the control chamber D as shown in (ii). When the steam pressure in chamber D acting over the full area of the disc exceeds the incoming condensate/steam pressure acting on the much smaller inlet area, the disc snaps shut over the orifice. This snap action is important. It removes any possibility of wire-drawing the seat, while the seating itself is tight, ensuring no leakage. As shown in (iv) the incoming pressure will eventually exceed the control chamber pressure and the disc will be raised, starting the cycle all over again.

The rate of operating will depend on the steam pressure and on the ambient air temperature. In practice, the trap will usually open after 15–25 s, the length of time open depending on the amount of condensate to be discharged. If no condensate has been formed, then the trap snaps shut immediately. From the foregoing it will be seen that the trap is never closed for more than 15–25 s, so condensate is removed virtually as soon as it is formed.

Vacuum (pumping) traps

The layout of drain systems can often be improved by the use of automatic pumps, sometimes referred to as vacuum traps. Condensate can be drained by gravity more readily to local receivers and then pumped back to the engine room as required. Similarly, engine room drains can be relieved of much back pressure by taking them to a low level hot well before pumping to a high level boiler feed tank. A high level feed tank will enable the pumps to handle warm feed water and can improve the overall efficiency of the plant. A typical pumping trap is shown in Figure 2.23.

When the trap is empty the exhaust valve C is open and the steam valve D closed. Water flowing into the trap, through the non-return valve A raises the float E until it compresses the spring H, exerting a force which plucks the spindle J away from the magnet G so that the exhaust valve is projected into the closed position and the steam valve D is opened through the movement of the lever. Steam is therefore admitted to the trap and drives out the water through the non-return valve B. The float falls with the level of the water inside the

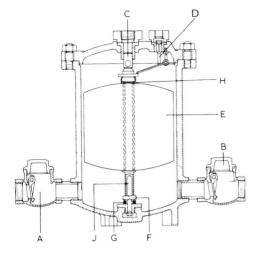

Figure 2.23 Automatic pumping
or vacuum trap
A, B. Non-return valves
C.   Exhaust valve
D.   Steam valve
E.   Float
F.   Collar
G .   Magnet
H.   Spring
J.   Spindle
(Royles Ltd.)

trap and engages the collar F, pulling the spindle J down so that the exhaust valve opens and the steam valve closes. The cycle of operation is then repeated. Being float controlled, the trap operates only when water is flowing into it.

Maintenance of traps

A defective trap wastes steam. To aid detection a sight glass or test cock should be fitted after each trap so that its performance can be observed. If these are not provided then the only alternative is to open up traps regularly for inspection. The only exception is the thermodynamic type. When operating correctly this gives a characteristic 'click', usually at intervals of between 20 and 30 s so that its performance can be checked simply by listening.

Before inspecting any trap, it is advisable to check the strainer which should always be provided at the inlet of the trap. The contents of the strainer screen will give an indication of the cleanliness of the system, while fine dirt passing through the strainer is one of the chief causes of defective steam traps. Pieces of pipe scale or dirt jammed across the seat can prevent proper closure and if left for a period of time can give rise to wire drawing.

In most cases cleaning and reassembling should be sufficient to ensure satisfactory operation. If the thermo-dynamic type fails to operate after cleaning, then it could be that the disc and seat require lapping. Mechanical traps should be checked for defective floats and buckets and wear on any linkages, while the valves and seats should be renewed if necessary. In the case of thermostatic traps, the elements should be checked to ensure that they are sound. Particular care should be taken with the elements of bi-metallic steam traps. Incorrect adjustment can give rise to excessive waterlogging or cause the traps to blow steam, while the elements themselves can sometimes assume a permanent 'set'.

STRAINERS (Filters)

Strainers, sometimes called filters, are devices designed to prevent the passage of unwanted solids into or further along a system. The simplest strainer consists of a box with a removable lid in which a flat perforated plate is inserted such that the fluid must pass through the perforations. Such strainers will be found close to sea water suction valves and immediately before bilge valves. In the latter application they are known as strum boxes. Perforation sizes vary according to duty and manufacture but are invariably in the range 3–12 mm.

Plates corrode/erode and, when checking strainers, due attention should be paid to the condition of the plate. If a gap has formed at the top or bottom of the plate, or if it is in danger of breaking up, then it should be renewed. Many strainer covers are hinged and are particularly prone to poor seating. The state of the gasket should be checked and care should be taken when closing the cover. Since these covers are frequently of cast iron and are secured by lugs and thumb screws care must be taken not to use undue force when closing them.

In many installations basket strainers are used. These consist of a cylindrical container in which a perforated metal or wire basket is suspended (Figure 2.24). Flow through these units is from the top, into the basket and out from the outside of the basket. They are

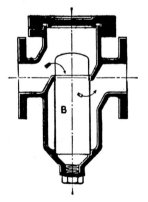
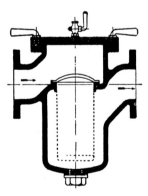

Figure 2.24 Single strainers for (left) high pressure water, steam or oil service and (right) low pressure water or oil service

frequently used in duplex units with three way cocks at inlet and outlet so that one or both baskets can be in use, or shut down for cleaning or for standby.

Lubricating oil systems are fitted with a wide variety of strainers some of which can be cleaned in situ while others have replaceable cartridges. Among the former type is the knife edge strainer, Figure 2.25. This consists of a series of discs ganged to a shaft. Interspaced between the discs are a number of thin fingers. The solid particles are trapped on the edges of, and between, the discs. By rotating the disc shaft the particles are cleared by the fingers and fall to a sump, which is drained periodically. It is essential to operate these strainers regularly since if they are allowed to clog up it is difficult to free the shafts.

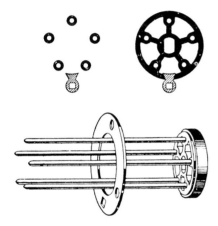

Figure 2.25 The Auto-Klean strainer showing circular straining plates, separation washers and spider
(Auto-Klean Strainers Ltd.)

Another type of edge strainer is the Stream-line (Figure 2.26). This also removes a certain amount of moisture. To clean the Stream-line strainer compressed air is blown through in the reverse direction to normal flow. The accumulation of solids is discharged from the sludge outlet.

Drinking water treatment

To remove fine particulate matter and the larger micro-organisms from drinking water, filter stacks, loaded with a very fine earth power are frequently used, usually as part of a complete treatment plant incorporating a chlorination system. Figure 2.27 shows schematically the layout of a Metafilter water treatment plant. This uses two filter stacks consisting of a number of metal rings (Figure 2.28) screwed onto a rod. The rings are scalloped so that when clamped together the element forms a stainless steel cylinder having a

46

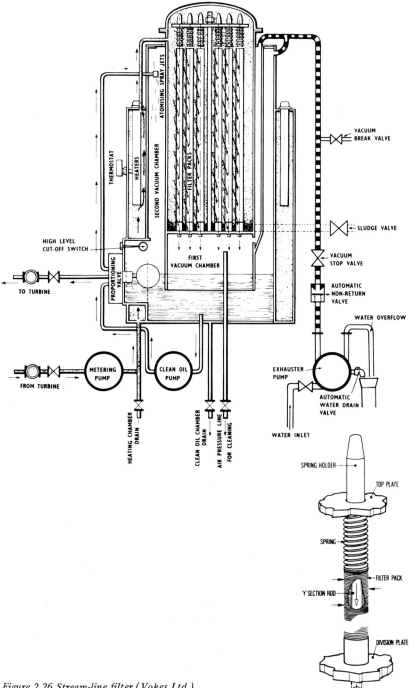

Figure 2.26 *Stream-line filter (Vokes Ltd.)*

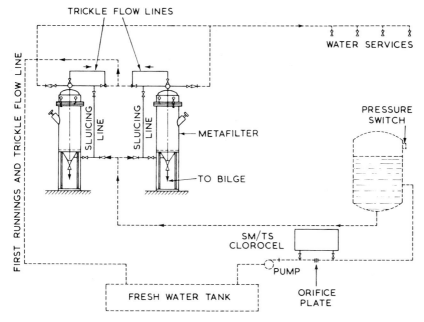

Figure 2.27 Layout of water filtration plant

Figure 2.28 Metafilter 'ring pack' element (Stella-Meta Filters Ltd.)

large number of 0.08 mm deep grooves. These are precoated with diatomaceous earth powder (Kieselguhr).

To bring the unit into operation a quantity of powder, the amount depending on the size of the unit, is mixed with water to form a slurry. This is then poured into the filter and the inlet and 'first running' valves are opened to circulate the contents back to the fresh water tank. When the precoat bed has formed (after about five minutes) the service outlet and trickle flow valves are opened and the 'first running' valve is closed. Filtration continues until the contaminant builds up on the outside surfaces of the precoat bed, blocking the pores of the powder and increasing the pressure drop across the filter. When the life of the filter bed is complete the stand by unit can be pre-coated and the original filter can be back-flushed ready for future use.

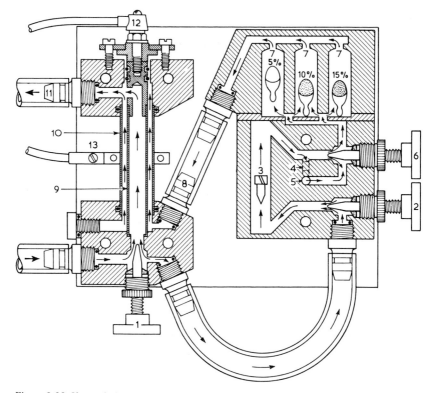

Figure 2.29 Clorocel electrolytic cell and brinometer
1. Cooling water return valve.
2. Flow control valve.
3. Flow indicator.
4. Water inlet to salt saturator.
5. Saturated salt solution outlet.
6. Salt control valve.
(Stella-Meta Filters Ltd.)

8. Clorocell salt solution nipple
9. Cathode.
10. Anode.
11. Hypochlorite outlet.
12. Cathode connection.
13. Anode connection.

Chlorination in the Metafilter system is effected by a Clorocel electro-chlorinator (Figure 2.29). This unit produces sodium hypochlorite by the electrolysis of a solution of sodium chloride.

BIBLIOGRAPHY

Holmes, W.E., *Industrial Pipework Engineering,* McGraw-Hill.
Steam at sea, Spirax-Sarco Marine Bulletin.
Valve Users Manual, British Valve Manufacturers Association.

3 Pumps and pumping

A pump is a device which adds to the energy of a liquid or gas causing an increase in its pressure and perhaps a movement of the fluid. A simple pumping system is shown schematically in Figure 3.1. The basic system consists of a suction branch, a pump and a discharge branch.

PUMP CHARACTERISTICS

Suction conditions

The pump only adds to the energy of the system. The energy required to bring the fluid to the pump is an external one and in most

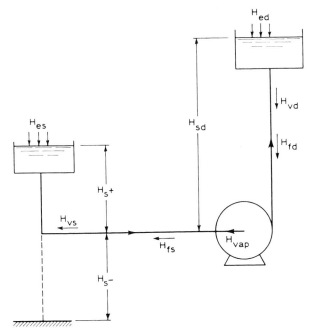

Figure 3.1 A simple pumping system

practical conditions is provided by atmospheric pressure. In discussing
Figure 3.1 it is assumed that the fluid being pumped is a liquid and
that it is thus incompressible. The case of pumping gases is slightly
more complicated and is only partially relevant to this chapter.

The diagram shows pressure head H_{es} acting on the liquid surface
at the suction inlet. The vertical distance of the pump centre H_s from
the surface of the liquid will affect the head available at the pump
and must be added algebraically to H_{es}. If the pump is below the
liquid level then H_s will be positive; if it is above the liquid level H_s
will be negative. The pipe will have some frictional resistance
resulting in a loss of pressure head H_{fs}. A further head loss H_v, due
to the velocity of the liquid will also occur but, except for very high
velocities, is negligible.

Providing that the sum of these head losses — $H_v + H_{fs} \pm H_s$ — is
less than H_{es}, the suction condition at the pump might be thought to
be adequate. There are two further factors to take into consideration,
however. These are the vapour pressure of the liquid being pumped
and the amount of remaining positive suction head required at the
pump suction to effect the designed delivery rate. This factor is
known as the required NPSH (net positive suction head).

Every liquid has a pressure at which it will vaporise and this
pressure varies with temperature. If the combination of pressure and
temperature within the suction pipe is such that vaporisation occurs,
the efficiency of the pump deteriorates and a condition can be
reached where the pump will cease to function. The vapour pressure
H_{vap} is thus usually shown as a suction head loss.

The summation $H_{es} \pm H_s - H_{fs} - H_{vs} - H_{vap}$ is known as the
available NPSH (net positive suction head). In application to systems
and neglecting the velocity head the expression becomes:

$$\text{available NPSH} = \frac{10.2}{\rho} \left[P_{bar} + P_{es} - P_{vap} \right] - H_{fs} \pm H_s$$

where:

ρ = density of liquid at max operating temp, kg/litre.

P_{bar} = barometric pressure at the pump, bar.

P_{es} = minimum pressure on the free liquid level at the suction
inlet (negative when under a vacuum), bar gauge.

P_{vap} = vapour pressure of the liquid at the maximum operating
temperature, bar abs.

H_s = height of liquid free surface above centre line of pump
(negative when level is below pump), m.

H_{fs} = friction head losses in suction piping system, m.

In application, the available NPSH must always be greater than the required NPSH. The former may be calculated knowing the details of the suction piping while the latter may be obtained from the pump manufacturer.

The significance of vapour pressure is most easily seen when considering a pump drawing from a negative suction head (usually referred to as a suction lift).

The theoretical suction lift of a pump at sea level with water at 15°C is 1.013 × 10.2 = 10.3 m, where the barometric pressure is 1.013 bar (1 atm) and 10.2 m is the head of water equivalent to 1 bar (1 bar = 10^5 N/m² = 14.51 lb/in²). In practice the suction lift will exceed 7 m only under very favourable conditions. This is because of friction losses in the suction pipe and because of the limitations of the pump design. Any increase in water temperature above 15°C will have a detrimental effect on the vapour pressure; e.g. at 50°C water will boil at an absolute pressure of 0.14 bar, so that the lift reduces to 10.2(1.013 − 0.14) = 9 m, drastically reducing the available NPSH. It follows that suction lift should be as small as conditions allow and that for water temperatures above about 75°C the suction head must be positive or if this is impossible the suction pipe must be short, straight, free from interference and the speed of flow must be low, say less than 1 m/sec.

Discharge conditions

Some of the energy fed into the pump will be dissipated as heat due to mechanical inefficiencies, the remainder will be converted into pressure rise and fluid velocity. Some of the pressure head generated will be lost in overcoming the friction of the discharge pipe H_{fd}, some in the static head of the pipe system H_{sd}, and some in the pressure head acting on the free surface at the terminal point H_{ed}. There will also be a velocity head loss but as in the case of the suction line, for most practical purposes this can be neglected.

Pump power

The total work done by the pump, neglecting losses within the pump itself will be proportional to the equivalent head difference between the points of suction and discharge. This is known as total head H_{tot}:

$$H_{tot} = H_{fs} + H_{fd} + H_{vap} + H_{sd} \pm H_s$$

The power absorbed by the pump, P_a, then becomes:

$$P_a = \frac{Q \times H_{tot} \times w}{K}$$

where:

P_a	=	Power absorbed (kW).
Q	=	Quantity delivered in litres/s.
H_{tot}	=	Total head in metres
w	=	Density of liquid in gm/ml (1 for fresh water)
K	=	101.9368 (102)

The input power P_i to the pump required from the prime mover is

$$P_a \times \frac{1}{\text{pump efficiency}}$$

For an electrically driven pump, the power consumed is

$$P \times \frac{1}{\text{pump efficiency}} \times \frac{1}{\text{motor efficiency}} \quad (\text{kW}).$$

Where the head available is small, the suction line, passages and valves are specially designed and of large area to reduce the suction losses to a minimum. This increases the cost of the pump and installation and reduces efficiency, but is unavoidable for duties such as the extraction of condensate from a condenser where the head available is frequently a matter of millimetres.

Generally speaking, suction heads require to be greater for high speed or large capacities than for low speed or small capacities. Condensate pumps, heater-drain pumps and feed pumps operating with direct-contact feed heaters, must be arranged below the water level as the static head of water is the only force available to cause the water to flow into the pump, because the water and the steam on the surface are at the same temperature.

Before liquid can flow into a pump, the air or vapour in the suction line must be evacuated sufficiently to cause the liquid to flow into the suction chamber.

Some pumps (known as self-priming pumps) do this automatically when they are started. In others special priming devices must be used to withdraw the air and lower the pressure in the pump sufficiently to cause flow.

Friction losses

The sum of these losses depends upon the sectional area and the internal condition of the pipes and fittings, the velocity and viscosity of the liquid being pumped and the friction caused by bends, valves and other fittings. Frictional resistance to the flow of water varies approximately as the square of the velocity. Thus, if the frictional resistance of a condenser and system of piping is equivalent to a head of 5 m when 800 litres/s are passing, the frictional resistance would rise to 11.25 m with 1200 litres/s and to 20 m with 1600 litres/s. Table 3.1 gives the head in metres required to overcome the friction of flow in every 30 m of new straight cast-iron pipe.

The general law of frictional resistance due to the flow of water in a straight circular pipe running full of water may be expressed accurately enough for practical purposes as

$$Hm = \frac{KLV^2}{2GR} \quad \text{if } R = \frac{\text{area of pipe bore}}{\text{wetted perimeter}} = \frac{D}{4}$$

or $\quad Hm = \dfrac{KLV^2}{2GD}$

where:

Hm = head loss (m)
L = length of pipe (m)
V = speed of flow (m/s)
D = bore of pipes (m)
G = Gravitational constant = 9.81 m/s^2 = 9.81N/kg

To this must be added the loss due to bends each equivalent to from 3 to 6 m of straight pipe, depending upon the radius and due to fittings (Table 3.2) from which it can be seen that suction pipes, if not bell-mouthed, T-pieces and elbows give rise to the greater losses.

Example. Water flow 26.25 litres/s, static lift 20 m. The suction is through a strainer with bell-mouthed entry, a foot valve, 2 m of straight pipe 125 mm bore and a similar bend.

Suction loss (equivalent length) (Table 3.2)
Strainer	0.58 m
Foot valve	1.43 m
Straight pipe	2 m
Bend	4.27 m
	8.28 m

Table 3.1 Head loss in metres for each 30 m of straight pipe

Flow Litre/sec	Bore of pipe, mm							
	25	38	50	65	75	100	125	150
0.75	4.05	0.517	0.121	—	—	—	—	—
1.50	15.5	1.99	0.487	0.151	—	—	—	—
2.25	—	4.50	1.07	0.350	0.143	—	—	—
3.00	—	7.92	1.83	0.610	0.246	—	—	—
3.75	—	12.10	2.83	0.915	0.35	—	—	—
4.50	—	17.36	4.1	1.310	0.548	0.121	—	—
5.25	—	23.80	5.49	1.80	0.75	0.167	—	—
6.00	—	—	7.3	2.28	0.94	0.216	—	—
6.75	—	—	9.1	2.83	1.185	0.241	0.0884	—
7.50	—	—	11.3	3.65	1.22	0.338	0.11	0.048
9.00	—	—	15.5	5.05	2.07	0.486	0.151	0.067
10.50	—	—	21.0	6.7	2.77	0.650	0.2	0.09
12.00	—	—	—	8.84	3.65	0.840	0.265	0.121
13.50	—	—	—	10.95	4.54	1.03	0.35	0.151
15.00	—	—	—	13.70	4.90	1.295	0.41	0.181
18.75	—	—	—	21.30	8.40	1.96	0.62	0.274
22.50	—	—	—	—	12.20	2.84	0.91	0.396
26.25	—	—	—	—	16.15	3.8	1.21	0.53
30.00	—	—	—	—	—	5.37	1.58	0.7
33.75	—	—	—	—	—	6.10	1.99	0.85
37.50	—	—	—	—	—	7.6	2.42	1.03
45.00	—	—	—	—	—	10.3	3.45	1.46
52.50	—	—	—	—	—	—	4.69	1.98
60.00	—	—	—	—	—	—	5.9	2.53

Table 3.2 Loss for fittings in equivalent lengths of straight pipe (m)

Bore of pipe mm	Vel. head for ordinary pipe	Vel. head for bell-mouthed entry	Bend	Foot valve	Non-return valve	Delivery valve full open	Strainer	Tees and elbows
25	1.37	0.82	0.76	0.24	0.305	0.24	0.091	0.83
38	2.2	1.31	1.13	0.36	0.49	0.36	0.152	1.31
50	3.0	1.8	1.52	0.52	0.7	0.52	0.214	1.8
65	3.8	2.31	1.95	0.64	0.85	0.64	0.275	2.31
75	4.75	2.86	2.44	0.79	1.06	0.79	0.305	2.87
100	6.4	3.96	3.3	1.10	1.43	1.10	0.427	3.96
125	8.5	5.2	4.27	1.43	1.86	1.43	0.58	5.2
150	10.7	6.4	5.27	1.76	2.31	1.76	0.70	6.4

From Table 3.1, the equivalent head loss is

$$\frac{1.21 \times 8.28}{30} = 0.33$$

Delivery loss (equivalent length) (Table 3.2)

Bell-mouthed entry	5.2 m
Delivery valve	1.43 m
Non-return valve	1.86 m
Straight pipe	15.25 m
Bend	4.27 m
	28.01 m

From Table 3.1: $\dfrac{1.21 \times 28.01}{30} = 1.13$ m

Total loss is 0.33 + 1.13 = 1.46 m. Add say 25% for future roughening of surfaces, 1.46 + 0.38 = 1.84, to which must be added the static lift of 20 m or 21.84 m in all.

The figures in Tables 3.1 and 3.2 show the resulting power losses if a pipe system is complicated, tortuous and not generously dimensioned.

The pressure corresponding to 1m of water is 0.098 bar or 98 mbar: conversely, the head corresponding to 1 bar is 10.17 m. It will be apparent that for practical purposes, these figures can be rounded to 0.1 or 100 mbar and 10 m respectively.

Drawings or prints are supplied with pumps by the manufacturers, giving sizes and particulars of flanges, positions of foundation bolts and other information necessary for the arrangements of pipe connections. These must be exactly adhered to; it is little use installing a highly efficient pump if the power is dissipated and increased by the use of unsuitable pipes and fittings, or by poor layout.

TYPES OF PUMP

Marine pumps fall into two broad classes:
1. *Displacement.* The liquid or gas is displaced from the suction to the discharge by the mechanical variation of the volume of a chamber or chambers. They can be subdivided into two classes, reciprocating pumps, in which a plunger or piston is mechanically reciprocated in a liquid cylinder, and rotary pumps, where the liquid is forced through the pump cylinder or casing by means of screws or gears.

2. *Centrifugal.* Flow through the pump is induced by the centrifugal force imparted to the liquid by the rotation of an impeller or impellers.

Figures 3.2 and 3.3 show diagrammatically the various types of displacement and centrifugal pumps, the alternative methods of driving employed, their duties and the range of capacities for which the various types are built.

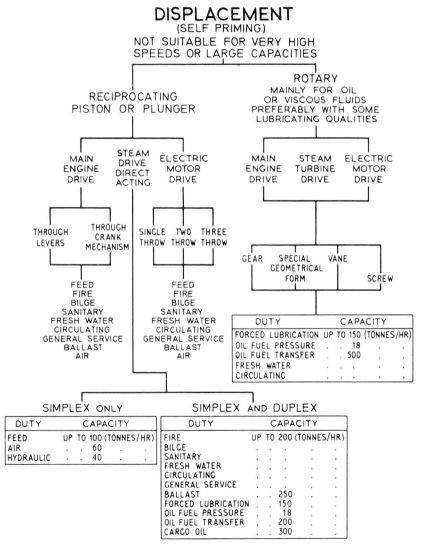

Figure 3.2 Types, duties and capacities of displacement pumps (Weir Pumps Ltd.)

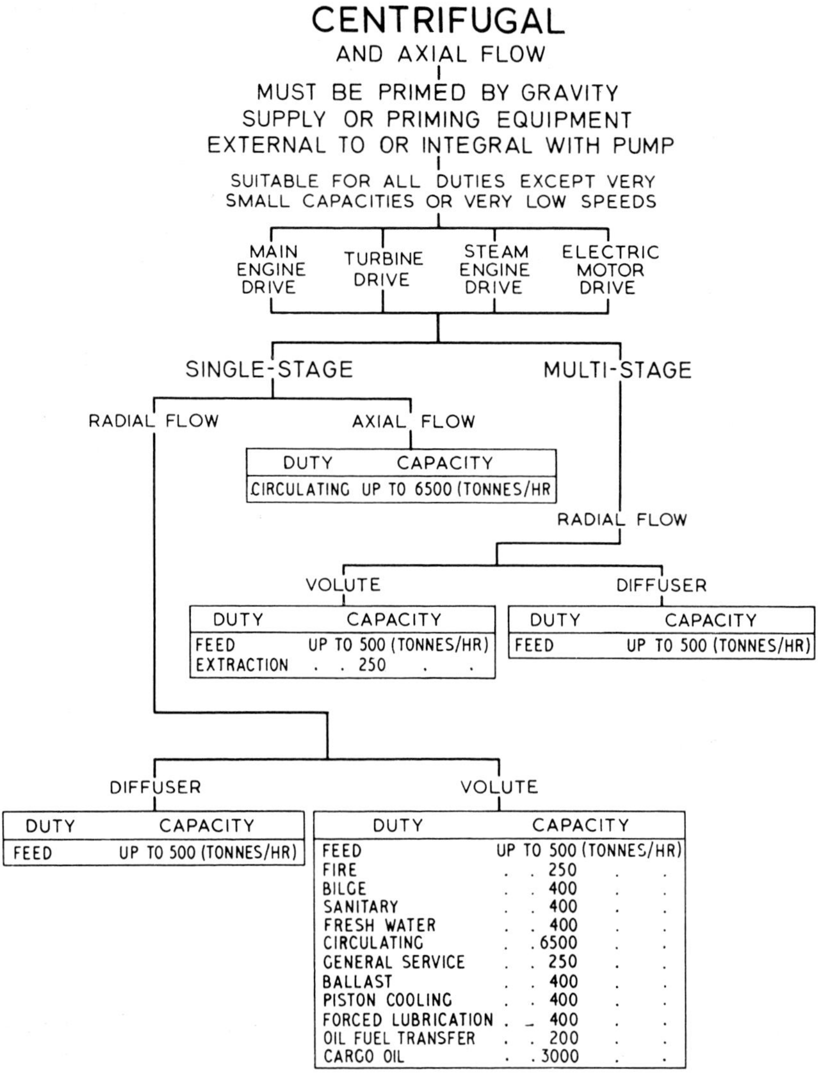

CENTRIFUGAL
AND AXIAL FLOW

MUST BE PRIMED BY GRAVITY
SUPPLY OR PRIMING EQUIPMENT
EXTERNAL TO OR INTEGRAL WITH PUMP

SUITABLE FOR ALL DUTIES EXCEPT VERY
SMALL CAPACITIES OR VERY LOW SPEEDS

| MAIN ENGINE DRIVE | TURBINE DRIVE | STEAM ENGINE DRIVE | ELECTRIC MOTOR DRIVE |

SINGLE-STAGE   MULTI-STAGE

RADIAL FLOW   AXIAL FLOW

DUTY	CAPACITY
CIRCULATING UP TO 6500 (TONNES/HR)	

RADIAL FLOW

VOLUTE   DIFFUSER

DUTY	CAPACITY
FEED	UP TO 500 (TONNES/HR)
EXTRACTION . . 250 . .

DUTY	CAPACITY
FEED	UP TO 500 (TONNES/HR)

DIFFUSER   VOLUTE

DUTY	CAPACITY
FEED	UP TO 500 (TONNES/HR)

DUTY	CAPACITY
FEED	UP TO 500 (TONNES/HR)
FIRE	. . 250 . .
BILGE	. . 400 . .
SANITARY	. . 400 . .
FRESH WATER	. . 400 . .
CIRCULATING	. . 6500 . .
GENERAL SERVICE	. . 250 . .
BALLAST	. . 400 . .
PISTON COOLING	. . 400 . .
FORCED LUBRICATION	. _ 400 . .
OIL FUEL TRANSFER	. . 200 . .
CARGO OIL	. . 3000 . .

Figure 3.3 Types, duties and capacities of centrifugal and axial-flow pumps (Weir Pumps Ltd.)

General applications

Reciprocating displacement units are the most suitable and efficient for dealing with small volumes and high differential pressure, and can handle any required viscosity. Rotary positive-displacement machines are suitable for the intermediate range of volumes, differential pressures and viscosities. This type of unit is covered by a wide range of designs, of which the gear and screw types are most commonly used, particularly for more viscous liquids such as oils. Centrifugal pumps find their widest application for dealing with large volumes with medium-to-low heads and viscosities, the circulating pump being an ideal example. Centrifugal or radial flow pumps give regular delivery, and are quiet in operation. Axial-flow and semi-axial-flow rotary pumps are frequently used for large volumes and low heads, as they are simple, occupy little space and are low in capital cost. Vertical pumps are in general use, because floor space is usually more valuable than height in engine rooms.

Reciprocating and rotary displacement pumps are self-priming, and can deal with liquid from a level below the pump. Centrifugal units require to be primed if the liquid level is below the pump. Small integral reciprocating displacement priming pumps are sometimes provided, but there is a wider use of water-ring pumps for priming purposes, because of their simplicity. Various other means, such as internal recirculation, steam ejectors, etc., are also used in appropriate cases.

Forms of drive

The size and class of the vessel and the main propulsion machinery must be considered before deciding on the type of pump drive. The alternatives available are

(a)   A.C. motor-driven pumps.
(b)   D.C. motor-driven pumps.
(c)   Turbine-driven pumps.
(d)   Direct-acting steam reciprocating pumps.
(e)   High-speed forced lubrication steam-engine-driven pumps.
(f)   Diesel-driven pumps.

Any comparison between the different methods of driving should take into consideration the efficiency of the power generation and transmission from fuel to pumping power, ease of maintenance, the fuel cost, the capital cost, flexibility, value of independent operation and the supervision required. It will be appreciated that factors vary widely and each case requires individual consideration.

Few engine rooms are completely fitted out with steam driven pumps. The usual arrangement is to use electrically driven units — either a.c. or d.c. depending on the electrical installation. The only real domain of the steam driven pump is for cargo pumping duties aboard tankers (page 109) and for boiler feeding duties (page 88).

The former application arises from safety considerations and the latter in order to ensure feed to a highly rated water tube boiler even in the event of a failure of electricity supply. It is common to find an electrically driven main feed pump installed with a turbine-driven standby unit, the turbine being arranged for automatic start-up in the event of failure of the electrically driven pump.

Main engine-driven lubricating oil and cooling water pumps have been favoured in the past but now rarely encountered apart from a few rotary pumps fitted to auxiliary diesel engines, which are usually chain driven from the crankshaft. All diesel engines have engine-driven fuel pumps but this is a special pumping application and is discussed where applicable in Chapter 6 of this book and in C.C. Pounder's 'Marine Diesel Engines', in the Butterworths marine engineering series.

When pumps are driven by d.c. electric motors, the pump output can be controlled by varying the pump speed. In the case of a.c. installations, however, this is not so easy to achieve and it is usual in such cases to find pump recirculating and cooler by-pass valves arranged to afford a measure of control. Some large pumps, notably main circulating pumps, have double-wound motors to give a choice of two speeds — say two thirds and full speed.

DISPLACEMENT PUMPS

The types of pumps found at sea under this general classification are listed in Figure 3.2. While their respective characteristics differ in detail they all bear the following main operating features if they are working under non-cavitating conditions:

1. Output is almost directly proportional to speed.
2. Output is marginally reduced at increased pressure — usually there is more slip with less viscous fluids.
3. The pump will develop a discharge pressure equal to the resistance to be overcome, irrespective of speed.
4. They are self-priming.

Figure 3.4 shows a simple single-acting reciprocating force pump, a hand-pump from which the water is forced by the action of the piston. This illustrates the principle of all reciprocating pumps. As the handle is forced down, water follows the plunger, and any water on top of the plunger is forced out. When the handle is raised, the bottom valve seats and water is forced through the valve in the plunger.

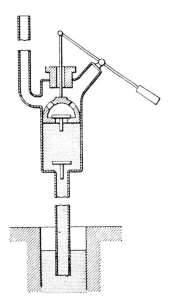

Figure 3.4 Principle of a single-acting reciprocating hand-pump

Capacity of the reciprocating pump

As the single-acting pump usually discharges water on its backward stroke through the pump cylinder, its theoretical displacement will be equal to the sectional area of the plunger or piston multiplied by the length of the stroke. Its theoretical capacity is therefore equal to the displacement multiplied by the number of strokes in a given period. This can be found by multiplying the area in square centimetres by the length of the stroke in centimetres and by the number of strokes per minute, and dividing by 1000 (1000 cm^3 = litre of water) giving the capacity in litres per minute.

For practical purposes the capacity of a single-acting pump can be found from the formula

$$L = \frac{D^2 SN}{1300}$$

where:

L = litres/min;
D = dia. of plunger in cm;
S = length of stroke in cm;
N = number of strokes per minute

The result obtained by this formula is about 3% below the theoretical displacement, but owing to possible leakage past the piston or valves, or slip past the suction valve, it is still about 3% greater than the actual discharge capacity under the best possible conditions.

The double-acting pump discharges water on both the forward and backward strokes. On the forward stroke the effective area of the piston is reduced by an amount equivalent to the area of the pump piston-rod. If this is neglected, then the double-acting pump displaces twice the amount of water of a single-acting pump of equal size.

Air and vacuum chambers

A reciprocating pump does not discharge a steady flow, but in a series of pulsations. This causes vibration and hammering at high speeds with shock to the pump and pipe fittings; reciprocating pumps are therefore usually fitted with air vessels on the discharge side of the pump in order to reduce this. At the peak of discharge the air is compressed and some water enters the air vessel. When the piston reaches the end of its stroke the air expands, discharges the water collected in the air chamber and so helps to keep to a steady rate of flow.

The internal diameter of an air vessel should be at least that of the pump bucket and its height determined by

$$\frac{\sqrt{[\text{internal diameter (cm)]}} \times \text{water pressure (bar)}}{0.65}$$

Example. If, water pressure = 7 bar and bore of vessel = 22.5 cm, then

$$\text{Height} = \frac{\sqrt{(22.5)} \times 7}{0.65} = 50 \text{ cm}$$

Similar vessels of comparable size may be fitted to the pump suction chest but since the water speeds are lower, they are often omitted if suction lines are short, particularly on bilge suctions.

It is now almost invariable practice to provide de-aerated water in order to avoid corrosion in boilers. For this reason, when a reciprocating boiler-feed pump is used in a system where the water has been de-aerated, the air vessel of normal construction cannot be used and, in general, boiler-feed pumps in such installations are without air vessels or the air vessel is provided with a float. The float minimises the absorption of the air into the feed water because it is made a fairly close fit and the water-line area subjected to the air is reduced to a small amount. The base of the float may be arranged as a valve so that when the pump is stopped, it settles on to a valve-seat and the air is trapped in the vessel.

To avoid shock in the pipelines, it is normal to design the pump steam-distribution chest to permit expansive working; in this way the water columns in the suction and discharge piping are gradually accelerated at the beginning and gradually decelerated at the end of each stroke.

Examination of Figure 3.62 will show that the steam auxiliary valve operating lever is not secured to the valve spindle but there is lost motion, i.e. the pump has moved part of its stroke before the auxiliary valve spindle is picked up. The greater the lost motion, the longer will be the stroke of the pump, up to the limit of travel.

In simplex pumps, reciprocating motion can only be obtained by introducing a free valve between the slide valve and the steam cylinder; steam reversal then takes place while the piston is still in motion, carrying the piston over the dead centre, ready for the return stroke. The essentials are shown in Figure 3.5 in which the auxiliary valve operates the shuttle which in turn distributes steam to the cylinder.

Applications of direct-acting steam pumps

Direct-acting steam-driven pumps have the advantage of extreme flexibility in operation; they have an infinitely variable capacity from no load to full load, and can be controlled manually or automatically with ease. They are self-priming, and can handle mixtures of liquid and gas when necessary. Figure 3.6 shows a simplex direct-acting pump as used for boiler feed.

ROTARY DISPLACEMENT PUMPS

Positive displacement rotary pumps have largely supplanted reciprocating pumps; they are self-priming and capable of producing high vacua. A number of types have been developed having rotors of special geometrical form. They give a steady flow but are less efficient than reciprocators because of the large areas with running clearance exposed to the differential pressure between suction and discharge. Wear increases the clearances with consequent loss of efficiency, especially when handling low viscosity fluids.

All rotary displacement pumps show the same loss characteristic if the pump is not operating under cavitation conditions.

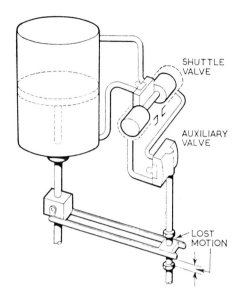

*Figure 3.5 A simplex pump shuttle
and auxiliary valves*

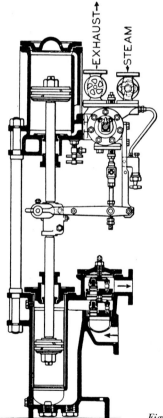

Figure 3.6 Simplex direct-acting pump (Weir Pumps Ltd.)

Pumped volume

It will readily be seen that at zero pressure the volumetric efficiency should be 100% but as the differential pressure increases the amount of leakage (slip) past the clearance will increase. This slip (the terminology normally used) will naturally be less the more viscous the pumped liquid is for any given pressure. It should also be noted that the slip, being a function of the clearance, viscosity and differential pressure is constant irrespective of running speed. In practice, changes in flow conditions affect this statement slightly.

In theory the pressure is limited only by the slip becoming so great that it equals the pump displacement and so delivery ceases. In practice overall efficiency and/or scantling strength is the usual commercial limitation.

Power

The power requirement of the pump may be split into two components, namely hydraulic power and frictional power. The hydraulic power is that required for the pumped medium. Since slip is only 'slip' because it has been pumped and then leaked back via the clearances it is only necessary to consider the pump displacement, at its running speed, and the differential pressure through which the liquid is being raised. Theoretically the rotors are not in contact, so that the frictional loss is minimal.

Acceleration forces

These are by far the greatest losses in a rotary pump and again the liquid properties are beyond the designers' control. This leaves only the distance of movement of the dynamic unit and speed in the designers' control (acceleration is a function of both).

Depth and form of tooth influence this in a gear pump. With the screw pump the pitch of the screw is the major facet. Thus high helix angle screws can be used up to a relatively high speed on small pumps and lower helix angles on larger pumps to keep the pitch within limits set by the field requirements for suction performance.

Types of rotary pump

A variety of types can be found but can be generally sub-classified as follows:

1. Screw pumps.
2. Rotary vane pumps.

3. Lobe pumps.
4. Special geometric forms.

Screw pumps

Both double-screw pumps in which the screws are driven in phase by
timing gears (Figure 3.7) and triple screw pumps, in which the centre
screw is driven and the outer screws idle (Figure 3.8) are used at sea,
especially for pumping high viscosity liquids such as oil and a variety
of other liquid cargoes. Since they are self-priming and able to pump
liquid and vapour without loss of suction they are particularly useful

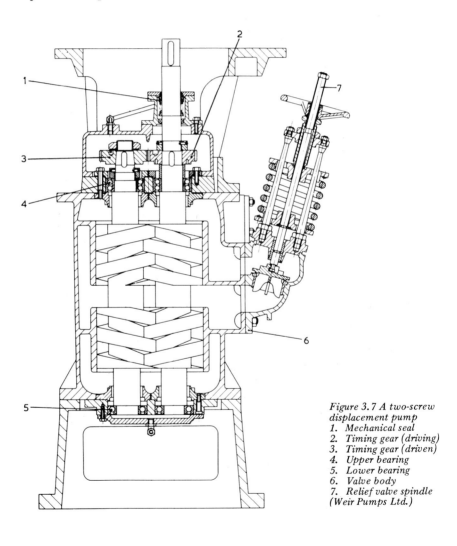

Figure 3.7 A two-screw
displacement pump
1. Mechanical seal
2. Timing gear (driving)
3. Timing gear (driven)
4. Upper bearing
5. Lower bearing
6. Valve body
7. Relief valve spindle
(Weir Pumps Ltd.)

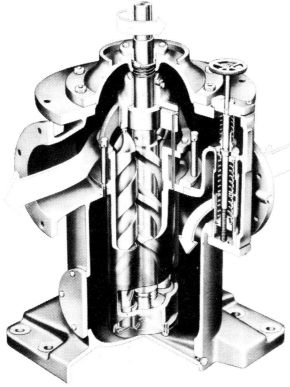

Figure 3.8 A triple-screw displacement pump (IMO Industri)

when draining tanks of high vapour pressure liquids. They are suitable for operation at high rotational speed (units are in operation with speeds of 3500 rev/min, delivering over 1000 litres/min) and can thus be easily matched with standard electric motors. Performance characteristics of screw pumps are illustrated in Figure 3.9.

In the IMO triple screw pump only the centre screw is driven mechanically, via a flexible coupling. The two outer screws are driven by the fluid pressure and act purely as seals. The screws work in a renewable cast iron sleeve mounted in a cast iron pump casing. When the screws rotate, their close relation to each other creates pockets in the helices; these pockets move axially and have the same effect as a piston moving constantly in one direction. These pumps work well at high pressure and with high viscosity fluids (up to 4000 centistokes). The axial thrust on the power rotor is balanced hydraulically, that of the idlers by thrust washers.

Double-screw pump with timing gears

This type of pump can be mounted both horizontally and vertically. Pumping is effected by two intermeshing screws rotating within a

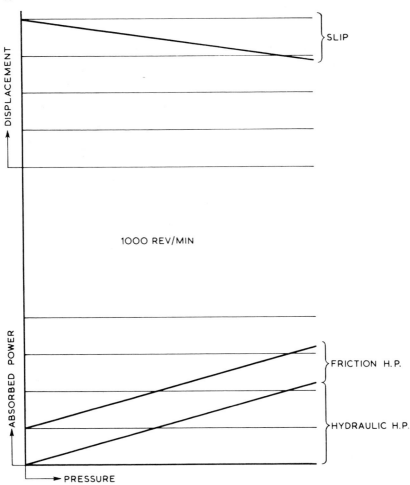

Figure 3.9 Screw pump performance characteristics

pump casing. Each screwshaft has a right and a left hand screw which ensures axial hydraulic balance, there being no load imposed on the location bearing.

Metal contact is avoided by driving the screwshaft through hardened and ground timing gears. Once the casing has been filled with liquid, the pump is self-priming and ready for operation. Displacement on pumping takes place when the screws are rotated and liquid is drawn into the screws at the outer ends and pumped inwards to discharge into the pump outlet branch, located about mid-length of the rotors, without pulsation.

When pumps are installed for handling non-corrosive liquids of

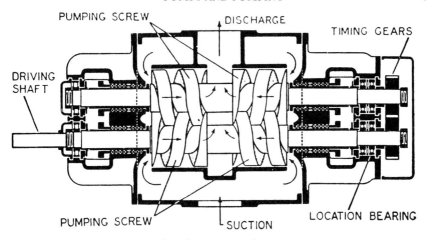

Figure 3.10 Counterscrew pump for oil or water service

reasonable lubricity, it is normal to find units with internal bearings. For handling more onerous liquids e.g. corrosive or abrasive liquids, chemicals, and liquids with a lack of lubricity and/or high viscosity, designs incorporating outside bearings are required. The latter can then be independently lubricated. (Figure 3.10).

Inside bearing pumps are to be preferred when possible, to their outside bearing counterparts as they are shorter, lighter and have only one shaft seal as against four; a separate timing gearbox is required also in the outside bearing type.

Shaft sealing

In the double-screw pump shown in Figure 3.7, shaft sealing can be effected by either mechanical seals or by packed stuffing boxes. In both cases, sealing is effected at the suction end of the pump, so that the seals are subjected only to low pressure or vacuum.

Selection of mechanical seals, sometimes with higher extra cost, is necessary to ensure leak-free operation of the pump, even under the most arduous service conditions. This is of paramount importance when handling toxic or aggressive liquids. Provision is made for the mechanical seals to be cooled and lubricated generally by the pumped liquid. Sometimes, however, external means are necessary.

Materials

For the general range of liquids handled, cast iron is suitable for the casing and bearing housings; the screwshafts are normally of high grade carbon steel. Sea-water pumps with bronze casings and stainless steel screwshafts have a longer life. Materials having high corrosion-resistant qualities such as stainless steel EN 58 J and Hastelloy may also be used in addition to the more common materials. Since the screws are not in direct contact they will not be scuffed.

Since screw pumps are essentially displacement pumps and will produce increasing pressure until rupture or drive failure occurs it is necessary to safeguard the pump, prime mover and its associated pipework in the event of a discharge line valve closure. Thus the pumps are equipped with full flow relief valves capable of bypassing the entire throughput of the pump. This is for safety purposes only, however, and should only operate for a short time otherwise excessive liquid/pump temperatures will result. The valve may be fitted with either manual or automatic control to facilitate starting up under no-load condition. This is necessary where the discharge system is under pressure, to avoid excessive starting torque (electrical load) and long run-up times.

Relief valves are also often fitted with automatic volume control valves, which control the output of the pump in order to maintain either a constant pressure or vacuum at a specific point in the system, as in diesel engine lubricating oil supply, to ensure constant pressure at engine inlet irrespective of oil viscosity.

When hot or viscous liquids are handled it may be necessary to preheat the casing of a pump already filled with liquid. Means of heating are available, e.g. electric immersion heaters or a coil through which low pressure steam or hot oil is circulated.

Stothert and Pitt Lobe pump

The Stothert and Pitt rotary pump has two types known as the Three-Four and Seven-Eight (Figure 3.11). The pumping elements consist of an inner and an outer rotor, which rotate in a renewable liner fitted in the pump body. The inner rotor is eccentric to the outer and is fitted to a shaft located by bearings in the pump covers. Rotation of the inner rotor creates a pocket of increasing capacity between the rotors on the rising side and a corresponding decrease on the falling side; displacement is thus effected, and the pumping action draws the fluid into the pump through ports in the outer rotor while the pocket is increasing in size and forces it out of the opposite side of the pump when the pocket is decreasing.

The normal range of these pumps covers pressures rising to 21 bar and capacities up to 400 tonne/hr.; special designs have been made for pressures rising to 83 bar for use in connection with hydraulic

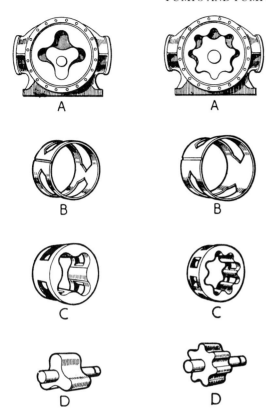

A

A

B

B

C

C

D

D

Figure 3.11 Lobe displacement pumps
Pumping elements of (left) Three-Four type for high-viscosity fluids and (right) Seven-Eight type for low viscosity fluids
(Stothert & Pitt Ltd.)

control. The Three-Four types are particularly suited for handling high-viscosity fluids and are set to run at comparatively slow speeds which usually call for the introduction of a gear between the prime mover and the pump. The Seven-Eight types are to meet the requirements of high-speed machinery, and operate efficiently at speeds around 720 rev/min when handling fluids in the lower viscosity range.

CENTRIFUGAL PUMPS

The centrifugal pump converts energy (external driving source) into kinetic energy in the fluid by giving impulses to the fluid by impellers. A diffuser or volute converts most of the kinetic energy into pressure, i.e. these machines use the basic law (Bernoulli's) of converting potential energy into kinetic energy and vice versa. The equation being:

Total energy (referred to unit weight of liquid) is

$$H = Z + \frac{P}{\gamma} + \frac{V^2}{2g}$$

where:

H =   total head at a given point.

Z =   height above datum.

$\dfrac{P}{\gamma}$ =   pressure head.

$\dfrac{V^2}{2g}$ =   dynamic or velocity head.

γ =   specific gravity (1 for fresh water)

Figure 3.12 shows the main types of centrifugal pump and whilst there are many other kinds of centrifugal pumps, they are usually hybrids of those shown.

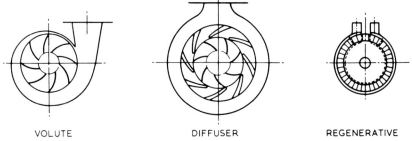

VOLUTE   DIFFUSER   REGENERATIVE

Figure 3.12 Types of centrifugal pump (Hamworthy Engineering Ltd.)

The volute pump is the most common, being found in large numbers whilst the diffuser pump is met mainly on high pressure pumps, usually multi-stage, such as boiler feed.

The regenerative pump is used where a relatively high pressure and small capacity are required.

From a mathematical consideration of the action of a centrifugal pump it can be shown that the theoretical relationship between head H and throughput Q is a straight line (Figure 3.13), with minimum throughput occurring when the head is maximum. Because of shock and eddy losses caused by impeller blade thickness and other mechanical considerations there will be some head loss, increasing slightly with throughput. These losses, together with friction losses due to fluid contact with the pump casing and inlet and impact losses, result in the H/Q curve shown in the figure. The final shape of this curve will vary according to the design of the pump. Depending on application, centrifugal pumps can be designed with relatively flat H/Q curves or if required the curve can be steep to give a relatively large shut-off head.

From Figure 3.13 it can be seen that minimum power occurs when there is no flow and when the discharge head is at its highest — in other words when the discharge valve is closed. Since throughput

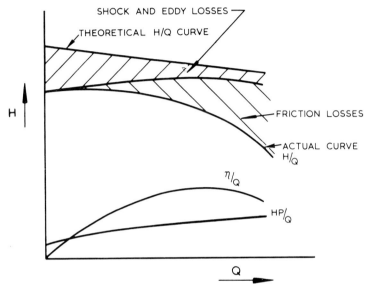

Figure 3.13 Showing centrifugal pump action by mathematical relationship (Hamworthy Engineering Ltd.)

decreases as the discharge head is increased there is no necessity to fit a relief valve to centrifugal pumps. It will also be noticed that the efficiency curve for the pump is convex which means that maximum efficiency occurs at a point somewhere between maximum and minimum discharge head and throughput conditions.

In the case of a variable speed pump:

1. Head varies as the square of the speed.
2. Capacity varies directly as the speed.
3. Power varies as the cube of the speed since it is a function of head and capacity.

In the case of a constant speed pump:

1. Head varies as the square of the diameter.
2. Capacity varies as the diameter.
3. Power varies as the cube of the diameter.

Where the head in a given installation is known, the following formula will be found of use in calculating the speed of the pump:

$$N = \frac{95 \ H.C}{D}$$

where:

N = rev/min.
D = dia. of impeller over blade tips in m.
H = total head in m.
C = constant.

The value of C varies considerably according to the hydraulic design of the pump but is generally between 1.05 and 1.2, the higher value being taken for pumps working considerably beyond their normal duty, or for pumps with impellers having small tip angles.

CONSTRUCTION AND INSTALLATION

Marine pumps are usually vertical with the motor above the pump to keep it clear of moisture. This gives the best possible NPSH conditions and takes up the least floor space.

Shaft sealing.

It is preferable on vertical pumps to have shaft sealing at the pump upper end only. This allows for:

(a) Adjustment; if soft packed,
(b) Observance of conditions (i.e. leakage) if mechanically sealed,
(c) Containment of liquid in the pump, over long idle periods, for wetting of bearings etc. on start.

Soft packed stuffing boxes predominate and are short with only four or five rings. Except in feed and circulating pumps, the stuffing box is exposed to low head, if situated appropriately. If however, the pump suction line is working under vacuum conditions, it is usual to have a grease packed sealing lantern beneath the stuffing box. Mechanical seals are used increasingly. Figure 3.14 shows some typical arrangements of both types.

It is important that cooling/lubricating liquid is led to mechanical

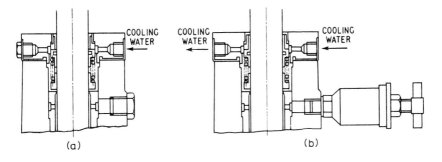

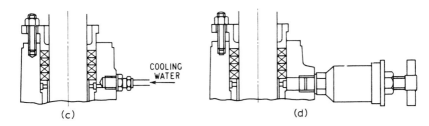

Figure 3.14 Typical arrangements of pump seals
(a) Water lubricated bearing   (c) Water lubricated bearing, soft packed
(b) Grease lubricated bearing   (d) Grease lubricated bearing, soft packed
(Hamworthy Engineering Ltd.)

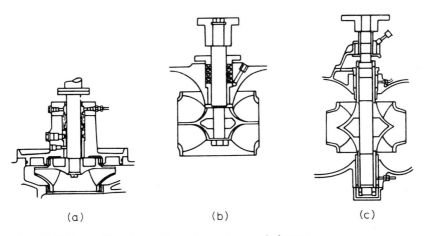

Figure 3.15 Types of bearing configuration used on vertical pumps
        (a) Rigid coupled pump with top bearing
        (b) Rigid coupled pump without bottom bearing
        (c) Flexible coupled pump with top and bottom bearing
(Hamworthy Engineering Ltd.)

seals from the lowest point on the pressure side of the pump, to ensure that some liquid reaches them, even when priming. They must not run in an air pocket and care must be taken to prevent ingress of foreign matter. Also, most mechanical seals incorporate a carbon face and there is a possibility of electrolytic action. For these reasons, soft packing has advantages in sea-water pumps.

Figures 3.15(a), (b) and (c) show some of the bearing configurations used on vertical pumps, the trend being to eliminate the bottom bearing. However, where design dictates a bottom bearing an internal one is preferable to eliminate the stuffing box.

It is possible to have internal bearings lubricated/cooled by the pumped liquid where liquid is always available when the pump is running and the same materials are successful with controlled applications of grease when used on pumps running under dry conditions. Should the dry running period be short, as for a fresh water cooling pump no external lubrication is necessary.

Construction

The construction of the pump varies according to the purpose for which the pump is to be used i.e.:

(a) *Non-salt water pumps (potable, engine cooling water etc. but excluding boiler feed pumps).* High grade cast iron is used for the casings with bronze internals, the shaft material being either bronze or stainless steel (EN57. 18 Cr/2Ni) the latter material giving the better wear life.

(b) *Seawater pumps.* (These must, of course, also handle harbour, river and canal water.) It is normal to use all-bronze pumps (zinc free), the casing being gunmetal, the impeller aluminium bronze (BS 1400 AB2 9: 5A1/5 Fe/5.5 Ni) and the shaft material either stainless steel (EN57) for soft packed stuffing boxes, or EN58J (18/10/3 Cr. Ni. Mo) under mechanical seals or bearings. Stainless steel shafts and stainless steel impellers for certain duties are becoming more appreciated.

(c) *Boiler feed pumps.* Because of the high pressures and temperatures involved these invariably have cast steel casings (0.25% carbon steel) with stainless steel shafts and impellers. Highly rated turbo-driven pumps have 3% Ni-Cu Mo forged alloy steel shafts.

AIR HANDLING

Because of roll and pitch, marine pumps at times have to handle very highly aerated water, even under flooded suction conditions in only moderate weather; the amount of air can be sufficient to air-lock a non-self priming pump if the water inlets or suctions are not well placed. Expansion of the air at the pump entry and its subsequent compression in the pump, gives rise to noise similar to cavitation, especially in positive displacement pumps, where compression is rapid. Also, it can be very destructive of pump and pipe materials, by corrosion, erosion or both.

Pumps may be mounted above the level of the liquid to be pumped even though placed low in the ship, and they must be equipped with means to create a vacuum in the pipeline. Some others must be similarly equipped so that the maximum amount of liquid can be extracted from the tanks, bilges, etc. To achieve this, the air handling facilities must be good.

As the velocity of the outer tips of the impeller of a centrifugal pump is relatively low, the suction effort of the pump, when empty, rarely exceeds 12 mm water gauge and a centrifugal pump must be primed with water as it cannot exhaust the contained air as a displacement pump can. When the pump is placed below the level of the water, as in marine circulating sets, the filling is effected by opening the injection valve on the ship's side and the air-cock on the top of the pump casing. The pump is fully primed when solid water emerges from the air-cock.

Air handling methods

(a) *Recirculation of discharge.* This is a very inefficient method.

(b) *Liquid ring primer* (see Figure 3.16). This type is most frequently used. Its air handling capacity is good, the extracted air being vented to atmosphere, although it can be used to pump against pressure. It is used as a gas compressor and as an air exhauster. Air gulps are quickly cleared, small air leakages and aerated water are continuously handled without fall in pump performance.

The liquid ring air pump consists of a bladed circular rotor shrouded on the underside, rotating in an oval casing. Sealing water is drawn into the whirlpool casing through a make up supply pipe. The water follows the periphery of the casing due to the centrifugal force imparted to it by the rotor and the 'water-ring', revolving eccentric to the blade recedes from and re-approaches the rotor boss twice in one revolution, thus producing in effect a series of reciprocating water pistons between the blades. The inner edge of the water-ring forms the boundary of two eccentric cores round the rotor boss, while the blades run full of water — A and B (Figure 3.16).

Assuming the space between each blade to be a cylinder, then in one-half revolution the water is thrown from F out to G and back again to F, constituting one suction and one discharge stroke and this occurs twice in one revolution. It will be understood therefore, that

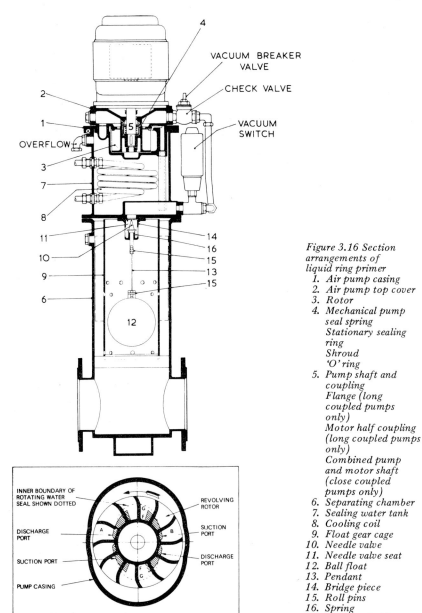

Figure 3.16 Section
arrangements of
liquid ring primer
 1. Air pump casing
 2. Air pump top cover
 3. Rotor
 4. Mechanical pump
    seal spring
    Stationary sealing
    ring
    Shroud
    'O' ring
 5. Pump shaft and
    coupling
    Flange (long
    coupled pumps
    only)
    Motor half coupling
    (long coupled pumps
    only)
    Combined pump
    and motor shaft
    (close coupled
    pumps only)
 6. Separating chamber
 7. Sealing water tank
 8. Cooling coil
 9. Float gear cage
10. Needle valve
11. Needle valve seat
12. Ball float
13. Pendant
14. Bridge piece
15. Roll pins
16. Spring
(Weir Pumps Ltd.)

if shaped suction and discharge ports are provided in way of the path
of the eccentric cores formed by the rotating water, air will be drawn
through the suction ports and expelled through the discharge ports,
as each blade passes the ports. Such ports are arranged in the
stationary rotor plate fitted in the cover above the rotor.

In each revolution, therefore, the water recedes from the rotor
boss, drawing air through the suction ports in the rotor plate into the
eccentric cores of the water-ring, from whence it is forced through
the discharge ports in the rotor plate after the points of maximum
throw-out at G have been passed and the water re-approaches the
rotor boss. A continuous supply of sealing water is circulated from
the reservoir to the whirlpool casing, and is discharged with the air
back to the reservoir. (The air passes to atmosphere through the overflow pipe.) This circulation ensures that a full 'water-ring' is maintained, and the cooling coil incorporated in the reservoir limits the
temperature rise of the sealing water during long periods of operation.
The supply for the cooling coil can be taken from any convenient
seawater connection. About 0.152 litres/sec is required at a pressure
not exceeding 2 bar.

(c) *Ejector.* If sized correctly these are effective; efficiency is low.

(d) *Eccentric vane primer.* Has good air handling but life can be
short due to wear of vane tips and jamming in slots.

Central priming system

This system is often used when more than four pumps require
priming facilities. It gives a large air exhausting reservoir as well as a
capacity greater than individual pumps can carry and the pump
casings can be filled with liquid before starting. The air exhausting
units can be of any type but are usually of the liquid ring type. A
typical schematic arrangement is shown in Figure 3.17.

Air interceptor vessels

The main purpose of these is to close off the passage to the air
exhausting unit when the water has risen to a pumpable height. This
height is preferably a minimum of 300 mm above the impeller eye.
The air passage valve orifice must be sufficiently large to avoid
restricting the air flow. A top atmospheric breaker valve may be
fitted to carry out two functions. The operation is as follows:

(a) The moving spindle of the lower assembly contacts the upper
     valve assembly just before seating. Because of atmospheric

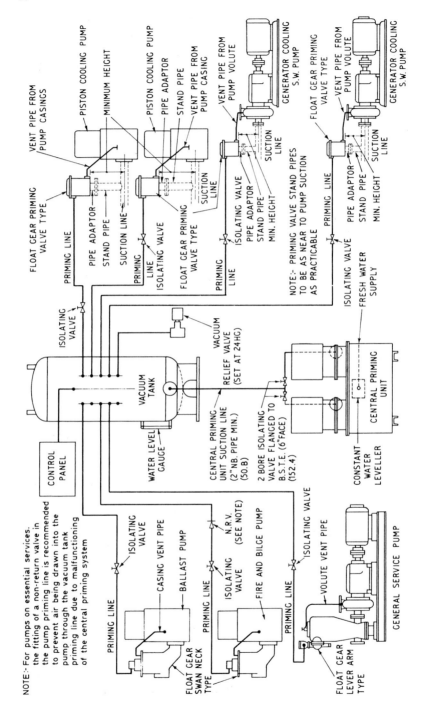

Figure 3.17 Layout showing central priming system (Hamworthy Engineering Ltd.)

pressure, the upper valve assembly resists the rise of the lower valve float assembly until the liquid level rises further up the float and creates extra buoyancy. This lifts the upper valve off its seat and the lower valve is then pushed firmly home by the excess buoyancy.

(b)   Once the upper valve is off its seat, atmospheric air flows easily to the priming unit and puts it on its lightest load condition. The upper valve must not be used with a central priming system.

GENERAL PURPOSE PUMPS

Single entry general purpose pumps

Used for salt and fresh water circulating, bilge and ballast etc., a typical configuration is shown in Figure 3.18. It will be noted that the impeller is suspended from the shaft and is not supported from below. A neck bush provides lateral support. The eye of the impeller faces downwards, i.e. the inlet is below the impeller. A renewable bush or wearing ring is located around the impeller boss, the design clearance between boss and bush being such that liquid in the discharge volute does not return to the suction side. In this design internal access for maintenance is via the top cover. A distance piece is arranged in the shaft which, when removed, permits impeller, shaft and cover to be lifted out without disturbing the motor or the pipework.

Figure 3.19a shows a different design of pump intended for the same duties, giving a throughput of 425 m^3/hr against heads of up to 54 m. The impeller is arranged with its eye uppermost, the suction branch being elevated. This arrangement is claimed to give better venting to eliminate any possibility of vapour locking. Another significant design difference is that the casing is split vertically so that the impeller and shaft can be removed sideways. Again wearing rings are fitted to prevent leakage back to the suction. To prevent the wearing rings and neck bush slipping around in the casing lips are provided which abut with the removable part of the casing.

Where a single entry pump is to be used to supply a greater pressure head an impeller of bigger diameter is used (Figure 3.19b). In this particular model which delivers up to 260 m^3/hr at heads up to 84 m, the casing is again split vertically and the impeller eye faces upwards. Because of the greater diameter of the impeller however it has been considered necessary to add a lower guide bush. To ensure a flow of water to the bottom bush a service pipe is arranged from the pump suction chamber to the bush housing. Holes drilled close to

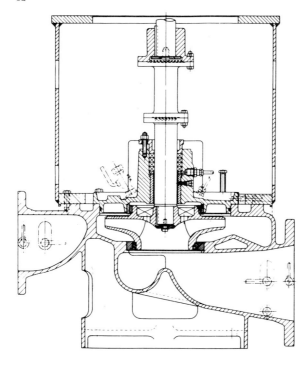

Figure 3.18
Typical single stage
centrifugal pump
(Hamworthy
Engineering Ltd.)

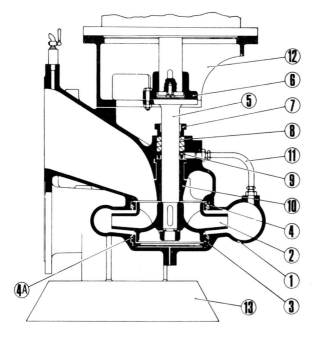

Figure 3.19a Single-
entry pump giving
throughput of 425
m³/hr against heads
of up to 54m
   1. Pump casing
       and cover
   2. Impeller
   3. Casing ring
       (bottom)
   4. Casing ring
       (top)
   4A. Locking pins
   5. Pump spindle
   6. Coupling
       (motor half)
   7. Gland
   8. Packing
   9. Lantern ring
       (split)
  10. Neck bush
  11. Water service
       pipe to stuff-
       ing box
  12. Motor stool
  13. Pump foot
(Wier Pumps Ltd.)

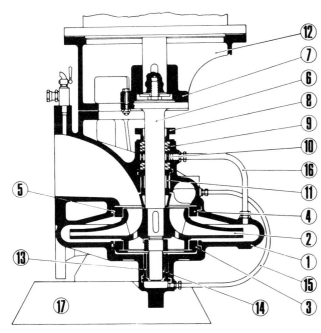

1. Pump casing
   and cover
2. Impeller
3. Casing ring
   (bottom)
4. Casing ring
   (top)
5. Locking pins
6. Pump spindle
7. Coupling
   (motor half)
8. Gland
9. Packing
10. Lantern ring
    (split)
11. Neck bush
12. Motor stool
13. Bottom bush
    housing
14. Bottom bush
    liner
15. Water service
    pipe to
    bottom bush
16. Water service
    pipe to stuff-
    ing box
17. Pump foot
(Weir Pumps Ltd.)

Figure 3.19b Large diameter impeller for large pressure head

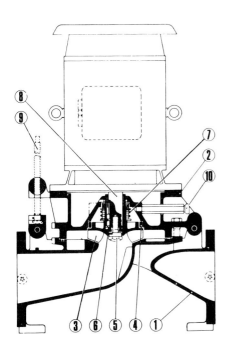

Figure 3.20 Single-entry pump with
open-sided impeller
   1. Pump casing
   2. Casing cover
   3. Impeller
   4. Casing ring
   5. Impeller locking screw
   6. Shims
   7. Mechanical seal
   8. Combined pump and motor shaft
   9. Screw jack
10. Air release plug
(Weir Pumps Ltd.)

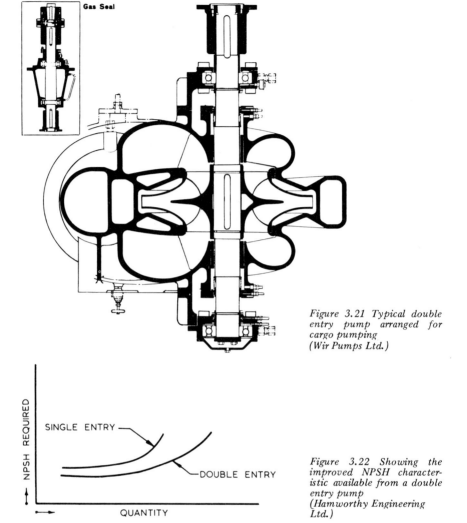

Figure 3.21 Typical double entry pump arranged for cargo pumping (Wir Pumps Ltd.)

Figure 3.22 Showing the improved NPSH characteristic available from a double entry pump (Hamworthy Engineering Ltd.)

the impeller boss connect the space above the bush housing with the suction chamber.

A somewhat different design of single entry pump is shown in Figure 3.20. In this pump the impeller eye faces downwards but the impeller is open-sided, the bottom of the pump casing effectively shrouding the vanes. This design allows the motor and cover of the pump to be hinged so that operation of a simple screw jack exposes all internal parts. A mechanical seal prevents water leakage or air ingress. Such pumps are available for capacities of up to 260 m³/hr at heads of up to 91 m.

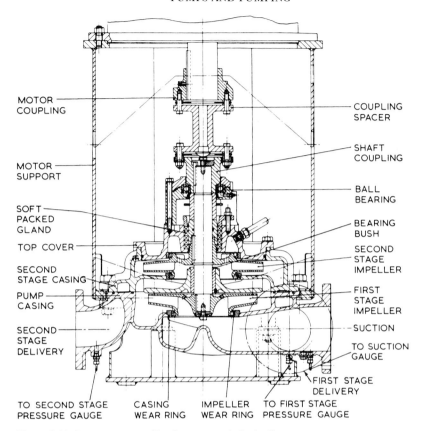

MOTOR
COUPLING

COUPLING
SPACER

MOTOR
SUPPORT

SHAFT
COUPLING

BALL
BEARING

SOFT
PACKED
GLAND

BEARING
BUSH

TOP COVER

SECOND
STAGE
IMPELLER

SECOND
STAGE CASING

PUMP
CASING

FIRST
STAGE
IMPELLER

SECOND
STAGE
DELIVERY

SUCTION

TO SUCTION
GAUGE

FIRST STAGE
DELIVERY

TO SECOND STAGE   CASING   IMPELLER   TO FIRST STAGE
PRESSURE GAUGE   WEAR RING   WEAR RING   PRESSURE GAUGE

Figure 3.23 A two-stage centrifugal pump, typical of a fire pump
(Hamworthy Engineering Ltd.)

A double entry pump may be used for larger capacities. This pump
is, fundamentally, two single stage impellers placed back to back
(Figure 3.21). Figure 3.22 shows the better NPSH available from a
double entry pump.

Figure 3.23 shows a two-stage pump, typical of a fire pump. It can
readily be used for a double duty of, say, bilge/fire, i.e. a lower head
duty by pumping through the first stage impeller only and a higher
head duty when pumping through both impellers.

Axial flow pumps

An axial flow pump is one in which a screw propeller is used to
create an increase in pressure by causing an axial acceleration of
liquid within its blades. The velocity increase is then converted

into pressure by suitably shaped passages in the propeller and in the outlet guide vanes.

Axial flow pumps are often placed in the same class as centrifugal pumps although centrifugal force plays no part in the pumping action. As can be seen from Figures 3.24 and 3.25 the H/Q and working efficiency characteristics of these pumps are somewhat different from those of centrifugal pumps.

Drawn in each case for constant speed the curves show:

1. The axial pump power decreases as the delivery increases while the centrifugal pump has a rising absorbed power with delivery although there is a marginal fall off beyond normal duty.
2. The axial pump retains reasonable efficiency over a wider head range.
3. At zero discharge the absorbed power of the axial pump is much greater than at normal duty — the pump needs to be unloaded at lower than normal throughputs or the pump motor must be over-rated.

There are three other features of the axial flow pump not indicated by the graph but of particular importance in their application. These are:

1. Under low head, high throughput conditions — e.g. 2.5–6.2 m head and 2800–9500 m³/h as commonly required by main condensers — an axial flow pump will run at a higher speed than an equally matched centrifugal pump, i.e. it can be driven by a smaller motor.
2. The pump will idle and offer little resistance when a flow is induced through it by external means.
3. The pump is reversible.

This combination of characteristics makes the axial flow pump ideal for condenser circulating duties, especially in conjunction with a scoop injection (a system whereby the motion of the ship under normal steaming conditions is sufficient to induce a flow through the idling pump and the condenser).

In addition its reversability and high throughput make it ideal for heeling and trimming duties.

When used for sea water circulation the pump will normally have a gunmetal casing; for heeling and trimming applications it is more usual to find pumps with cast iron casings. Impellers are of aluminium bronze, guide vanes of gunmetal and the shaft is of stainless steel, with a renewable stainless steel sleeve in way of the bush. A typical axial flow pump is shown in Figure 3.26.

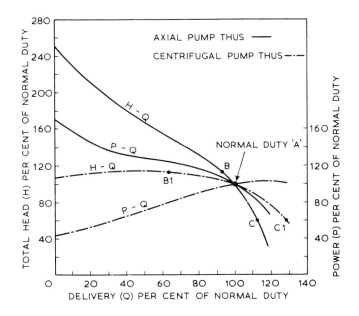

Figure 3.24
Head/quantity
curves, at con-
stant speed, of
axial-flow and
centrifugal
pumps

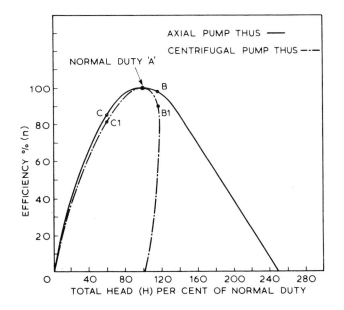

Figure 3.25
Working efficiency
of axial-flow and
centrifugal pumps
where the head
varies within large
limits

Figure 3.26 812/812 mm inclined axial flow pump
(Figures 3.26 to 3.48 are reproduced by courtesy of Weir Pumps Ltd.)

PUMPS FOR SPECIAL PURPOSES

The pumps so far described above each have their own range of application but there are some specific applications on board ship which require special pump designs. These include boiler feed, condensate extraction, lubrication, cargo-discharge and de-ballasting.

Boiler feeding is a very important pump duty. For small capacities the reciprocating pump is still widely fitted, because of its flexibility in capacity and efficiency, but for large capacities and high powers, the turbo-feed pump is used. All reciprocating pumps have an intermittent discharge and reversal of flow into and out of their chambers, although a multiplicity of chambers gives a comparatively even flow. Nevertheless, the intermittent flow limits the speeds which can be used without troublesome hammer action arising.

In centrifugal pumps the flow of water through the pumping passages is unidirectional and continuous, thereby enabling much higher speeds to be used. In reciprocating pumps the speed of water following the buckets is from 0.6 to 0.9 m/sec depending on the duty and suction conditions. In two-throw or three-throw pumps speeds of 1.1 m/sec can be used, whereas in a high-speed centrifugal pump the water passes through the impeller at speeds up to 15.1 m/sec. The dimensions of the respective types of pump to deal with the same volume of water are roughly proportional to the speed of water passing through the pump.

Weir reciprocating boiler feed pump

A sectional view of a Weir reciprocating boiler feed pump is shown in Figure 3.6. In pumps 125 mm dia and under, Kinghorn-type flat disc valves are fitted. For 150 mm dia pumps, Weir patent flat-faced group valves are fitted, both valves and seats being of special stainless steel. For 175 mm dia pumps and over, similar patent flat-faced group valves are fitted, with stainless-steel seats expanded into gun-metal decks.

Temperature and suction head. At temperatures up to 79°C the direct acting feed pump will draw from a source below its level, but for temperatures above that figure it is necessary that there should be a head of water over the suction branch of the pump to ensure satisfactory working. Table 3.3 gives particulars of the suction lift and head for various temperatures:

Table 3.3 Suction lift and head

Temp. °C	Suction lift (m)	Temp. °C	Head over suction valve (m)
54	3.04	87	1.5
65	2.1	93	3.04
76	0.61	98	4.56
79	0	104	6.7

Weir steam slide-valve chest

The Weir steam slide-valve chest is shown in section (Figures 3.27 and 3.28). The chest contains a main and an auxiliary valve. The main valve A distributes steam to the cylinder. The auxiliary valve B has two functions: it distributes steam to work the main valve; and its outer edge cuts off steam entering the main ports C and D leading to the top and bottom of the cylinder.

Both valves are simple slide valves; but the main valve is cylindrical. On the back of the main valve a flat face is formed upon which the auxiliary valve works. On this face the ports C, D, E, F are cut, with the exhaust ports H in the centre (Figure 3.27). The ends of the main valve are fitted with loose bells or cylinders in which the valve

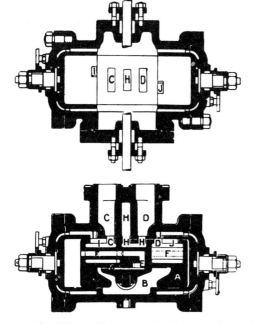

Figure 3.27 Steam slide-valve chest, sectional elevation showing bypass ports I and J

Figure 3.28 Plan view of Figure 3.27

works. These bells are held in position by the end covers and by faces cast on the chest for the purpose.

The operation of the valves during a double stroke of the pump is as follows: When the piston is at the bottom of the stroke the main valve is in the right-hand position (Figure 3.28), the auxiliary valve B is also at the bottom of its travel, and the port C leading to the bottom of the cylinder is open to the steam pressure. This port remains open until the piston reaches half-stroke, when the auxiliary valve B begins to move in the same direction as the piston, and at about three-quarters stroke the auxiliary valve closes the port C leading to bottom of cylinder. The remainder of the stroke is completed by the expansion of the steam already shut in the cylinder, or by more steam admitted through the bypass, which will be described later. Port E leads to the left-hand end of the main valve, and port F to the right-hand end. When the piston reaches the top end of its stroke, port E is opened to exhaust by the auxiliary valve B, and at this position port F is open to steam; the main valve is therefore thrown over until the exhaust from the left-hand end is cut off; the steam remaining after cut-off acts as a cushion and prevents the main valve from hitting the end cover. The main valve is now at the opposite end of its stroke, and the port C, which admitted steam to move the piston on the up stroke, is open to exhaust. The port D leading to the top of the piston is now open to steam.

The action described above also takes place on the down stroke; the main piston moves half its stroke before beginning to move the auxiliary slide which again cuts off steam at about three-quarters of the stroke. The stroke is completed by the expansion of the steam in the cylinder, or by additional steam admitted through bypasses.

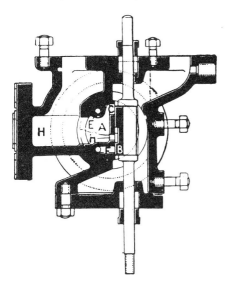

Figure 3.29 (left) Sectional side elevation of steam slide-valve chest

Figure 3.30 Auxiliary valve

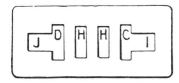

Figure 3.31 Main valve face

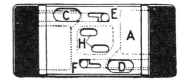

Figure 3.32 Main valve back face for auxiliary valve

Bypasses. Under certain conditions the pump will not complete its stroke by the expansion of the steam in the cylinder; for instance, if the pump were started with the cylinder cold, the steam would rapidly condense and fall below the pressure necessary to move the piston. In such circumstances it is necessary to admit steam after the auxiliary valve is closed to main ports C and D on the face of the main valve, and this is done by means of the bypass on each end of the chest. These bypasses I and J (Figure 3.27 and 3.31) are made by cutting a port on each of the bells and a corresponding port on the back of the main valve; and provision is made on each end of the steam chest for opening and closing the ports by hand.

Water-stop valve-chest

As an example of the construction and arrangement of Weir water-stop valves, Figure 3.33 shows, partly in section, a standard suction-stop valve-chest for one feed pump arrangement.

The valve-chest and covers are of cast iron, fitted with stainless-steel seats, securely fixed, gunmetal valves, brass nuts and glands with bronze spindles. The suction valves are of the screwdown non-return type.

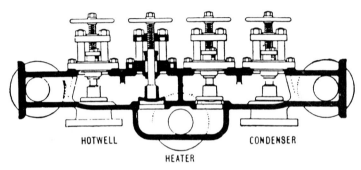

Figure 3.33 Suction stop valve-chest partly in section

In the suction chest shown for this arrangement there is a division, and it should be noted that each pump can draw only from the valves on its own side of this division. The discharge valves are of the screwdown type.

Buckets

Figure 3.34 shows the Weir Express bucket. The bucket A is of gunmetal, in which two grooves are turned. Rings BB, of specially manufactured ebonite, are fitted into these grooves, and are cut at an angle CC.

Automatic control of direct-acting feed pump

Figure 3.35 shows the design of the Weir automatic steam controller. The discharge pressure of the pump acts on a diaphragm against the bias of the spring until the spring load and the pressure load are in equilibrium. As shown, the valve is in the full-open position, and the pressure applied under the diaphragm causes the valve to move upwards, partially closing the passage for steam flow, thereby controlling the steam supply to the feed pump.

For any given boiler load, the discharge pressure of the pump finds a position of equilibrium against the spring load with an

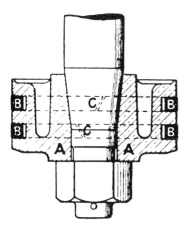

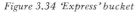

Figure 3.34 'Express' bucket

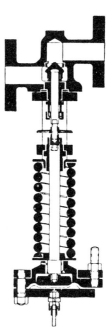

Figure 3.35 (right) Automatic steam controller for reciprocating pump

opening of the steam valve, which passes the quantity of steam required to drive the feed pump to give the discharge pressure corresponding to the given characteristic at the required feed flow.

Weir multi-stage turbo feed pump

The water tube boiler with its low storage capacity in relation to its steaming capability, demands a steady rate of feeding and pure feed water. The centrifugal pump, turbine driven, meets the requirements and is robust and reliable. Cycle efficiency demands the use of exhaust in feed heating and the turbine gives uncontaminated exhaust.

The successful development of water lubricated bearings and their use in the Weir turbo feed pump permitted the turbine and pump to be close-coupled in a very compact unit, Figure 3.36. During normal running, a multiplate restriction orifice allows feed water from the first stage impeller discharge to flow through a two-way non-return valve and a strainer to the bearings. A relief valve is incorporated. A secondary supply of lubricating water is introduced through the two-way non-return valve from an outside source such as the main condensate extraction pumps, to protect the bearings from damage during starting, stopping, or standby periods of duty. Figure 3.37 shows the arrangement of the lubricating water system.

94

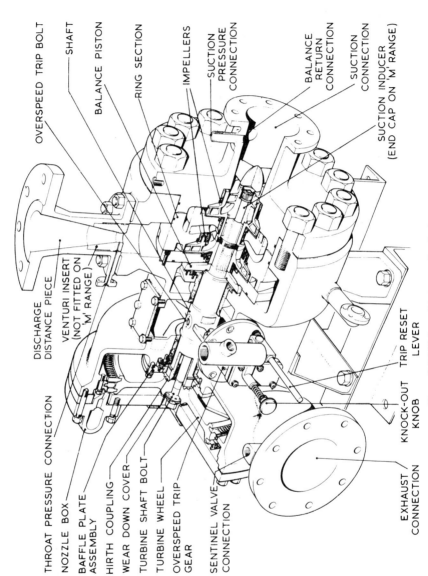

OVERSPEED TRIP BOLT

SHAFT

BALANCE PISTON

RING SECTION

IMPELLERS

SUCTION PRESSURE CONNECTION

BALANCE RETURN CONNECTION

SUCTION CONNECTION

SUCTION INDUCER (END CAP ON 'M' RANGE)

DISCHARGE DISTANCE PIECE

VENTURI INSERT (NOT FITTED ON 'M' RANGE)

THROAT PRESSURE CONNECTION

NOZZLE BOX

BAFFLE PLATE ASSEMBLY

HIRTH COUPLING

WEAR DOWN COVER

TURBINE SHAFT BOLT

TURBINE WHEEL

OVERSPEED TRIP GEAR

SENTINEL VALVE CONNECTION

EXHAUST CONNECTION

KNOCK-OUT KNOB

TRIP RESET LEVER

Figure 3.36 Weir multi-stage turbo-feed pump

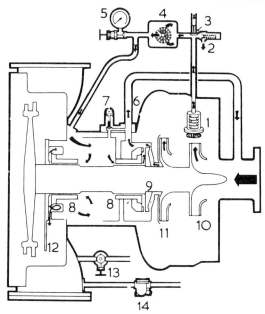

Figure 3.37 Lubricating water system

1. *Multi-plate restriction*
    orifice
2. *Relief valve*
3. *Two-way non-return*
    valve
4. *Strainer*
5. *Pressure gauge*
6. *Balance chamber*
    leakage

7. *Leak-off control valve*
8. *Bearing*
9. *Balance piston*
10. *1st stage impeller*
11. *2nd stage impeller*
12. *Baffle plate*
13. *Drain valve*
14. *Drain trap*

(Weir Pumps Ltd.)

The overspeed trip is triggered by a spring loaded unbalanced bolt mounted in the shaft between the two journals.

Pressure governor. A discharge-pressure operated governor (Figure 3.38) and a rising head/capacity curve from full load to no load gives inherent stability of operation. The main feature of the governor is that if the pump loses suction the steam ports are opened wide, allowing the pump to accelerate rapidly to the speed at which the emergency trip acts.

An adjusting screw, the collar of which bears against a platform in the casing, is threaded into the upper spring carrier and allows the compression of the spring to be altered by varying the distance between the upper and lower spring carriers.

The piston, fitted with an O-ring and a spiral back-up ring, slides in a close fitting liner. A flange on this liner locates in a recess in the governor casing cover and is sealed by an 'Armco' iron joint ring.

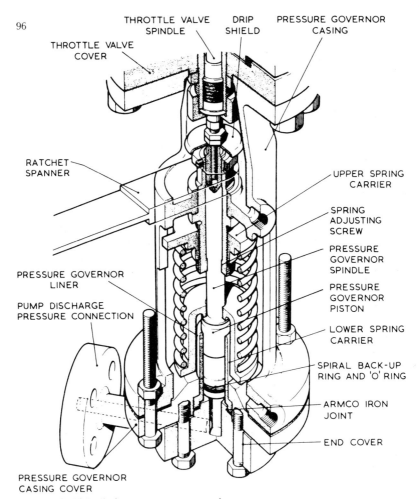

THROTTLE VALVE SPINDLE

DRIP SHIELD

PRESSURE GOVERNOR CASING

THROTTLE VALVE COVER

RATCHET SPANNER

UPPER SPRING CARRIER

SPRING ADJUSTING SCREW

PRESSURE GOVERNOR SPINDLE

PRESSURE GOVERNOR LINER

PRESSURE GOVERNOR PISTON

PUMP DISCHARGE PRESSURE CONNECTION

LOWER SPRING CARRIER

SPIRAL BACK-UP RING AND 'O' RING

ARMCO IRON JOINT

END COVER

PRESSURE GOVERNOR CASING COVER

Figure 3.38 Weir discharge-pressure operated governor

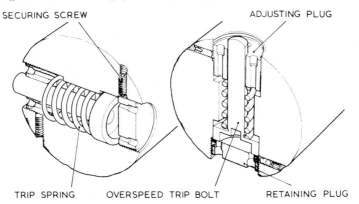

SECURING SCREW

ADJUSTING PLUG

TRIP SPRING   OVERSPEED TRIP BOLT

RETAINING PLUG

Figure 3.39 Overspeed trip (bolt type)

When the pump is started, the throttle valve moves upwards due to the increasing discharge pressure under the piston until the desired pressure is reached. At this point, the upward force exerted on the governor spindle by the piston is equal to that being applied downwards by the spring, and the throttle valve is admitting the correct quantity of steam to the turbine to maintain the desired discharge pressure. With an increased demand on the pump the discharge pressure falls, allowing the pressure governor piston to move down under compulsion of the spring until the throttle valve opens to provide steam to satisfy the new demand. The reverse action takes place when the demand decreases.

Safety (overspeed) trip. Figure 3.39 shows a 'Bolt' type overspeed trip. This consists essentially of a spring-loaded stainless steel bolt which, due to its special design, is heavier at one end than the other. The rotary motion of the turbine shaft tends to move the bolt outwards, while the spring retains it in its normal position until the turbine speed reaches a predetermined safety level. At this speed the centrifugal force exerted by the heavier end of the bolt overcomes the spring opposing it and the bolt moves outwards to strike the trip trigger. This in turn disengages the trip gear, allowing the steam stop valve to shut.

Turbine-driven oil-lubricated pump

Prior to the introduction of the water lubricated turbo-feed pump an oil-lubricated pump with a relatively long horizontal shaft was commonly used. In this particular pump the overspeed trip was of the 'Ring' type as shown in Figure 3.40. The pump was mounted on a taper at the end of the turbine shaft adjacent to the trip gear, and was secured by a mild steel set bolt tapped into the end of the shaft. This set bolt was locked in place by a copper lock-washer. The overspeed mechanism consisted of a case-hardened steel ring which, bored eccentrically, was weighted off-centre. The ring, however, was spring loaded to maintain concentricity with the shaft until the speed of the turbine reached a predetermined safety limit. At that speed the centrifugal force exerted by the ring would overcome the force of the opposing spring and would then move outwards to strike the trip trigger.

Hydraulic balance mechanism

To control the axial movement of the rotating assembly, a balance piston is arranged to counteract the effect of the thrust of the

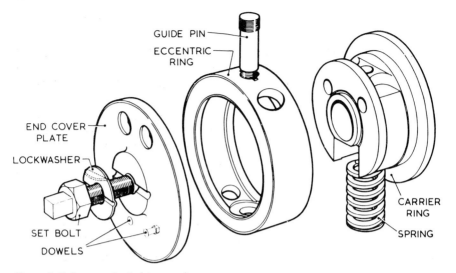

GUIDE PIN

ECCENTRIC
RING

END COVER
PLATE

LOCKWASHER

SET BOLT

DOWELS

CARRIER
RING

SPRING

Figure 3.40 Overspeed trip (ring type)

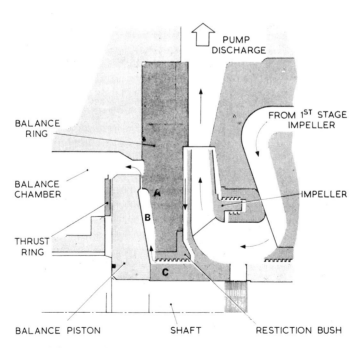

PUMP
DISCHARGE

BALANCE
RING

BALANCE
CHAMBER

THRUST
RING

FROM 1ST STAGE
IMPELLER

IMPELLER

BALANCE PISTON

SHAFT

RESTICTION BUSH

Figure 3.41 Hydraulic balance

turbine and impellers. Figure 3.41 shows a typical balance arrangement.

The arrangement keeps the rotating assembly in its correct position under all conditions of loading. Water at the approximate pressure of the pump discharge passes from the last stage of the pump between the impeller hub and the balance restriction bush C into the annular space B dropping in pressure as it does so. The pressure of water in the chamber B tends to push the balance piston towards the turbine end. When the thrust on the balance piston overcomes the turbine and the impeller thrust the gap A between the piston and balance ring widens and allows water to escape. This in turn has the effect of lowering the pressure in chamber B allowing the rotating assembly to move back towards the pump end.

Theoretically this cycle will be repeated with a smaller movement each time until the thrust on the balance piston exactly balances the other axial forces acting on the assembly. In practice the balancing of the forces is almost instantaneous and any axial movement of the shaft is negligible.

Weir electrofeeder

The Weir electrofeeder (Figure 3.42) is a multi-stage centrifugal pump mounted on a common baseplate with its electric motor; the number of stages may vary from two to fourteen depending upon the size of the pump and the required discharge pressure. The pump body consists of a number of ring sections fitted with diffusers and

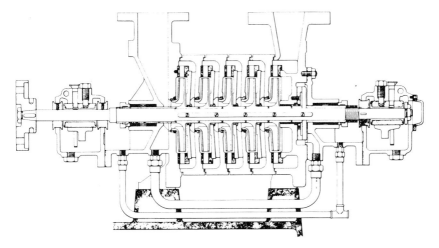

Figure 3.42 Electrically-driven multi-stage feed pump

held in position between a suction and discharge casing by virtue of the 'crush' imposed on them by a ring of steel tie bolts.

The unit is supported on pads on the baseplate by two feet on each of the end casings, these feet being drilled to accommodate the holding down bolts. Tapered dowels are used to maintain the correct alignment, and the driving torque from the electric motor is transmitted through a flexible coupling.

The shaft assembly is supported on two ring lubricated white metal lined journal bearings bedded into plummer blocks, the lower sections of which form oil sumps. An internal hydraulic balancing arrangement similar to that found in the turbo-feed pump automatically maintains the shaft assembly in its correct axial position at all loads during running. To avoid excessive wear on this balancing arrangement when starting up the pump, it is essential that the discharge pressure be built up quickly, and for this purpose, and to eliminate the possibility of reverse flow, the pump is fitted with a spring loaded non-return discharge valve.

Condensate cooled stuffing boxes packed with high quality graphited asbestos packing are used for shaft sealing and these can be additionally cooled by water-circulated cooling jackets in the suction casing and the balance chamber cover. A pressure-operated cut-out switch may be fitted which will automatically isolate the driving motor from its supply if the first stage discharge pressure falls to a predetermined value due to loss of suction pressure, cavitation, etc.

Centrifugal condensate extraction pump

Removal of condensate from a condenser imposes very difficult suction conditions on the pump. The available NPSH is minimal because the condenser is situated low in the ship permitting a static suction head of only 450–700 mm and the condensate is at, or near, its vapour pressure. It is necessary therefore to ensure that the pump's required NPSH is correspondingly low and to this end suction passages and inlets are given ample area. The pumps used for the duty are two stage units (Figure 3.43), the first stage impeller being arranged as low as possible in the pump with an upward facing eye. This impeller feeds a second impeller via suitable passages in the pump casing.

Where extraction pumps are fed from de-aerators or drain coolers the pump suction level is maintained constant by a specially designed float control; in instances where the pump is drawing from the main condenser however it is growing practice to operate the pump on a free suction head. This means that the pump must operate with a

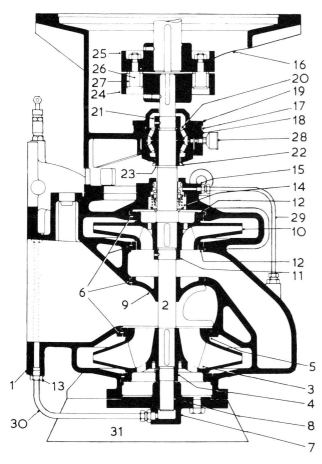

1. Pump casing (split)
2. Pump spindle
3. Impeller (1st stage)
4. Impeller nut and tag washer (1st stage)
5. Casing ring (1st stage)
6. Dowel pins
7. Bottom cover
8. Bottom bush liner
9. Intermediate bush
10. Impeller (2nd stage)
11. Impeller nut and lock screw (2nd stage)
12. Casing ring (2nd stage)
13. Plug
14. Mechanical seal
15. Seal clamping plate
16. Motor stool
17. Thrust bearing housing
18. Thrust bearing cover
19. Thrust bearing end cover
20. Thrust bearing
21. Thrust nut and lock screw
22. 'V' ring
23. Distance piece
24. Flexible coupling (pump half)
25. Flexible coupling (motor half)
26. Coupling bolt and nut
27. Coupling pad
28. Grease lubricator
29. Water return pipe
30. Water supply pipe
31. Pump feet

Figure 3.43 Weir two-stage extraction pump

variable NPSH at varying flow rates. Figure 3.44 shows the H/Q curve for a two stage extraction pump operating under a variable NPSH. The system resistance curve is interposed on the diagram.

When operating at the specified maximum capacity, the flow rate corresponds to that at the intersection of the natural H/Q characteristic of the pump and the system resistance curve. As the level of condensate in the suction sump falls below the point where the 'available NPSH' intersects with the minimum 'required NPSH' the pump starts to cavitate and its output is regulated. The natural H/Q characteristic is modified as the discharge pressure is reduced to that required to overcome the resistance of the system at the reduced

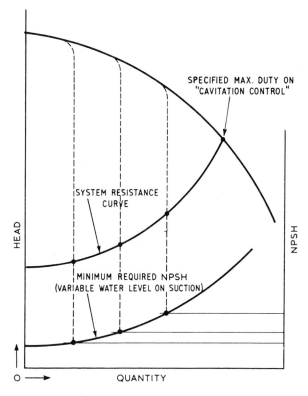

Figure 3.44 Extraction pump cavitation control curves

flow rate. When the suction level increases the flow rate increases. Therefore, an inherent control of the flow rate is achieved without the use of a float controlled regulator. Such a system is usually referred to as 'cavitation control'.

Air ejector

A steam-jet ejector is frequently used to withdraw air and dissolved gases from the condenser. In each stage of the steam-jet ejector, high pressure steam is expanded in a convergent/divergent nozzle. The steam leaves the nozzle at a very high velocity in the order of 1220 m/sec and a proportion of the kinetic energy in the steam jet is transferred, by interchange of momentum, to the body of air which is entrained and passes along with the operating steam through a diffuser in which the kinetic energy of the combined stream is re-converted to pressure energy. The maximum pressure ratio that can be obtained with a signal stage is roughly 5:1 and consequently it is necessary to use two or even three stages in series, to establish vacua in the order of 724 mmHg, with reasonable steam consumption.

There are a variety of styles in service but all work on the same principle. Older units have heavy cast steel shells which serve as vapour condensers and also contain the diffusers. These are arranged vertically, the steam entering at the top (Figure 3.45). More recent designs have the diffusers arranged externally and the vapour condenser shell is somewhat lighter in construction. Horizontal and

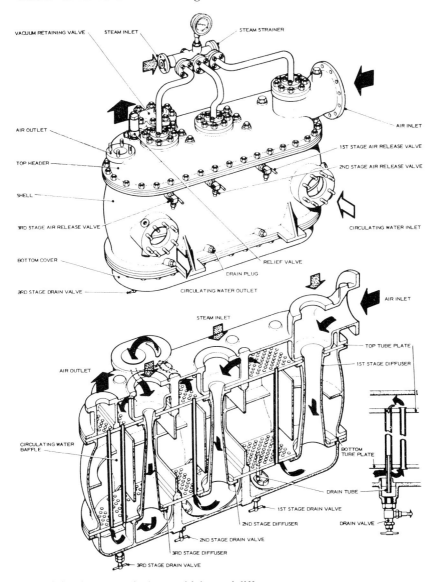

Figure 3.45 Three-stage air ejector with internal diffusers

vertical arrangements can be found and some units are arranged as combined air ejectors and gland steam condensers.

Horizontal single element two-stage air ejector

An air ejector now in common use is shown schematically in Figure 3.46. The unit comprises a stack of U-tubes contained in a fabricated mild steel condenser shell on which is mounted a single element two-stage air ejector.

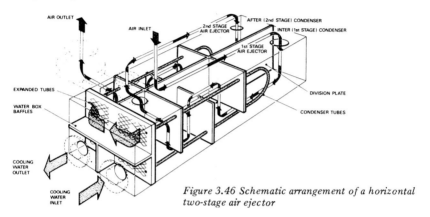

Figure 3.46 Schematic arrangement of a horizontal two-stage air ejector

The condensate from the main or auxiliary condenser is used as the cooling medium, the condensate circulating through the tubes whilst the air and vapour passes through the shell. The high velocity operating steam emerging from the first stage ejector nozzle entrains the non-condensables and vapour from the main condenser and the mixture discharges into the inter (or 1st stage) condenser.

Most of the steam and vapour is condensed when it comes into contact with the cool surface of the tubes, falls to the bottom of the shell and drains to the main or auxiliary condenser. The remaining air and water vapour are drawn into the 2nd stage ejector and discharged to the after (or 2nd stage) condenser. The condensate then passes to the steam drains tank and the non-condensables are discharged to the atmosphere through a vacuum retaining valve.

The vacuum retaining valve is shown in Figure 3.48 which is fitted as a safety device to reduce the rate of loss of vacuum in the main condenser is the air ejector fails. It is mounted on a pocket built out from the second stage condenser, and consists essentially of a light stainless steel annular valve plate which covers ports in a gunmetal valve seat. When the pressure inside the after condenser exceeds atmospheric pressure the valve lifts and allows the gases to escape to

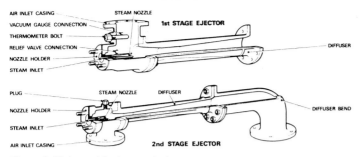

Figure 3.47 1st and 2nd stage ejectors

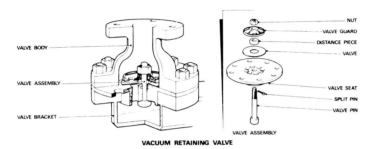

VACUUM RETAINING VALVE

Figure 3.48 Vacuum retaining valve

atmosphere. A relief valve is fitted on the first stage condenser shell of the twin element unit.

The ejector stages, Figure 3.47, consist of monel metal nozzles in mild steel holders discharging into gunmetal diffusers. Expansion is allowed for by sliding feet at the inlet end.

Nash rotary liquid ring pumps

Nash rotary liquid ring pumps, in association with atmospheric air ejectors may be used instead of diffuser-type steam ejectors and are arranged as shown in Figure 3.49. The pump, discharging to a separator, draws from the condenser through the atmospheric air ejector, creating a partial vacuum of about 600 mm Hg. At this stage the ejector, taking its operating air from the discharge separator, which is vented to atmosphere, comes into action: sonic velocity is attained and vacua of the order of 725 mm Hg maintained in the condenser. The liquid ring pump is sealed with fresh water recycled through a seawater cooled heat exchanger.

The Nash pump using recycled fresh water for sealing may be also used to provide oil-free instrument air at a pressure of 7 bar, Figure 3.50.

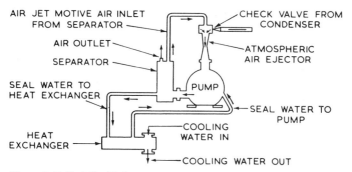

Figure 3.49 Nash liquid ring pump

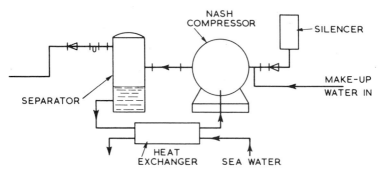

Figure 3.50 Using recycled fresh water for sealing
(Nash Engineering Co. (G.B.) Ltd.)

Centrifugal pumps for lubricating oil duties

Because of their self-priming abilities positive displacement pumps
are widely used for lubricating oil duties. This practice is completely
satisfactory in installations where the pump speed is variable but
when the pump is driven by a constant speed a.c. motor it is necessary
to arrange a bypass which can be closed in to boost flow. By using
a centrifugal pump with an extended spindle, such that its impeller
can be located at the bottom of the oil tank, the H/Q characteristics
of the centrifugal pump can be used without the priming disadvantages. Known as the tank type pump this pump has a small open
impeller. It can be driven directly by a high speed alternating current
motor without capacity restrictions, whereas the permissible
operating speed of the positive displacement type, pumping an
incompressible fluid, decreases as the capacity increases.

The pump will deliver an increased amount of oil as the system
resistance is reduced, when the oil temperature rises and viscosity
falls. In practice this is what the engine requires. A conventionally-
mounted self-priming centrifugal pump is satisfactory for this duty

if the suction pipe is short, direct and generously dimensioned. In a positive displacement pump the output varies little, if at all. The latter point is readily appreciated on reference to Figure 3.51 in which, for simplicity, the effect of change in viscosity of the oil on the pump characteristics has been omitted, although this would favour the centrifugal and enhance the comparison still further.

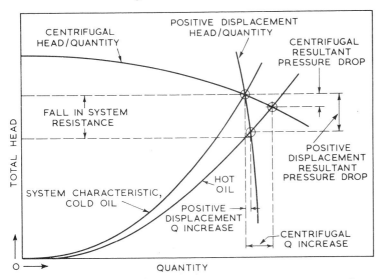

Figure 3.51 Comparison of centrifugal and positive displacement pump characteristics with respect to lubricating oil duties

Metering pumps

A metering pump is a pump that has been designed primarily for *measuring* and dispensing a liquid, rather than merely transferring it from one site to another. It has to be a precisely designed piece of equipment, manufactured to close engineering limits. It is frequently used to dispense concentrated and highly corrosive liquids and for this reason considerable care has to be taken to ensure that the materials used in construction, particularly those of parts in direct contact with the liquid, will resist attack.

Most metering pumps are of the positive displacement type. They consist of a prime mover, a drive mechanism and a pumphead. Pumpheads are usually of the piston, or plunger type where the pump is to be used against high pressures; for lower pressure duties the diaphragm version is generally used (Figure 3.52). In comparing the two types, it should be realised that the plunger model is more exact in its performance than the diaphragm, whilst the diaphragm type, which requires no glands, is completely leakproof.

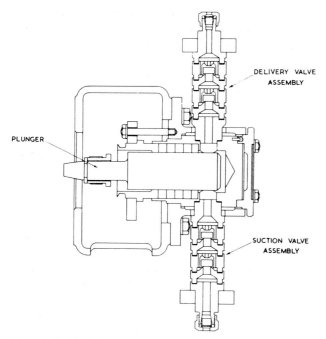

Figure 3.52a Typical plunger head for MPL Type Q pump

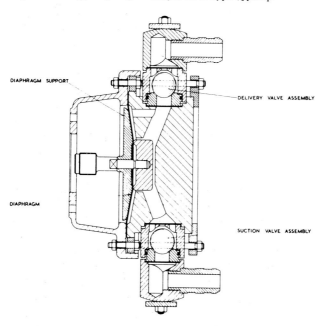

Figure 3.52b Typical diaphragm head for MPL Type Q pump

Pump output is controlled by varying the length of the pump stroke; in this way it is possible to adjust output progressively from zero to maximum whilst the pump is in operation as well as when it is at rest.

A further feature of many metering pumps is that arrangements are made in design so that more than one pumphead and mechanism can be accommodated, each with its own stroke control. Similarly each pumphead mechanism can be equipped with a different gear ratio, thus enabling one pump to meter various liquids in different ranges of flow rates. The metering pump is thus highly versatile, capable of adaptation to a wide range of duties.

Cargo pumping

In large tankers, very high pumping rates for cargo discharge and for de-ballasting are required; consequently, the reciprocating pump has been superseded by large rotary-displacement and for really high rates, say above 600 m^3/hr., by centrifugal pumps, electrically or turbine driven.

Within its capacity, a pump will discharge at the rate at which fluid will flow to its suction orifice; this will be determined by viscosity, specific gravity, line friction, entrained air (as the tank level falls), and in the case of gassed oils, the release of volatile light fractions if the pressure in the suction pipe falls below the vapour pressure of the oil. This is the highest practicable rate; it will fall under adverse pumping conditons but a useful rate will be obtained from a displacement pump while enough oil passes to keep it sealed. The centrifugal pump however, will not regain a lost suction until the pump fills with liquid after the vapour is released. This problem disappears if the non-liquid elements are separated from the fluid before it reaches the suction orifice and if the discharge rate is limited to that of flow to the suction. Figure 3.53 shows a typical system of this kind, for one pump.

A separator 10 fitted with a vapour outlet pipe, a sight glass and a level controller 1, is placed near the pump in the suction line. The vapour outlet is led to an interceptor tank 11 having a moisture eliminator, a drain to 10 and an outlet to a vapour extraction pump 7 through an exhaust control valve 2. This pump, (a water-ring type, described on page 85) controlled by the pressure switches A and B, draws from the interceptor tank and discharges to atmosphere at a suitable height through a tank 8 having an internal cooling coil; the small quantities of water carried with the vapour drop out in the tank and return to the pump by the pipe shown. This arrangement, by means described later, ensures that the pump is always fully primed. The discharge rate is regulated by a butterfly valve 6 in the cargo discharge line, controlled by a pneumatic positioner; a thermostat guards against overheating when the pump is working against a restricted or closed discharge.

As non-liquid elements separate out, the liquid level in the separator tank will fall, so long as the exhaust control valve is closed. The level controller is supplied with clean, dry air (this is a prerequisite of satisfactory operation) at a suitable pressure and is so arranged that its outlet pressure increases progressively from say, 200 mbar to 1000 mbar, as the level falls. At 240 mbar the pressure switch A will make but the extraction pump 7 will not start; at 270 mbar the exhaust control valve 2 will open and if there is a head on the suction line, 10 will vent to atmosphere through 1, 2, 7 and 8; the level in 10 will rise, the level controller outlet pressure will fall, 2 will close and A will break.

As pumping continues and the entrained air, etc., increases, the level in 10 will fall again, the outlet pressure of 1 will rise, 2 will open, A and B will both make, 7 will start and again 10 will be

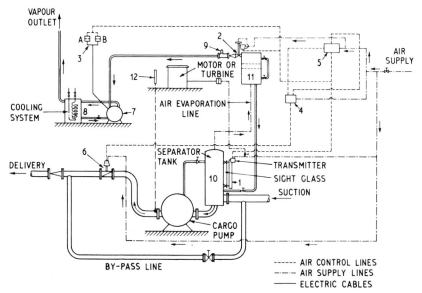

Figure 3.53 Diagram of 'Vac-Strip' cargo system
1. *Leveltrol (level controller)*
2. *Exhaust control valve*
3. *Pressure switches*
4. *Auto/manual selector (panel loader)*
5. *High selector relays*
6. *Butterfly valve*
(Worthington-Simpson Ltd.)

7. *Vacuum pump*
8. *Reservoir tank*
9. *Vacuum breaker*
10. *Separator*
11. *Interceptor tank*
12. *Thermostat*

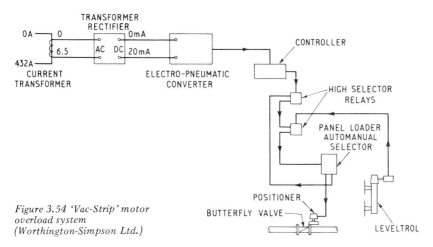

*Figure 3.54 'Vac-Strip' motor
overload system
(Worthington-Simpson Ltd.)*

vented; the level in 10 will rise, 7 will stop, 2 will close, A and B will break. This sequence will be repeated until frequent snoring at the suction brings in so much air that 7 runs continuously. As pumping goes on, the situation becomes one in which the cargo pump is discharging at a faster rate than oil can flow to the suction and the level in the separator tank will fall continuously. As a result, the level controller outlet pressure will be high and will act to increase the discharge head, so reducing the pump capacity by closing the butterfly valve 6. The level controller outlet is led to the high-selector relay 5 and thence to the butterfly valve positioner, through the auto-manual selector 4. This may be set by hand to limit the pump discharge pressure to that desired in the receiving system, or to operate in automatic response to the level controller outlet pressure fed to it from 5, i.e. the manual setting will over-ride the automatic if the discharge pressure desired is below that consequent upon restricting the pump discharge rate to a figure not above the rate of flow to the suction. The butterfly valve will therefore be either fully open or partly closed, either to limit discharge pressure or to limit pump capacity. The pump speed may be reduced in a similar manner before the butterfly valve begins to close. It is important to bear in mind the difficulties of working centrifugal pumps in parallel at different speeds — the slower-running pump may cease to pump at all.

The above description refers to a basic automatic system for one turbine-driven pump. Separators, interceptors, etc. and their fittings will increase in number with the pumps; one or more vapour extraction pumps may serve all the interceptors. It is possible to overload the motor of an electrically-driven cargo pump and this may be prevented by the means shown in Figure 3.54. The motor armature current taken through transformer, rectifier and electro-pneumatic converter, produces pressure variations in the feed to a controller, between no-load and overload, of the same order as those arising from variations in separator tank level; the controller has an adjustable set point. These pressure variations are fed to the auto-manual selector through two high-selector relays, to vary appropriately the opening of the butterfly valve. Two high-selector relays are necessary because these instruments can only select the higher of two impulses transmitted to them; here the highest of three pressures has to be selected, from the level controller, the auto-manual selector manual setting and the electro-pneumatic converter.

The system described pre-supposes the absence of stripping pumps. One or more are sometimes fitted, either as a safeguard against unforeseen difficulty with the main pumps or to deal with tank washings.

The level controller comprises a vessel, common to the separator tank, in which is suspended a body connected by an arm to a torque tube in such a way that variations in liquid level (i.e. variations in the immersion of the suspended body) give rise to proportionate variations in the torque applied to the rotary shaft, see Figure 3.55.

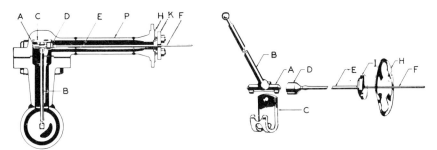

Figure 3.55 Level controller (left) exploded view (right) plan view
A.   Torque tube driver
B.   Arm
C.   Torque tube driver bearing
D,E. Torque tube assembly
F.   Rotary shaft
H. Positioning plate
I. Torque tube flange
K. Retaining flange
(G.E.C.-Elliott Control Valves Ltd.)

Figure 3.56 shows the instrument which translates the degree of rotation of the rotary shaft into the controller outlet pressure operating the exhaust control valve, the pressure switches and the pump discharge butterfly valve. Air at 1½ bar is fed to the inlet valve 0, to the upper diaphragm chamber L and to nozzle D through orifice J. D is large enough to bleed off all the air passing J if it is not restricted by the beam G (i.e. L is free of pressure and O is closed

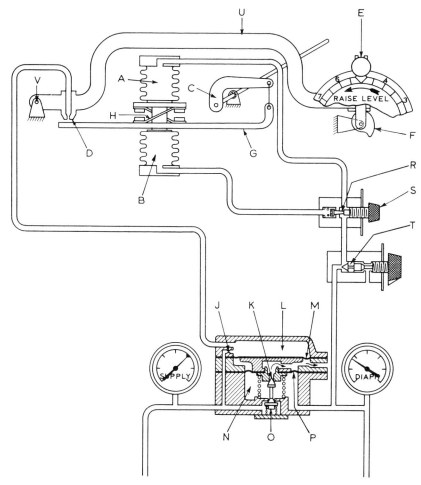

Figure 3.56 Air pressure instrument

A, B. *Bellows*
C.   *Torque arm*
D.   *Nozzle*
E.   *Level adjustment*
F.   *Level adjustment arm*
G.   *Beam*
J.   *Air supply orifice*

K.   *Exhaust valve*
R.   *Reset valve*
T.   *Proportional valve*
M,P. *Diaphragm assembly*
L,N. *Upper and lower*
      diaphragm chambers
O.   *Inlet valve*

(G.E.C.-Elliott Control Valves Ltd.)

unless D is wholly or partly closed by G as a result of a fall in the liquid level). The double diaphragm assembly M and P is fully floating and pressure balanced. When the pressure in L rises, the assembly moves downwards, opens the inlet valve O and so admits pressure to the lower diaphragm chamber N; the assembly then rising, returns to its original position and O closes. When the pressure in L falls,

114

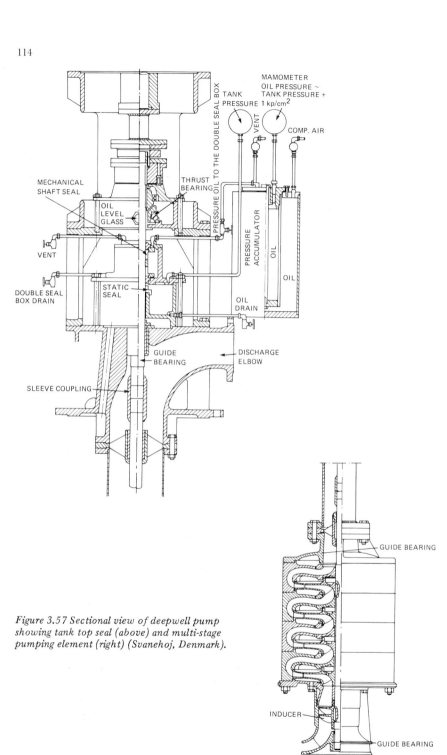

Figure 3.57 Sectional view of deepwell pump showing tank top seal (above) and multi-stage pumping element (right) (Svanehoj, Denmark).

because of the rise in liquid level, the assembly moves upwards, opens exhaust valve K and so reduces the pressure in N until the assembly returns to its original position again and K closes. It will be seen that the pressure at the controller outlet is the same as that in N, varying with the degree of restriction of D by G and that there is a lead to the bellows A through the proportional valve T so that, as the outlet pressure rises, A extending, moves G away from D and limits the pressure rise. M and P are so proportioned that the pressure in N and therefore at the controller is always thrice that in L.

Deepwell pumps

It is sometimes beneficial to site the pump in the bottom of the cargo tank. Reasons for this could be to facilitate the simultaneous discharge of different cargoes (e.g. chemical carriers) or in the case of particularly volatile cargoes, where NPSH becomes critical but where the design or duty of the ship precludes the use of cargo pump rooms below the weather deck (see IMCO Gas Code Chap. 3.3.1). In such cases deepwell pumps will frequently be found. These pumps are vertical centrifugal pumps driven, via a long shaft, by a motor mounted above the cargo tank top.

A multi-stage deepwell pump is shown in Figure 3.57. This particular pump is designed for LPG cargoes and features an oil-filled gas seal at tank top level. The oil within the seal chamber is maintained constantly 1 bar above cargo tank pressure by an accumulator containing a plunger which is operated by air or nitrogen pressure.

The drive shaft is located within the pump discharge pipe and is supported in carbon bearings. It is essential that this type of pump is not allowed to run dry or with a throttled discharge valve.

CLEANING AND MAINTENANCE OF PUMP SYSTEMS

Air handling has already been covered, and in the pipelines themselves all that need be mentioned is that commonsense must be used to provide natural venting and to avoid arrangements which permit pockets of air to collect (such as upward loops). Frictional losses of the system must be calculated correctly so as not to create cavitation at the pump (excessive frictional loss in suction pipework) or excessive losses in the system generally which will reduce the delivered capacity of the pump. With a centrifugal pump, system losses much below those expected may allow excess capacity to be pumped and so overload the pump.

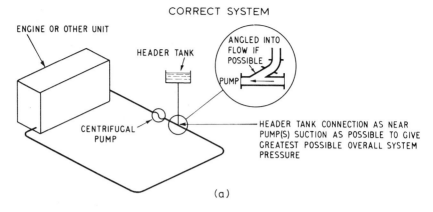

Figure 3.58 Closed pumping systems (Hamworthy Engineering Ltd.)

Referring to overloads, it is important not only to check the differential head across the pump, but to check the generating frequency since the power, when running at 62½ Hz instead of the correct 60 Hz, will increase by 8%. Figure 3.58 shows how a closed system, with header tank, should be arranged.

Connecting pipes of the correct bore should be fitted; it is specially important that the bore of the suction pipe should not be curtailed. Pipe connections should be as direct as possible, sharp bends must be avoided and loops such as that shown in Figure 3.59 will probably cause a direct-acting pump to knock at the end of its stroke. See that there are no leaks in the suction pipe, as it is undesirable that the pump should draw air. If there are many bends in the suction pipe, or if this pipe is very long, an air vessel on the suction branch of the pump may be necessary.

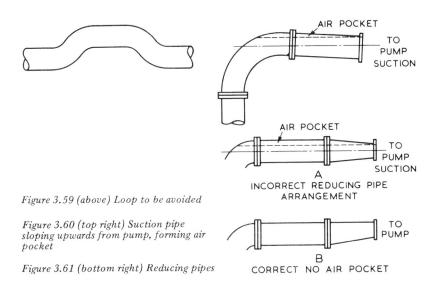

Figure 3.59 (above) Loop to be avoided

Figure 3.60 (top right) Suction pipe sloping upwards from pump, forming air pocket

Figure 3.61 (bottom right) Reducing pipes

A common fault, causing a great deal of trouble, is shown in Figure 3.60. Here the suction pipe connected to the pump slopes upwards away from the pump, with the result that an air pocket is formed, which gives rise to loss of suction, irregular discharge and hammer.

A similar fault is found when reducing pipes are used to connect an oversize pipe to the pump inlet. The type of reducing pipe shown at A in Figure 3.60 can often be usefully employed, but when used in conjunction with a 90° bend, it is difficult to obtain sufficient downward slope to prevent a bad air pocket forming in horizontal pipes. Such a pipe is quite satisfactory on the discharge side of the pump when used to increase the diameter of the piping, but for suction purposes the type of reducer shown at B, should be used.

Cleaning the system

It is often found, particularly in new ships, that the bilges and bilge systems have not been thoroughly cleaned with the result that pieces of wood, waste, nuts, bolts, etc., are found inside valves and pipes. These choke the valve-chests, prevent the valves from being closed properly and also block strainers. It is, therefore, most important that steps should be taken to see that the whole system is thoroughly clean.

The system should be tested to ensure that all suction piping and joints, and the packing glands of valves, are free from air leaks.

Suction pipes

Suction tail pipes in tanks, etc., should be bell-mouthed, as the increased diameter assists in reducing the loss of head at the inlet and allows the pipe end to be placed nearer to the bottom. Suction bells should always be cross-baffled and great care should be taken in the tank construction to ensure maximum ease of flow to the suction bell.

Where strainers or weed-boxes are fitted on the suction side, it may happen (even when the pump has a suction head) that, due to choking of the strainer or weed-box, the area becomes so restricted that a vacuum is created in the pump suction; if air enters through leaky joints or badly-packed glands, it may accumulate at the impeller eyes, and pumping will cease.

Alignment

If pumps are designed so that the driving motor or turbine is mounted upon an extension to the pump casing proper, tendency to mal-alignment, due to pipeline stresses, is practically eliminated. Nevertheless, it is essential that the pipe systems and heavy valve chests, etc., are separately supported and stayed during installation, the flanged connection to the pumps being the last to be coupled up, after the faces are correctly aligned. This can contribute materially to the life of the unit.

Horizontal pumps should be laid down upon steel or hardwood chocks, accurately fitted to ensure that the couplings, with the bolts removed, are in correct alignment and with their faces parallel. This alignment should be checked after tightening down the holding-down bolts and again after the pipes are coupled and preferably full of liquid.

In pumps in which the spindle is not entirely supported by two external bearings, but receives some support from rigid coupling to the prime movers, it is essential that special half-bushes, without running clearances, be fitted into the stuffing-boxes to ensure that the spindle is centralised while adjusting the alignment.

If main circulating and other pumps are connected to freely floating condensers through bellows connections, it is essential that the pipes are firmly anchored on the pump side of the bellows.

Alignment of pumping engines

When installing pumping engines, care should be exercised when erecting the pipes to avoid distortion of the pump casing. It is quite easy to pull a strong pump casing out of true alignment by the bolts on the suction and delivery pipe joints, this is especially true with large gunmetal pump casings.

To set up pump and engine in correct alignment the following procedure is recommended:

1. Level up the engine approximately to the position in which it is to be finally bolted down and loosely bolt up the suction and discharge pipes.
2. Remove the bolts connecting the engine crankshaft to the pump shaft and check the fairness of the flanges of the couplings at their outside diameter with straight-edge and with feelers between the faces.
3. Adjust the pumping engine on its seating so that the closing lengths come into their fairest position with regard to the pump flanges.
4. Bolt up the closing lengths to the pump, checking from time to time the alignment of the pump and engine coupling which is still unbolted.
5. Wedge up the engine bed-plate and bolt down again, checking the alignment once more.
6. If the alignment is found to be exactly correct, insert the coupling bolts, which should go into place when tapped lightly with a hammer.

STARTING

Reciprocating pumps

When started for the first time, the pump will work irregularly for several strokes because the whole pipe system will be full of air.

Pumps should therefore be flooded with the liquid they are to handle, which prevents seizure or wear, especially when priming a long suction line.

To start the pump, proceed as follows. Check lubrication. Open the water-end valves, suction and discharge; open the isolating valves in the exhaust and steam lines; open the steam-cylinder drain cock to rid the steam cylinder of water of condensation (close the cock when the pump is running and heated through); open the cushioning valves, if fitted; open the exhaust valve. Open the throttle valve slowly and warm up gradually until the pump is brought up to the proper speed. If the pump is controlled by a pressure governor, open the throttle gradually until the governor takes control of the pump, and then open the throttle valve fully.

Never start the pump with the feed check or other stop valve shut.

Centrifugal pumps

A centrifugal feed pump must not be operated unless it is fully primed. The pump casing should be filled before starting, the suction pipe and pipe branch to the discharge stop valve must also be full. If the water enters the pump suction by gravity, priming is unnecessary and the pump will remain full of water when shut down. To fill the pump, open the small air-valve on the top of the pump casing until water commences to flow, then shut the air-valve.

If the pump is operating with a suction lift, it may be primed either from an independent water supply or from the discharge line, or by means of a vent connection or exhauster which will evacuate the pump and suction piping of air. A foot valve must, of course, be fitted in these cases. The discharge valve and air-valves should be kept closed during the priming of the pump if done by the latter methods.

To start a centrifugal pump, proceed as follows: Check lubrication, and see that pump glands are properly packed and adjusted. Open steam and exhaust casing drain cocks of the driving engine; open steam and exhaust valves; open suction valves and air-cocks. Open turbine throttle valve sufficiently to free the pipelines, steam chest and exhaust casing of water, running turbine at a very low speed. When free of water, close the drain cocks and bring the unit up to speed.

For reciprocating engines, ease the throttle to allow the cylinder to warm slowly; close the drain cocks when the engine is free of water. After all air has escaped, close the pump air-cock. When the pump has reached the correct speed, open the discharge valve.

For propeller-type pumps, it is preferable to start with the discharge valve open, as otherwise an excessive discharge head may be developed. Check all gauges to see that a proper pressure is being developed, and check the speed of the driving engine.

Electrically driven pumps

Before putting power on to the motor for the first time, or after long periods of idleness, it is advisable to test insulation resistance and to raise it if necessary, by drying out.

When starting, note by the first movement of the shaft if the direction of rotation is correct. If it is not, the wiring to the motor must be altered.

CURING TROUBLES – CENTRIFUGAL PUMPS

Failure to deliver water

First, make sure the pump is primed. The fault may be that the discharge head is too high, or the suction lift is too high. It should not be more than 4.7 m at 29.4°C. There may be insufficient speed; it was explained earlier in this chapter that the pressure or head at the periphery of the impeller depends upon the tip speed. Other faults to look for are an air leak in the suction line, or a broken or plugged-up impeller. The direction of rotation of the impeller should also be checked.

Pump will not prime

If the pump will not prime, the most probable cause is an air leakage of some sort. If there is a leakage at the pump gland, the gland should be adjusted and the recess filled with oil or the stuffing-box re-packed. All joints should be checked for a leak in the pump or suction pipe. Make sure that the delivery valve is not open.

The priming pump float gear, if fitted, should be removed and examined to ensure that the filter protecting the float valve has not become choked, nor the ball disconnected, allowing the spring to close the valve.

In bilge applications a frequent reason for failure to prime is a faulty bilge suction valve.

Failure to build up pressure

If the pump fails to build up adequate pressure, or to discharge water when the discharge valve is opened and the speed brought up to normal then the following checks should be made.

Stop the driving unit. Make sure that the pump is primed, that all air has been expelled through the air-cocks on the pump casing, and that all valves in the suction line are open. Start the pump again, and if the discharge pressure is still not normal, stop the pump and find the exact cause of failure. It may be that the speed is too low, or that there is air in the water. The impeller may be damaged or the wearing rings worn, or some other mechanical defect may require attention.

Insufficient capacity

As with complete failure to deliver water, check first the whole pump arrangement. The total dynamic head may be higher than that for which the pump is rated, or the suction lift may be too high. Check also the temperature of the water and the speed of the pump. The foot valve may be too small or may have become obstructed, or the foot valve or suction pipe may not be immersed deeply enough.

If the above are correct, then the most probable cause is that the impellers have become partially obstructed or choked with dirt from the bilges or pipes. Other possible causes are air leaks in the suction or stuffing-boxes, defective packing or worn wearing rings.

If there is low output with abnormally high vacuum reading, the probable causes are that there is an obstruction in the suction line, such as the blind joint, choked strainer, valve or mud-box filter, or the requisite valve in the piping system has been inadvertently left closed, or others left open.

Pump loses water after starting

Check the suction lift and the temperature of the water. A leaky suction line may be suspected.

Pump vibrates

If the alignment of the pump is correct and the foundations secure, it is probable that the impeller has become partially clogged so that the balance is disturbed. There may also be mechanical faults, such as worn bearings, a bent shaft or an eroded impeller.

Pump overloads driving engine

It may be that the speed is too high, or that the pump is pumping too much water because the total dynamic head is lower than that for which the pump is rated. Check also that the liquid is of the correct specific gravity and viscosity for the rating of the pump.

Maximum vacuum test

Where pumps have suctions to positive suction heads, e.g. a sea suction on a bilge pump or general service pump.

1. Open sea-suction valve.
2. Start the pump and raise to normal speed.
3. Open sea-discharge valve.
4. Close sea-suction valve.
5. Open suction valve to an empty compartment until the vacuum has fallen away and then close the valve. The pump vacuum should rise rapidly to 64.5 mbar or more, if the air pump is in satisfactory condition and the pump is free from air leakage.
6. Stop the pump. The vacuum should be retained if the pump is airtight.

If the pump is *not* airtight, stop the pump, readjust the pump gland and fill the recess with oil. Make a further test, and if no difference in performance occurs the gland is not at fault.

If the pump-casing joints are quite airtight and the suction valve properly shut, the fault is then attributable to a faulty air-pump, a leaky n.r. valve, a choked basket strainer or a detached float, which can be easily checked by removing the float-gear distance piece. Remove the n.r. valve-cover plug and grind in the valve with fine abrasive paste.

With the pump running, and the air-pump suction pipe disconnected, note the pull exerted by the air-pump upon a hand placed over the suction branch of the air-pump. This will indicate if the air-pump is pulling up the required vacuum. If the air-pump is found to be in order, open the pump suction valve and the discharge valve, and, with all other individual bilge or tank sections closed on the line, ascertain the maximum vacuum which can be obtained in the main-line suction when the pump is operating. The pump should be watched for overheating in the water end under a prolonged run when not pumping, and if necessary, a fresh supply of water admitted by opening the sea-suction valve for a few minutes.

If the vacuum fails to rise, then the defect is due to excessive air leaks in the suction piping through bad joints, badly packed valve spindles or valves left open or leaky. If the vacuum rises satisfactorily, open the suction valve of the tank or bilge to be pumped and watch the vacuum gauge. If the vacuum falls away suddenly and fails to recover, look for a leak in the tail pipe above the water level in the tank. If the vacuum remains higher than that required to overcome the static lift, and the pump fails to deliver, there is a chokage in the suction pipe, which can be readily ascertained. A choked impeller will show little or no increase in vacuum after the pump is primed and a low motor current value.

It should be remembered that, although the air-pump will continue to operate pulling water from the suction system via the float valve, it is not desirable to operate the unit under these conditions. Should the water be constantly discharged by the air-pump, it is probable that the float valve is dirty or requires to be lightly ground in.

OVERHAUL OF RECIPROCATING PUMPS

Group valves

In many pumps brass Kinghorn valves are fitted. These should be examined to see whether any cupping has taken place. If it has, the valves should be inverted or replaced.

If the seats have been eroded, they should be either scraped by hand, or if the erosion is too deep, skimmed in a lathe. Where group valves are of the mitre type, the valves and valve-seats should be examined and the valves ground in as necessary to make them perfectly tight. In general, the lift of valves should be the minimum which will give satisfactory operation. Greater lifts than necessary increase wear and tear.

The steam cylinder, piston and piston rings must be in good condition, to avoid leakage and increased steam consumption. Pistons are commonly iron castings or steel forgings having two piston ring grooves in each of which two rings are fitted: the rings are usually hammered cast iron, i.e. hammered on the inner surfaces to give spring. The pistons have running clearances varying with diameter, from 0.5 to 1.5 mm; ring clearances in the grooves should be as small as will allow them to float freely and the gaps, again varying with diameter, from 0.25 mm to 1.0 mm, the gaps being diametrically opposite. These dimensions are not critical but excessive increases are undesirable. In time, ridges form at the limits of ring travel, making it impossible to fit piston rings with a reasonable gap. Reboring is then necessary and the fitting of oversize pistons and rings.

Piston valve-chest

Although the following paragraphs apply to the Weir piston valve-chest, they can be taken as typical of all types of steam-distributing valve-chests.

The Weir piston valve-chest comprises a main and auxiliary valve. These valves are simple slide valves, although the main valve is cylindrical with a flat face formed at mid-length along one side on which the auxiliary valve works. As these valves are simple slide valves, they are normally held up to their respective faces by steam pressure, but wear or corrosion and erosion will allow steam to leak to exhaust, causing increased steam consumption.

If this condition arises, the valve face should be rebedded, but only for the length of the valve travel. The portion on the main valve into which the bells fit should not be touched, as this would impair the shooting of the main valve. The auxiliary valve should be filed and scraped to bring up a bearing between the two faces and liners, as necessary, fitted between the valve spindle and the back of the auxiliary valve, to hold it up to position when the steam is shut off. After lining up, it should be possible to push the main valve to and fro by hand. The Weir chest is designed to allow expansive working. The expansion pointers on the end covers should be closed down as far as possible without impairing the speed at which the pump operates. This will ensure the maximum expansive working and the minimum steam consumption.

If the wear is too great to enable the chest to be refitted easily at site, it should be returned to the manufacturer for overhaul. Corrosion is a serious cause of wear. Oil should be injected into the piston valve chest and cylinder just before a pump is shut down, to oil the working surfaces and minimise corrosion when idle.

Gland packing

The packing in the stuffing-boxes in water and steam ends should be carefully fitted and maintained in good condition.

Buckets and ebonite rings

The Weir pump bucket is gunmetal, is fitted with two rings of a special grade of ebonite manufactured to stand up to the action of hot feed water. The ends of the rings are cut at an angle of about 60°, and in use the rings expand and make a good watertight fit in the pump chamber.

In overhauling it may be found that the ebonite rings do not entirely fill the pump barrel. This does not mean that the rings must be badly worn, and it is not necessary to replace them on this account, as the water pressure behind the rings forces them outwards, so that while working they are quite tight.

When the gaps in the rings become too wide, indicating considerable wear, it is necessary to fit new rings.

Fitting ebonite rings

When fitting new rings to the pump bucket the following procedure should be followed. It is assumed that the pumps are recent types, when the gaps stated are correct; in these pumps the depth of the ring grooves in the buckets allows 1.5 mm clearance all round behind the ring, as indicated in the diagram at X, Figure 3.61.

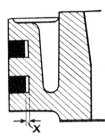

Figure 3.61 Clearance under ebonite rings

First, test the ebonite ring for lateral clearance by inserting it in its groove. The lateral clearance should be 0.25 mm for pumps up to 150 mm bore, and 0.4 mm for pumps over 150 mm bore. Move the ring round until it has all passed completely within the groove. If all is in order, saw through the ring at an angle of 60°, using a fine hacksaw, so that the gap is approximately 0.8 mm. When fitting oversize rings, or when the feed water to be pumped is above 100°C, the gap width should be adjusted by filing if necessary, to give the following clearances when the ring is placed in position in the pump barrel:

   0.8 mm gap for pumps up to 150 mm bore;
   1.6 mm gap for pumps 175 mm bore;
   2.4 mm gap for pumps 250 mm bore and upwards.

If the gaps are much greater, there will be a tendency to leakage.

The pliable ring is now ready for springing over the bucket into its groove, and must be made pliable by submerging in boiling water for several minutes. Spring the ring into position, and working from the side opposite to the gap, press the ring into the groove; push it well home, until the gap is 10 mm. Hold the ring firmly in this position for one minute while it cools and sets hard; it will then have regained its spring. On inserting the pump bucket into the barrel, the gap will close to its proper size, and the ring will be given sufficient spring to ensure freedom from leakage.

If pump barrels are considerably worn, oversize ebonite rings should be fitted. These are stocked in sizes 3.2 mm larger in diameter than standard but special sizes can also be obtained.

The quality of ebonite is important, as the material is subjected to considerable wear and works in high temperature water; the quality recommended by the makers is Triple H which gives excellent results. When the water end is opened up, the bucket rod should be checked for necking and, if this is significant the rod should be skimmed and a new neck bush fitted to the water end cover.

Pump barrels

In examining a pump for overhaul, the condition of the pump barrel itself is important. Some wear is inevitable and in time this leaves ridges at the limits of ring travel. In order to avoid excessive wear and breakage of the ebonite rings, and to maintain pump efficiency, the ridges should be removed.

Where the wear in the pump barrel is less than 5 mm, the unworn end portions should be bored out parallel. Should the wear exceed this, the complete pump barrel should be bored out, and a new bucket fitted to support new ebonite rings of the new diameter. If the wear is bad enough to warrant a new liner, the makers should be consulted, as the work is difficult and requires a press for inserting the liner.

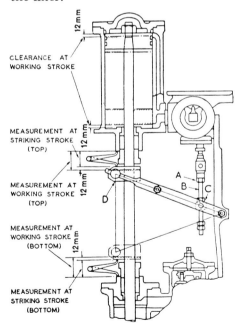

Figure 3.62 Method of adjusting stroke on Weir reciprocating pump (Weir Pumps Ltd.)

Stroke adjustment

To ensure the maximum steam economy with a direct-acting pump, the pump must run the normal length of stroke. If the stroke is short the clearance volume will be correspondingly increased, and a greater amount of steam will be used for a correspondingly smaller amount of water pumped. The following is a method of adjusting the stroke, as described in relation to the Weir pump, but it is a method which can be adopted for most makes of simplex reciprocating pump, see Figure 3.62.

While the pump is working on normal duty, slacken the locknut A and gradually raise the bottom portion of the valve spindle B by means of the hexagon C until the piston strikes the cylinder cover. Measure the distance between the steam-cylinder gland and the top of the crosshead D; screw the spindle B in a downward direction till the distance between the gland and the crosshead is 12 mm more than when previously measured. This completes the operation for the top part of the stroke. The process should be repeated for the bottom end of the stroke.

Slip test

The condition of the water end of the pump can be assessed by carrying out a slip test. For this a pressure gauge on the discharge chest of the pump is necessary.

Open the pump suction and exhaust valves. The pump discharge valve is then closed, preferably blanked off and the steam valve opened up slowly until the pressure gauge on the discharge chest is showing normal discharge pressure. The pump will work slowly because of leakage past the bucket rings and the valves. If, under these conditions the pump makes more than two double strokes/min, it requires further examination. Note that the normal discharge pressure must not be exceeded.

OVERHAUL OF CENTRIFUGAL PUMPS

Impeller clearance

In order to keep the water under pressure in the volute from returning to the suction, the impeller of a centrifugal pump has to be fitted close to the casing. Both wear and erosion may occur and the clearance increase. In order to return to the original clearance, renewable wearing rings or bushes are fitted. The side clearance of the impeller should be checked, and if it is found to be excessive, it is time to fit new wearing rings.

The casing and impeller should be examined to see that no portion has worn unduly thin. Diffuser blades showing wear or erosion at the tips should be cut back until they are at least 2.4 mm thick at the points, and dressed to give a slightly rounded nose. Facing rings, if eroded where adjacent to diffusers, should be replaced.

Shaft

The shaft should be carefully examined, especially the stuffing-box, and if excessive wear has taken place, the quality of the packing used should be investigated and a larger amount of lubricant used while running. A temporary repair with plastic white metal tinned on can be made to a shaft worn at the stuffing-box.

Bushes and bearing

Whenever the pump casing is opened the clearance of the various bushes which prevent leakage from one element to another should be measured for excessive wear. If the bearings are worn down excessively, renewal of the bushes is necessary. They should be re-metalled and bored out.

Stuffing-boxes

As the life of the shaft depends to a very great extent on the quality and treatment of the packing, it pays to maintain the stuffing-box in good order and to re-pack from time to time, especially where there is grit in the water. It is advantageous to run with a full stuffing-box, as by this means the intensity of pressure on the packing is reduced and consequently the frictional resistance and wear on the shaft.

In re-packing a stuffing-box, see that the packing is fitted so as to give uniform thickness all around the shaft sleeves. An excess of packing on one side of the shaft will result in deflection of the shaft and frequently in shaft breakage. In fitting new packing, the stuffing-box should be packed loosely and the gland set up lightly, allowing a liberal leakage in the case of stuffing-boxes subject to pressure above atmospheric. Then, with the pump in operation, tighten the gland in steps so as to avoid excessive heating and possible scoring of shafts or shaft sleeves.

A slight drip is necessary from the gland in order to provide

Table 3.4 Pump fault finding chart

| Reason for failure | Type of failure | | | |
	Failure to deliver liquid	Pump does not deliver rated capacity	Pump does not develop rated pressure	Pump loses liquid after starting
Wrong direction of rotation	X		X	
Pump not primed/filled with liquid	X			
Suction line not filled with liquid	X			X
Air or vapour pocket in suction line	X	X		X
Inlet or suction pipe insufficiently submerged	X	X		X
NPSH available too low	X	X		X
Pump not up to rated speed	X	X	X	
Total head greater than design	X	X		
Air leaks in suction line or stuffing box		X		X
Foot valve too small		X		
Foot valve clogged		X		
Viscosity greater than rated		X	X	
Wear rings worn*†		X	X	
Impeller damaged*†		X	X	
Internal leakage† (gaskets)		X	X	
Gas or vapour in liquid			X	X
Liquid seal to lantern ring plugged				X
Lantern ring not properly located				X
Speed too high				
Total head lower than rated (low Ns)		X†		
Total head higher than rated				
Viscosity and/or specific gravity				
Starved suction				
misalignment				
Worn or loose bearings				
Rotor out of balance				
Impeller blocked or damaged*				
Bent shaft				
Improper location of discharge valve				

* Centrifugal pumps only   †Mechanical defects
‡ Positive displacement pumps only

Pump overloads driver	Vibration	Bearings oveheat	Bearings wear rapidly	Motor heating up	Seized pump	Irregular delivery	Pump does not prime	Noisy pump
							X	
					X		X	
	X					X		X
	X							X
	X					X		X
						X	X	X
								X
X								
X								
X								
	X						X	X
	X							
X								
X								
X‡								
			X					
X			X			X		X
	X					X		X
	X	X	X					X
	X							X
	X		X					X
	X	X	X					X
	X							X

Table 3.4 (continued)

Reason for failure	Type of failure			
	Pump does not deliver rated capacity	Pump does not develop rated pressure	Pump loses liquid after starting	Pump overloads driver
Foundation not rigid				
Packing too tight				
Packing not lubricated				
Wrong grade of packing				
Insufficient cooling water				
Box badly packed				
Oil level too low/high				
Wrong grade of oil				
Dirt in bearings				
Moisture in oil				
Failure of oiling system				
Bearings too tight				
Oil seals too close fit				
Excessive thrust				
Lack of lubrication				
Bearings badly installed				
Too much cooling				
Too much grease in bearings				
Relief valve jammed ‡	X	X		
Relief valve spring badly adjusted‡	X			
If viscosity low, too low discharge pressure				
Pump does not deliver rated capacity (see type of failure)		X		
Pipes exert forces on pump				X
Vibration (see type of failure)				
Speed too high, check motor name plate				
Foreign matter in pump				
Viscosity lower than rated ‡	X	X		

‡ Positive displacement pumps only

Vibration	Stuffing box overheats	Bearings overheat	Bearings wear rapidly	Motor heating up	Seized pump	Irregular delivery	Pump does not prime	Noisy pump
X								X
	X							
	X							
	X							
	X	X						
	X	X					X	
		X						
		X						
		X	X					
		X	X					
		X						
		X						
		X						
			X					
			X					
			X					
			X					
		X						
							X	
					X	X		
								X
X	X	X	X		X			
			X					
				X				
					X			

(Reproduced by courtesy of Stothert and Pitt Ltd.)

lubrication for the packing. Excessive pressure should not be applied to the glands to prevent excessive leakage. If it is found that leakage from the gland is excessive, the correct procedure is to re-pack the stuffing-box at the earliest opportunity.

Hydraulic balance

Every 6 months and after an overhaul, it is advisable to check the balance piston clearance of the pump where such a device is fitted. As manufactured, the total clearance within the balancing device is 0.25 mm. The pump axial clearance may increase due to wear, which may be caused by the erosive action of solid matter in the feed water or by contact between the rotating and stationary faces of the balancing device during the first few revolutions when starting up.

As an aid to detecting faults in both centrifugal and positive displacement pumps Table 3.4 is reproduced by courtesy of Stothert and Pitt Ltd.

4 Various machines and appliances

AIR COMPRESSORS

Compressed air is used on board ship for starting diesel engines, in pneumatic control systems and for variuos pneumatic tools and cleaning equipment. By far the greatest use is for diesel engine starting using air pressures of 25 bar or more provided by reciprocating compressors.

This type of compressor will give a compression ratio of 7:1 in each stage depending on the size, cooling and speed of the machine.

Multi-stage units of various cylinder configurations and piston shapes have been used to produce the air pressures required. A selection of these are shown in Figure 4.1 but, because of its simplicity, accessibility and ease of maintenance, the two-stage two-cylinder, two-crank machine (type a) is now used almost invariably. Such machines are capable of compressing to 25—40 bar.

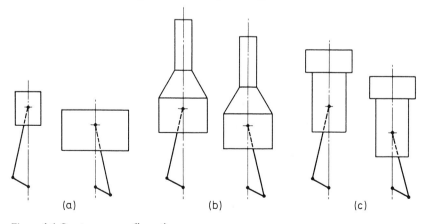

Figure 4.1 Compressor configuration

Cycle of operation

On the compression stroke the pressure rises to slightly above the discharge pressure. A spring-loaded non-return discharge valve opens and the compressed air passes through at approximately constant pressure. At the end of the stroke the differential pressure across the valve, aided by the valve spring, closes the discharge valve, trapping a small amount of high pressure air in the clearance space between the piston and the cylinder head. On the suction stroke the air in the clearance-space expands, its pressure dropping until such time as a spring-loaded suction valve re-seats and another compression stroke begins (Figure 4.2).

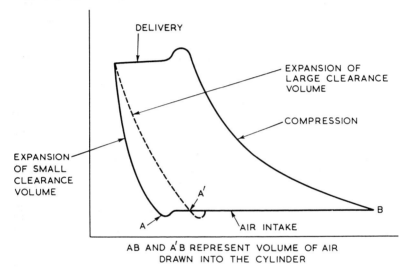

AB AND A'B REPRESENT VOLUME OF AIR
DRAWN INTO THE CYLINDER

Figure 4.2 Compressor indicator diagram
(Figures 4.2 to 4.8 by courtesy of Hamworthy Engineering Ltd.)

Cooling

During compression much of the energy applied is converted into heat and any consequent rise in the air temperature will reduce the volumetric efficiency of the cycle. To minimise the temperature rise, heat must be removed. Although some can be removed through the cylinder walls, the relatively small surface area available severely limits the possible heat removal and as shown in Figure 4.3 it is preferable to compress in more than one stage and to cool the air between the stages.

Most small compressors use air to cool the cylinders and the intercoolers, the cylinder outer surfaces being extended by fins and

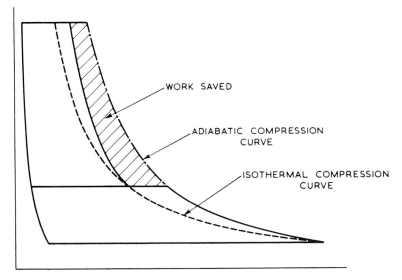

WORK SAVED

ADIABATIC COMPRESSION
CURVE

ISOTHERMAL COMPRESSION
CURVE

Figure 4.3 Multi-stage indicator diagram for 2-stage compressor with intercooling

the intercoolers usually being of the sectional finned-tube type over which a copious flow of air is blown by a fan mounted on the end of the crankshaft. In larger compressors used for main engine starting air it is more usual to use water-cooling for both cylinders and intercoolers.

Sea water is most commonly used for this purpose but fresh water ring mains serving compressors and other auxiliaries are not uncommon (see Chapter 1). In the former case the coolant can be circulated by a pump driven by the compressor or it can be supplied from a sea water ring main.

Two-stage starting air compressor

The compressor illustrated in Figure 4.4 is a Hamworthy 2TM6 motor-driven compressor typical of the two-stage, two-crank single-acting machines found at sea. This machine is available in a range of sizes to give free air deliveries ranging from 200 m^3/h at a discharge pressure of 14 bar to 274 m^3/h at 40 bar.

The crankcase is a rigid casting which supports a spheroidal graphite cast iron crankshaft in three bearings. The crankshaft has integral balance weights and carries two identical forged steel connecting rods.

Both the first and second stage pistons are of aluminium alloy with cast iron compression and oil control rings. The pistons are

1. Oil pump
2. Oil pump chain drive
3. Auxiliary drive cover
4. Water pump
5. Big end half bearing
6. Connecting rod
7. Drain valve
8. Fusible plug
9. Cooler delivery
10. Gudgeon pin
11. Small end bearing
12. Piston
13. Oil gauge
14. Suction filter and
    silencer
15. H.P. gauge
16. Cylinder head
17. Delivery valve
18. Suction valve
19. Suction valve
20. Delivery valve
21. L.P. cylinder head
22. Piston
23. Small end bearing
24. Gudgeon pin
25. Cooler return
26. Connecting rod
27. Drain plug
28. Flywheel
29. Key
30. Oil seal
31. Crankshaft
32. Crankcase
33. Breather
34. Main bearing halves
35. Crankcase door
36. Oil filter
37. Relief valve
38. Inlet strainer

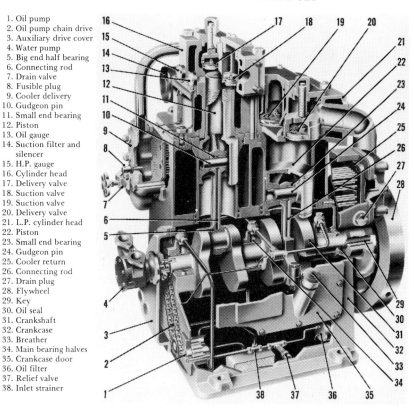

Figure 4.4 Sectional arrangement of Hamworthy 2TM6 compressor

connected to the connecting rods by fully floating gudgeon pins running in phosphor bronze bushes. The bushes are an interference fit in the connecting rods and are so toleranced that the collapse of the bore when fitting is allowed for, to provide the correct running clearance. Steel-backed white metal-lined 'thin shell' main and crank-pin bearings are used and all of the bearings are pressure lubricated by a chain driven gear pump. The lower cylinder walls are lubricated by oil mist.

The air suction and discharge valves are located in pockets in the cylinder heads. The valves are of the Hoerbiger type and are as shown in Figure 4.5. The moving discs of the valves have low inertia to permit rapid action. Ground landings are provided in the pockets on which the valve bodies seat: the bodies are held in place by set screws which pass through the valve box covers, capped nuts being fitted to the ends of the set screws. A combined air filter and silencer is fitted to the compressor air intake.

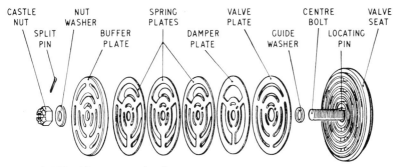

Figure 4.5 Hoerbiger automatic valve

The intercooler is of the single pass type. The shell forms an integral part of the cylinder block casting, the air passing through the tubes. The aftercooler is of the double pass U-tube type. Again the shell is integral with the cylinder block. Relief valves are fitted to the air outlets of each stage and are set to lift at ten per cent above normal stage pressure. The actual stage pressures vary according to the application. To protect the water side against over pressure in the event of a cooler tube failure, a spring loaded relief valve is fitted on the cylinder jacket. This valve has to be re-set by hand. Additional protection is afforded by a fusible plug fitted on the after-cooler discharge head.

Operation and maintenance

Compressors must always be started in the unloaded condition since pressures otherwise build up rapidly producing very high starting torques (Figure 4.6). During running the accumulated moisture in the separators must be drained off regularly. This is extremely important, firstly because the condensate, if allowed to pass from the first compression stage to a subsequent stage, may give rise to lubrication problems and secondly the oily emulsion so formed, combined with dirt picked up from the atmosphere (where no suction filters are fitted) can, if conditions are right, result in an explosion in starting air lines or in the reservoir. Moisture lying in the starting air lines will give rise to corrosion and can cause water hammer. It is also good practice to check air reservoirs regularly for the presence of liquid by giving the reservoir drains a good blow occasionally. A compressor should never be stopped without first unloading it. This is done by opening the 1st and 2nd stage drains. Failure to do so could result in a severe explosion.

The maker's instructions on the lubrication of the machine must be closely followed. Poor lubrication, or the use of an incorrect oil, can cause severe wear and sticky valve operation.

By far the greatest item of maintenance concerns the valves. The seats of these may need refacing from time to time due to their constant pounding. Whenever this is done care must be taken to see that the valve lift is not altered since increased wear will then occur. Poor selection of lubricating oil, dirt, or overheating, will give rise to valve sticking or seat pitting.

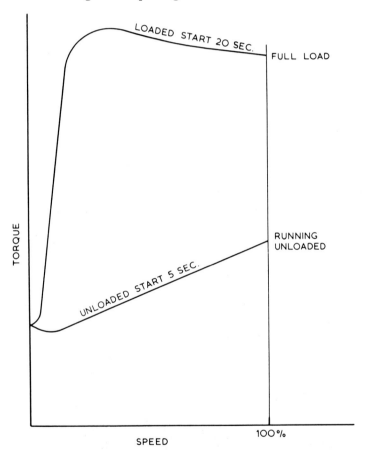

Figure 4.6 Compressor starting torque/speed curves

Bad valve operation can be detected by observation of the inter-stage pressures and is usually accompanied by excessive discharge temperature although the latter will also be a symptom of poor cooling.

Adequate attention must be given to the water cooling system. Overcooling can cause condensation on the cylinder walls, adversely affecting lubrication, while poor cooling, due perhaps to scale formation (especially in a sea-water cooled machine) will result in a fall-off in volumetric efficiency and rapid valve deterioration. In extreme cases compressor explosions can occur. Lloyds require that the compressor should be so designed that the air discharge temperature to the reservoir should not substantially exceed 93°C. Comments on the maintenance of heat exchangers appear elsewhere.

Automatic operation

When compressors are arranged for automatic or remote operation the cylinders can be unloaded, for starting and stopping, by a number of methods. These include:

1. Throttling the suction;
2. Speed variation;
3. Bypassing the discharge to suction;
4. Fitting depressors which hold the suction valve plates on their seats;
5. Changing the volumetric clearance of the cylinders;
6. Step unloading the cylinders in a multi-cylinder machine.

Methods 3 and 4 are commonly used. Method 4 requires slightly less power when running unloaded than method 3, but the latter can be effected by a modular unloader which is both robust and simple in construction.

Figure 4.7 shows a typical modular unloader which will vent the inter- and after-coolers to atmosphere via the inlet filter silencer, the initial blast of air helping to keep the inlet filter clean.

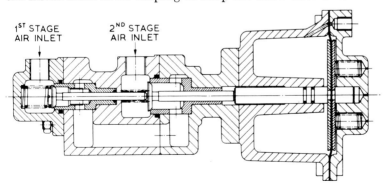

Figure 4.7 Modular unloader

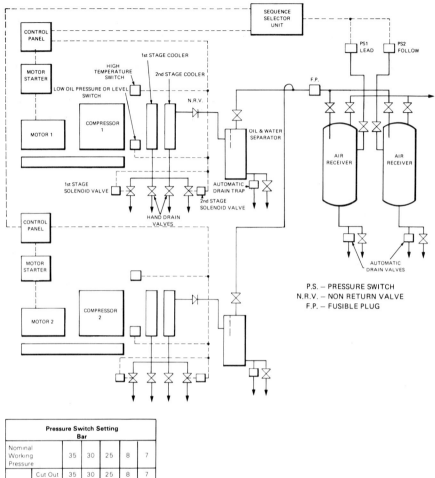

Pressure Switch Setting Bar						
Nominal Working Pressure		35	30	25	8	7
Lead	Cut Out	35	30	25	8	7
	Cut In	30	25	21	7	6
Follow	Cut Out	34	29	24	7 6	6 7
	Cut In	29	24	20	6 6	5 7

Figure 4.8 Automatic operation of compressors (Hamworthy Ltd.)

It should be noted that the unloader is always in the unloading position unless air is on the diaphragm (normally from the first stage) overcoming the actuating spring thrust.

To drain the coolers continuously during running, an automatic device must be fitted to each cooler.

Many manufacturers have now moved away from the modular unloader and are using solenoid-operated unloading valves on the first and second stage cooler drains (Figure 4.8). The solenoid valves are also opened briefly on an intermittently timed cycle, thus providing automatic draining as well as unloading.

The solenoid valves are normally open, venting the cooler and pressure spaces of the compressor whilst it is stationary and providing unloaded starting. Two timers are located in the control panel. One delays the closing of the cooler drains giving the compressor time to run up to speed; it also controls the time that the solenoid valve remains open during periodic draining. The other timer controls the frequency of this periodic draining. The unit gives a more positive draining sequence than the older float or thermodynamic drain valves.

Figure 4.8 also shows how two compressors may be arranged to provide automatic start and stop of both machines. Either machine can be selected as 'lead' machine. This will run preferentially during manoeuvring, automatically stopping and starting under the control of a pressure switch on the air receiver. The 'follow' machine is arranged to back-up the 'lead' machine during manoeuvring, cutting in after the 'lead' machine when the receiver pressure falls below a pre-set value (see table in Figure 4.8). When the pressure switch stops the compressor the solenoids holding the drain valves closed are de-energised, allowing the drain valves to open.

Delivery line non-return valve

A non-return valve in the air delivery line near to each compressor is essential. It must be of the low inertia type similar to the compressor discharge valve.

Automatic shut-off of the cooling water supply, where this is independent of the compressor, should also be fitted. This could be a valve, normally closed, which is opened by the first stage pressure acting on a diaphragm.

Monitoring

Monitoring of the following is necessary; additional monitoring however may often be applied.

1. Lubricating oil pressure (or level, if splash-fed).
2. The final stage air delivery temperature immediately after the delivery valve.

In the automatic operation of a battery of compressors the best arrangement is to have the lead machine running continuously during manoeuvring, automatically loading and unloading according to supply and demand, but stopping and starting automatically after 'full away'. A schematic arrangement is shown in Figure 4.8.

Instrument air

Instrument air must be clean and dry. Air leaving a compressor is 100% humid and has some free moisture present; the free moisture is easily removed by ceramic filters but an absorbent type drier is also required to give the desired dryness factor.

The source of the air can be from the main air reservoirs after reducing the air pressure to the 7 or 8 bar pressure required by the system. In this case it is likely that the dew point of the air will be such that an absorbent drier will not be needed and a ceramic filter will suffice.

Oil-free and non-oil-free rotary compressors

Both these machines deliver wet air which must be dried as described above but the non-oil-free machine passes over some oil which must also be removed. This is usually effected by a pre-filter followed by a carbon absorber, which removes the oil, followed by an after filter (to remove the remaining free moisture). Either system can be found at sea.

OIL/WATER SEPARATORS

Oil/water separators are necessary aboard vessels to prevent the discharge of oil overboard when pumping out bilges, deballasting or when cleaning oil tanks. The requirement to fit such devices is the result of international legislation. The need for such legislation is that free oil and oily emulsions discharged in a waterway can interfere with natural processes such as photosynthesis and re-aeration, resulting in destruction of the algae and plankton so essential to fish life. Inshore discharge of oil can cause damage to bird life and mass pollution of beaches. Ships found discharging water containing more than 100 mg/litre of oil or discharging more than 60 litres of oil per nautical mile can be heavily fined, as also can the ship's Master.

In consequence it is important that oil/water separators are correctly installed, used and maintained.

Principle of operation

The principle of separation by which all commercially available oil/water separators function is the gravity differential between oil and water.

In oily water mixtures, the oil exists as a collection of almost spherical globules of various sizes. The force acting on such a globule causing it to move in the water is proportional to the difference in weight between the oil particle and a particle of water of equal volume. This can be expressed as:

$$F_s = \frac{\pi}{6} D^3 (\rho_w - \rho_o) g \qquad (1)$$

where:

F_s = separating force.
ρ_w = density of water.
ρ_o = density of oil.
D = dia. of oil globule.
g = acceleration due to gravity.

The resistance to the movement of the globule depends on its size and the viscosity of the fluid. For small particles moving under streamline flow the relationship between these properties is expressed by Stoke's Law:

$$F_r = 3 \pi \mu v d \qquad (2)$$

where:

F_r = resistance to movement.
μ = viscosity of fluid.
v = terminal velocity of particle.
d = dia. of particle.

When separation of an oil globule in water is taking place, F_s will equal F_r and the above equations can be worked to express the relationship of the terminal (or in this case rising) velocity of the globule with viscosity, relative density and particle size:

$$v = \left(\frac{g}{18\mu} \right) (\rho_w - \rho_o) d^2 \qquad (3)$$

In general, a high rate of separation is favoured by a large size of oil globule, elevated temperature of the system (which affects both the specific gravity differential of the oil and water and the viscosity of the water) and the use of sea water. Turbulence or agitation should be avoided since it causes re-entrainment; conversely laminar (or streamline) flow is beneficial.

While heating coils are fitted to all types of separators found at sea (as much to enable high viscosity oils to be removed from the separator as to provide optimum separating conditions), a variety of means are used to encourage laminar flow and the formation of large oil globules, with associated increase in velocity. Units also exist which attempt a higher rate of separation by inducing cyclonic flow, the theory being that the centripetal force acting on the globule will effectively increase the velocity differential. It is extremely difficult however to induce this type of flow without causing considerable turbulence with resultant re-entrainment.

Pumping considerations

Since the rate of separation depends on the oil globule size it will be appreciated that any disintegration of oil globules in the oily feed to the separator should be avoided and this factor can be seriously affected by the type and rating of the pump used. Tests were carried out by a British government research establishment some years ago on the suitability of various pumps for separator feed duties and the results are shown in Table 4.1.

It follows that equal care must be taken to avoid turbulence occurring in pipework feeding the separator.

Placing a filter on the discharge side of a separator will 'polish' the discharge but overloading must be avoided.

Table 4.1 Pump suitability for oil/water separator duty

Type	Remarks
Double vane Triple screw Single vane Rotary gear	Satisfactory at 50 per cent derating
Reciprocating Hypocycloidal	Not satisfactory: modification may improve efficiencies to 'satisfactory' level
Diaphragm Disc and shoe Centrifugal Flexible vane	Unsatisfactory

Separator types

Figure 4.9 shows a Turbulo separator. This consists of a vertical cylindrical pressure vessel containing a number of inverted conical plates. The oily water enters the separator in the upper half of the unit and is directed downwards to the conical plates. Large globules of oil separate out in the upper part of the separator: the smaller globules are carried by the water into the spaces between the plates. The rising velocity of the globules carries them upwards where they become trapped by the under-surfaces of the plates and coalesce until the enlarged globules have sufficient rising velocity to travel along the plate surface and break away at the periphery. The oil rises, is caught underneath an annular baffle and is led up through the turbulent inlet area by risers to collect in the dome of the separator. The water leaves the conical plate pack via a central pipe which is connected to a flange at the base of the separator.

Two test cocks are provided to observe the depth of oil collected in the separator dome. When oil is seen at the lower test cock, the oil drain valve must be opened. An automatic air release valve is located in the separator dome. An electronically operated oil drainage valve is also frequently fitted which works on an electric signal given by liquid level probes in the separator. Visual and audible oil overload

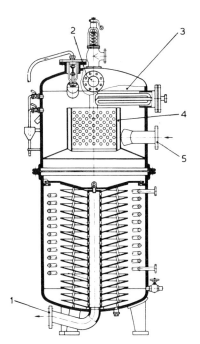

Figure 4.9 Sectional view of Turbulo separator
1. *Clean water run-off connection*
2. *Outlet*
3. *Oil accumulation space*
4. *Riser pipes*
5. *Inlet connection*
(Howaldtswerke-Deutsche Werft)

indicators may also be fitted. To assist separation steam coils or electric heaters are fitted in the upper part of the separator. Where high viscosity oils are to be separated additional heating coils are installed in the lower part.

Before initial operation, the separator must be filled with clean water. To a large extent the conical plates are self-cleaning but periodically the top of the vessel should be removed and the plates examined for sludge build-up and corrosion. It is important that neither this separator nor any other types is run at over capacity. When a separator is overloaded the flow becomes turbulent, causing re-entrainment of the oil and consequent deterioration of the effluent quality.

The Comyn oil/water separator

Figure 4.10 is a diagram of the *Comyn* oil water separator, and the following operating instructions are issued by the manufacturers.

General procedure. Before pumping oily water into the separator it is important that the separator is completely filled with clean water, either direct from the sea or through the filling valve provided for this purpose. It is essential to keep the lower internal surfaces and outer chamber from becoming coated with oil.

To fill the separator, slightly open scum valve 3, also high level test and air release valve 5 to vent the separator during filling. When water flows freely from these valves close and stop filling. Close filling valve. Before commencing to pump oily water through the separator, ensure that water discharge valve on ship's side is *open,* to allow the water effluent to be pumped overboard.

Slightly open both mixture inlet test valve and high level test valve. Leave these two valves slightly open during the working of the separator. Valve 16 will indicate the nature of the mixture entering the separator, while valve 5 will free the separator of air and also denote the presence of oil within the oil recovery dome.

Procedure using automatic control. Switch on the electronic control panel. If water completely covers the electronic probe the *green* indicator light will show. Alternatively, if the water level in the oil dome, after filling, is below the level of the electronic probe, the *red* indicator light will show, in addition to the green indicator light, and the alarm bell will ring. Note that oil has exactly the same effect on the electronic probe as air.

Check that the emergency hand control wheel fitted to the automatic oil outlet valve 2, is in the open position.

Check that the operating air or water pressure is 1.4—3.5 bar.

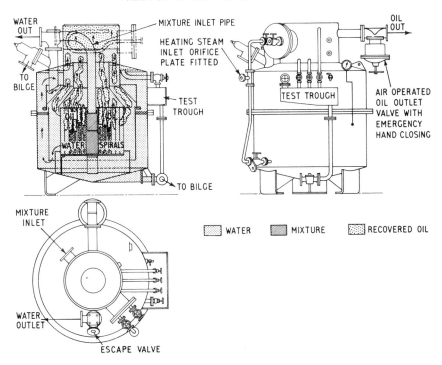

Figure 4.10 The 'Comyn' separator

1. *Escape valve*
2. *Oil recovery valve*
3. *Scum valve*
4. *Cleaning valve*
5. *High level test*
6. *Deep level test*
7. *Steam to coil*
8. *Exhaust from coil*
9. *Bosses for electronic probes*
(*Alexander Esplen & Co. Ltd.*)

11. *Drain to bilge*
12. *Test cock (water discharge)*
13. *Pressure gauge*
14. *Pressure gauge connection*
15. *Water outlet N.R. valve*
16. *Mixture inlet test valve.*
21. *Ring and blank plate*
22. *Coil steam header*

Open air or water supply to solenoid pressure control valve.

On commencing to pump oily water into the separator, the clean water surrounding the electronic probe causes the coil of the solenoid pressure controller to become energised and move the pilot valve to the *open* position, closing the outlet previously *open to vent*. Air is permitted to pass to the top of the diaphragm operated oil discharge valve 2 and immediately closes it.

As the pressure within the separator builds up to approximately 0.75 bar the spring-loaded water discharge valve opens.

A gradual build-down of oil takes place, which, in time, completely covers the electronic probe. At this juncture the control unit de-energises the coil of the pressure control valve, returning it to the *open to vent* position. This permits the spring assisted oil recovery valve 2 to open, and any oil present within the separator is automatically discharged back to the oil storage tanks. The *red* and *green* indicator lights will show on the control panel.

When valve 2 opens and — due to the fact that the recovered oil is being discharged against atmospheric pressure — the pressure within the separator falls to zero and the water discharge valve 15 automatically closes. Therefore it is impossible to discharge water while recovering oil.

The *red* indicator light on the control panel will go out when the oil recovery dome is free of oil and the electronic probe again immersed in water, the air or water pressure is again restored to the top of the diaphragm in the oil recovery valve 2, which causes it to close. The pressure is once more increased within the separator, which opens the water discharge valve 15.

General. Deep level test valve 6 is fitted to the centre oil chamber and indicates the level within the separator at which the oil must be recovered, in order to prevent overloading the separator.

Scum valve 3 is provided for cleaning purposes, and should be opened periodically to draw off any accumulated oil, particularly if it is suspected that the separator has become dirty or overloaded.

Test cock 12 on the side of the separator is fitted to allow samples of the water effluent to be taken. Such samples should be taken in a beaker, preferably glass, and should traces of oil be present, *immediately open scum valve 3.* Oil showing at 12 may also mean that the separator has become fouled, in which case it should be cleaned by steaming out.

It is important that the rated throughput is not exceeded.

Recovered oil valve 2. This is a diaphragm operated pressure closing type of valve, with a spring incorporated to assist in the opening.

When the pumping of oily water has been completed it is advisable to pump clean water through the separator until all oil has been recovered, thus leaving the separator clean for the next time. A full separator lessens the risk of corrosion.

The non-return spring-loaded water discharge valve 15 set at approximately 0.5 bar is fitted to the water outlet. Apart from its function previously described, this valve prevents syphoning, should the separator be positioned above the overboard discharge valve on the ship's side. Syphoning, if permitted once pumping has stopped, would allow any oil remaining in the separator to be pulled down and foul the lower internal surfaces of the separator, and possibly be carried overboard.

Where heating is fitted this is primarily for cleaning purposes, although it may sometimes be found advantageous to use the heating coil during separation in cold climates, or when dealing with oils of high viscosity. Note that on no account must the steam injection be used when the separator is in operation.

Cleaning the Comyn separator

To clean out the separator ensure that all oil possible has been drawn off through the oil recovery valve 2 and scum valve 3, and that the separator is full of water. Turn on steam to heating coil; also steaming out connection, and allow the separator to become thoroughly hot. Pump clean water slowly through the separator, and occasionally open oil recovery valve 2 to free the separator of any oil that is present. Use hand control cock during this operation to open and close oil recovery valve 2 as required.

Finally, open drain cock 11 to bilge and continue to pump clean water into the separator for a short period. Close valve 11 and leave the separator full of clean water at all times.

The automatic control system associated with all sizes of Comyn Separator consists of a transistorised level control unit, actuated by a probe mounted horizontally within the oil recover dome. The level control unit which incorporates indicator lights, is mounted, together with an 'on/off' fused switch, alarm bell and terminal box, on a suitable panel attached to the separator shell. The control panel is sometimes mounted at some position away from the separator. The actuation of the level control unit directly governs the opening and closing of the diaphragm operated pressure closing type oil outlet valve, via a suitable solenoid pilot control valve, which is also mounted on and pre-wired to, the control panel.

Air at approximately 3.5 bar actuates the oil outlet valve, which incorporates a handwheel for emergency manual operation of the separator. A secondary control system is sometimes fitted.

Procedure using hand control. In order to operate the separator manually the oil recovery valve is fitted with emergency handwheel control (on older models a three-way cock is fitted). The separator is filled with clean water as before, and the air is shut off the solenoid valve control. The oil recovery valve is shut manually. Pumping of oily water is allowed to continue until oil shows at the deep level test valve 6. When this is observed the hand control is used to *open* the oil recovery valve and the accumulated oil within the oil dome leaves the separator. When clear water again shows at the high level test valve 5, the hand control is used to close the oil discharge valve 2.

The pressure within the separator is once more restored to approximately 0.7 bar which automatically opens the water discharge valve 15.

Coalescing bed-type oily water separators

A coalescing bed type oil/water separator in common use is the Firtop. Figure 4.11 shows a standard 5 tonnes/hr throughput unit with automatic discharge of the collected oil and steam coils for heating the oil to facilitate separation and discharge.

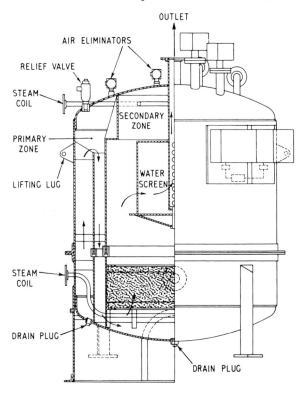

Figure 4.11 Example of coalescing bed separation (G. & M. Firkins Ltd.)

Vertical separators are available up to 200 tonnes/hr capacity in a single unit, but a horizontal arrangement with twin-tiered coalescing beds can be supplied to handle up to 500 tonnes/hr. The principle of operation is as follows (see Figure 4.11).

Oily water is pumped through a non-return valve in the inlet to the separator; it enters the outer annular space and flows vertically upwards towards the primary collection zone where the larger globules of oil, which separate easily by gravity, collect. The water then flows downwards through the inner annular space into the bottom of the separator, through ports into the coalescing bed at a very low speed when the oil adheres to the granular material in the bed and forms, in effect, a climbing film which moves upwards through the bed at a slower speed than the water. The top of the bed is covered by a perforated plate. When the oil reaches the underside of this plate it flows through the perforations and due to its viscosity and surface tension, begins to form globules which increase in size until they have sufficient buoyancy to float upwards at a faster rate than the water and collect in the secondary zone while the water flows through the water screen and into the outlet pipe.

The coalescer is capable of forming oil globules up to 12 mm diameter and the heavier fuel oils usually form long 'stalagmites' which grow to a considerable size before they are released. Because of the low velocities employed the collected oil can flow past the water screen without becoming re-entrained and carried through into the outlet.

The separators can be arranged for manual or automatic operation. Four test cocks are fitted, two to each zone with the internal pipes arranged so that the high and low levels of oil can be monitored. When oil reaches the high oil level test cock a manual discharge valve on the collection zone being monitored is opened to allow the oil to discharge to the oil collection tank. The low oil level test cock is then opened and when water appears the oil discharge valve is closed. The separator pressure valve is maintained at 1 bar either by having a high level overboard discharge or a spring loaded or manual discharge valve in the clean water outlet.

The fully automatic version employs electronic capacitance probe units to monitor the oil/water interface and to open automatically oil discharge valves — which can be of either the solenoid or electro-pneumatic type — when the oil reaches the high level. When the oil has been discharged a signal from the low oil level probe closes the valve.

To prevent the electronic probes from giving false signals due to the presence of any air the separator is fitted with air discharge valves which automatically vent any air in the primary and secondary zones.

It is advisable to fit a high capacity basket strainer with a 20 mesh screen between the oily water pump and the separator to prevent dirt and debris from fouling the underside of the coalescing bed. With a suitable strainer the bed will operate almost indefinitely without servicing and does not normally require cleaning by back flushing or renewal but without adequate filtration the bed is liable to become partially blocked resulting in a fall-off of effluent quality and a significant rise in back-pressure. In the event of this occurring the bed should be backwashed. This can be done by opening the drain valve in the bottom of the unit and allowing the contents to run to the bilge. For satisfactory operation it is essential that the bed is well charged with oil at all times; new separators are supplied with the bed already charged. When operating the separator for the first time it should be filled with clean water before use.

The separator is designed so that it is only necessary to lift the top section 75 mm to gain access to the interior and for full examination, the top half can be moved sideways. This facility enables the separator to be installed between the decks where the headroom is low. Steam coils or electric heaters may be fitted to reduce the specific gravity of the oil to aid collection and discharge. Drain and steaming out connections are provided so that the separator can be drained and steamed out before opening up for survey.

This design of separator produces an effluent quality of better than 20 mg/litre. Figures of 15 mg/litre with 25% oil concentration and 6 mg/litre with 5% oil concentration have been recorded during tests on board ship. When fitted with a clean water discharge valve which closes when the oil discharge valves are open, the separator can function adequately when handling 100% oil as for instance when the bilge water has been stripped off and only the oil is left.

FUEL AND LUBRICATING OIL TREATMENT

To ensure good combustion in diesel engines and reduce wear and corrosion in this type of engine and in turbines it may be necessary to remove certain impurities from fuel and lubricating oils. These include ash, various salts and water present in fuel oil and carbonaceous matter, metals, acids and water present in used lubricating oil.

When the impurities are heterogeneous i.e. suspended solids or immiscible liquids, they can be removed reasonably successfully by one or more of a number of methods. The main methods used at sea are centrifuging, filtration and coalescing. Of these three methods centrifuging is the most widely used.

Centrifuges

It was shown in the section on oil/water separators how liquids with a specific gravity difference can be separated by gravity. The process was expressed mathematically as

$$F_s = \frac{\pi}{6} D^3 \left(\rho_w - \rho_o \right) g \tag{1}$$

Clearly in a standing vessel the acceleration cannot be altered to enhance the separation force F_s, but by subjecting the operation to centrifugal force the above expression can be replaced by

$$F_s = \frac{\pi}{6} D^3 (\rho_w - \rho_o) \omega^2 r \qquad (4)$$

where:

ω = angular velocity
r = effective radius

Both the rotational speed and the effective radius are controllable within certain engineering limitations. Thus if our standing vessel is replaced by a rotating cylinder the separating force and hence the speed of separation can be increased. This, effectively, is what happens in a centrifuge.

For many years marine centrifuges were designed for batch operation, that is the machines were run for a period during which solids accumulated in the bowl. The machine was stopped when the accumulated solids began to impair its performance and the solids were removed. Batch centrifuging is still commonly used especially

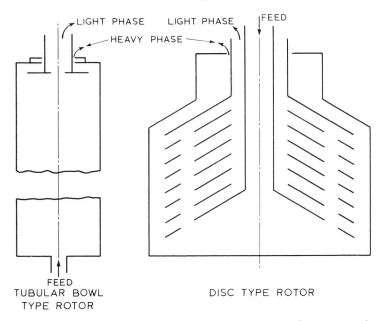

Figure 4.12 Comparison of narrow and wide bowl centrifuges (Pennwalt Ltd.)

for lubricating oil purification, but many machines capable of continuous or semi-continuous sludge discharge are now at sea. Two distinct types of batch operated machines have been used. These are illustrated in Figure 4.12 and classified in Table 4.2.

Table 4.2 Details of tubular and disc type centrifuges used for batch treatment of liquids

	Maximum centrifugal force	Bowl dia. (mm)	Length dia. (of bowl)	Drive
Batch tubular Bowl type	13 000 – 20 000 X g	Up to 180	Up to 7	Bowl suspended from above
Batch disc bowl Type	5 000 – 8 000 X g	Up to, say, 600	Generally < 1	Bowl supported from below

The tubular bowl machine is physically able to withstand higher angular velocites than the wide or disc bowl type, hence a higher centrifugal force can be applied. The heavier phase only has a short distance to travel before coming to the bowl wall where the solids are deposited and the heavy phase liquid is guided to the water discharge if the machine is so set up. On the other hand the sludge retention volume and liquid detention time for a given throughput can only be increased by lengthening the bowl. This gives rise to bowl balancing and handling problems.

The wide bowl type can retain more sludge before its performance is drastically impaired and can be cleaned in situ. Thus it has no handling problems. On the other hand settling characteristics in a wide bowl machine are relatively poor towards the bowl centre and the distance the heavy phase has to travel before reaching the bowl wall is great.

To overcome these problems a stack of conical discs, spaced about 2–4 mm apart is arranged in the bowl. The liquid is fed into the stack either from the outer edge of the discs or via distribution holes towards the periphery of the discs depending on the make of the machine, and flows along the spaces between adjacent plates. The plates then act as an extended settling surface, the heavy phase particles passing through the light phase as shown in Figure 4.13 and impinging on the under surfaces of the discs. Once the particles have impinged on the disc surface they are able to accumulate and eventually slide along the discs towards the periphery. At the periphery of the disc the water globules and solid particles leave the stack. The heavier solid particles pass to the bowl wall, and the water is sandwiched between the solids and the oil, which orientates itself towards the bowl centre. The boundaries at which substances meet are known as interfaces.

The oil/water interface is very distinct and is known as the e-line. To gain the fullest advantage from the disc stack the e-line should be located outside of it. On the other hand if the e-line is located outside the water outlet baffle (top disc) discharge of oil in the water phase will take place.

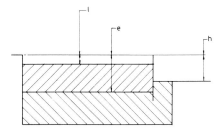

Figure 4.13a (above) Hydrostatic situation in a gravity settling tank

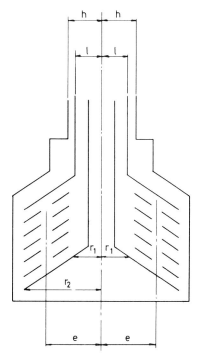

Figure 4.13b (right) Hydrostatic seal in a disc type centrifugal bowl (Penwalt Ltd.)

Referring back to gravity separation in a settling tank, if the tank is partitioned, as shown in Figure 4.13a, continuous separation will take place. Since the arrangement is a very crude U-tube containing two liquids of different specific gravities, the height of the liquid in the two legs will have the relationship

$$\rho_l \left(e - l \right) \doteq \rho_h \left(e - h \right) \qquad (5)$$

where:

ρ_l = density of oil.
ρ_h = density of water.

A very similar arrangement is found in the centrifuge. This is shown in Figure 4.13b and the equation now becomes

$$\omega^2 \, \rho_l \, (e^2 - l^2) = \omega^2 \, \rho_h \, (e^2 - h^2) \tag{6}$$

or $\dfrac{\rho_h}{\rho_l} = \dfrac{e^2 - l^2}{e^2 - h^2}$ \hfill (7)

The mechanical design of the centrifuge requires that the e-line is confined within certain strict limits. However variations in ρ_l will be found depending upon the port at which the vessel takes on bunkers. It is necessary therefore to provide means of varying h or l to compensate for the variation in specific gravity. It is usually the dimension h which is varied, and this is done by the use of dam rings (sometimes called gravity discs) of different diameters. Normally a table is provided with the machine, giving the disc diameter required for purifying oils of various specific gravities. Alternatively the disc diameter D_h may be calculated from the following formula which is derived from (7)

$$D_h = 2\sqrt{\left[l^2 \, \frac{\rho_l}{\rho_h} + e^2 \left(1 - \frac{\rho_l}{\rho_h} \right)\right]} \tag{8}$$

The dimension e can be taken as the mean radius of one thin conical plate and the heavy top conical plate (outlet baffle). If oil is discharging in the water outlet the gravity disc is too large.

It is important to realise that variation in oil temperature will cause a proportional variation in S.G. and for this reason the feed temperature must remain constant. To some extent the feed rate will have an effect on the e-line (because it can alter the overheight of the liquid flowing over the lip of the gravity disc): excessive feed should be avoided in any event since the quality of separation deteriorates with increase of throughput. To prevent oil passing out of the water side on start up it is necessary to put water into the bowl until the water shows at the water discharge. The bowl is then hydraulically sealed.

Alfa-Laval intermittent discharge centrifuge

Figure 4.14 shows a centrifuge bowl capable of being programmed for periodic and regular dumping of all or part of the bowl contents to remove the sludge build-up.

The sludge discharge takes place through a number of slots in the bowl wall(a). Between discharges these slots are closed by the sliding bowl bottom, which constitutes an inner, sliding bottom in the separating space. The sliding bowl bottom is forced upwards against a seal ring by the pressure of the operating liquid contained in the space below it. This exceeds the counter-acting downward pressure from the process liquid, because the underside of the sliding bowl bottom has a larger pressure surface (radius R_1), than its upper side (radius R_2). Operating liquid is supplied on the underside of the bowl via a device known as the paring disc under the bowl which maintains a constant operating liquid annulus (radius R_3) under the bowl, as its pumping effect neutralises the static pressure from the supply.

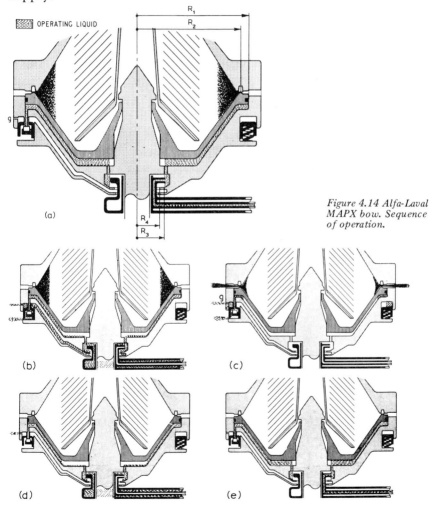

Figure 4.14 Alfa-Laval MAPX bow. Sequence of operation.

When the sludge is to be discharged, operating liquid is supplied through the outer, wider supply tube so that if flows over the lower edge of the paring chamber (radius R_4) and continues through a channel out to the upper side of a sliding ring, the operating slide. Between discharges, the operating slide is pressed upwards by coil springs. It is now forced downwards by the liquid pressure, thereby opening discharge valves from the space below the sliding bowl bottom so that the operating liquid in this space flows out (b).

When the pressure exerted by the operating liquid against the underside of the sliding bowl bottom diminishes, the latter is forced downwards and opens, so that the sludge is ejected from the bowl through the slots in the bowl wall. Any remaining liquid on the upper side of the operating slide drains through a nozzle g (c). This nozzle is always open but is so small that the outflow is negligible during the bowl opening sequence.

On completion of sludge discharge, the coil springs again force the operating slide upwards (d), thus shutting off the discharge valves from the space below the sliding bowl bottom. Operating liquid is supplied through the outer, wider tube, but only enough to flow to the space below the sliding bowl bottom and force the latter upwards so that the bowl is closed. (If too much liquid is supplied, it will flow into the channel to the operating slide and the bowl will open again.)

The outer, wider inlet is now closed while the inner, narrower one is open (e). The paring disc counter-balances the static pressure from the operating liquid supply, and the bowl is ready to receive a further charge of oil. The situation is identical with that shown in the first illustration of the series but for the difference that the sludge discharge cycle is now accomplished.

Bowls are available for four different types of sludge discharge:

Total discharge;
Rapid total discharge;
Partial discharge;
Combined programme.

In all these cases the operating liquid must be supplied under adequate pressure at the correct moment and in the right quantity during carefully calculated time intervals.

Periodically the bowl requires to be stripped and thoroughly cleaned. It is important to remember that this is a precision instrument, very carefully balanced and all parts should be treated with the utmost care. In dismantling it is necessary to note that a number of left hand threads are used in the machine and the full maintenance instructions should therefore be carefully read before stripping the equipment.

Treatment of detergent type lubricating oils

The main function of a detergent lubricating oil is to keep solid contaminations in suspension and prevent both their agglomeration and deposition in the engine. This function reduces ring sticking, wear of piston rings and cylinder liners, and generally improves the cleanliness of the engine. Other functions include reduction of lacquer formation, corrosion and oil oxidation. These functions are achieved by the formation of an envelope of detergent oil round each particle of solid contaminant. This envelope prevents coagulation and deposition and keeps the solids in suspension in the oil.

In engines of the trunk piston type with a combined lubrication system for bearings and cylinders, in addition to the deposition of the products of incomplete combustion which occurs on pistons, piston rings and grooves, some of these products can be carried down into the crankcase, contaminating the crankcase oil and causing deposition in oil lines, etc. Detergent oils are, therefore, widely used in this type of engine.

The detergent additives used today are, in most cases, completely soluble in the oil. There may, in certain instances, be a tendency for the detergent to be water soluble, or for emulsion to be formed, particularly if a water-washing system is incorrectly used during separation.

Manufacturers of centrifuges have carried out a considerable amount of research work in conjunction with the oil companies on the centrifuging of detergent lubricating oil using three different methods of centrifuging, i.e., purification, clarification and purification with water washing.

The following is a summary of the findings and recommendations based on results which were obtained:

When operating either as a *purifier* (with or without water washing) or as a *clarifier* all particles of the order of 3–5 microns and upwards are completely extracted, and when such particles are of high specific gravity, for example iron oxide, very much smaller particles are removed. The average size of solid particles left in the oil after centrifuging are of the order of only 1–2 microns. (One micron is a thousandth part of a millimetre.) No particles are, therefore, left in the oil which are of sufficient size to penetrate any oil film in the lubricating-oil system.

A centrifuge should be operated only with the bowl set up as a *purifier,* when the rate of contamination of the lubricating oil by water is likely to exceed the water-holding capacity of the centrifuge bowl between normal bowl cleanings.

When the rate of water contamination is negligible the centrifuge can be operated with the bowl set up as a *clarifier.* Any water separated will be retained in the dirt-holding space of the bowl.

Purification with water washing can be employed to remove water soluble acids from the oil, in addition to solid and water contaminants. This method can be used with most detergent lubricating oils but *it should not be used without reference to the oil suppliers.*

Normally it is not desirable to water wash detergent oils because one of their functions is to prevent the formation of acid in the oil. If it is decided to use water washing the quantity of wash water used must be strictly limited and should not be in excess of 1 per cent of the oil flow. If too much water is used emulsion troubles may occur.

Batch and continuous lubricating systems

For small or medium units without a circulatory lubricating system, the oil is often treated on the batch system. As large a quantity of oil as possible is drained from the engine at intervals and run into a heating tank, the system being replenished with clean oil at the same time. The heated oil is run through the purifier and pumped back to the main oil tank.

For removing soluble sludge, a system combining the batch and continuous systems is effective. The used oil drains to a tank, where it is allowed to rest for about ten days. The oil cools and any sludge settles. The oil is then drawn off, run through the purifier, and pumped to the clean oil tank. When the oil in use shows signs of sludging, it is drained off, the clean oil is pumped into the system and the cycle re-started. The purifier is used on the continuous system all the time, except for the short period when the batch of oil from the dirty oil tank is being purified.

Continuous bypass systems for diesel-engine and steam-turbine installations are illustrated in Figures 4.15 and 4.16. In Figure 4.15 (diesel engine) the bottom of the engine-sump tank is connected to the feed pump of the purifier by the pipe-line. A heater is fitted to the piping in order to make the oil sufficiently fluid for centrifugal treatment, especially when the engine is cold. The outlet for purified oil is connected to the top of the sump tank by means of the return pipe. The water and sludge outlet is connected to the waste by a pipe. The feed pipe should preferably be connected to a supply of hot, fresh water. This arrangement facilitates the priming of the purifier

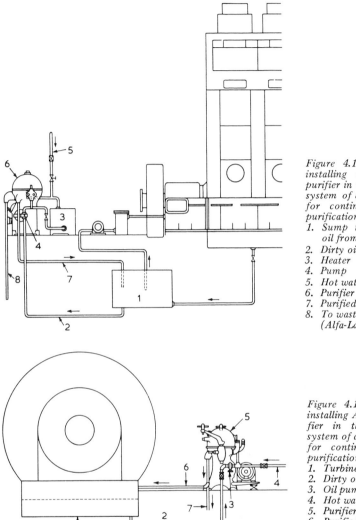

Figure 4.15 Method of installing an Alfa-Laval purifier in the lubricating system of a diesel engine for continuous by-pass purification
1. Sump tank for dirty oil from engine
2. Dirty oil to purifier
3. Heater
4. Pump
5. Hot water piping
6. Purifier
7. Purified oil
8. To waste
   (Alfa-Laval Co. Ltd.)

Figure 4.16 Method of installing Alfa-Laval purifier in the lubricating system of a steam turbine for continuous by-pass purification
1. Turbine oil tank
2. Dirty oil to purifier
3. Oil pump
4. Hot water piping
5. Purifier
6. Purified oil to turbine
7. To waste

bowl with water, and enables some of the dirt to be occasionally washed out from the bowl. In the case of small engines containing a rather small quantity of oil in their crankcases, a tank should be included in the system. The oil required for filling the purifier bowl and pipe-work, etc. is drawn from this tank, the oil level in the engine thus remaining constant.

In Figure 4.16 (steam turbine) the dirty oil is drawn from the bottom of the tank in the base of the turbine and taken by the pump into the purifier, whence the purified oil is returned direct to the turbine reservoir. Some hot condensate is let into the purifier, together with the oil, in order to wash out acid and sludge.

SEWAGE TREATMENT

The treatment of sewage on board ships is a relatively recent development resulting from legislation already imposed by the U.S. Coast Guard and the Canadian Government and in anticipation of the ratification of Annexe IV of the 1973 IMCO Conference on Marine Pollution.

The exact amount of sewage and waste water flow generated on board ship is difficult to quantify. European designers tend to work on the basis of 70 litres/man/day of toilet waste (including flushing water) and about 130—150 litres/man/day of wash water (including baths, laundries etc.). U.S. authorities suggest that the flow from toilet discharges is as high as 114 litres/man/day with twice this amount of wash water.

Some plants are designed so that the effluent is retained in the vessel for discharge well away from land, or to a receiving facility ashore; others are designed to produce an effluent which is acceptable to port authorities for discharge inshore. In the former type, the plant consists of holding tanks which receive all lavatory and urinal emptyings, including flushing water, while wash-basins, showers and baths are permitted to discharge overboard. Others are designed to minimise the amount of liquor such that it may be recyled for flushing. It is claimed that such a system only requires about one per cent of the retaining capacity of a conventional retention system.

To discharge sewage in territorial waters the effluent quality may have to be within certain standards laid down by the local or national authorities. These will usually be based on one or more of three factors, namely the biochemical oxygen demand (BOD), suspended solids content and e-coliform count of the discharge.

BOD is a measure of the total amount of oxygen which will be taken up by the chemical and organic matter in the effluent. Its importance is two-fold. Firstly if the waterway in which the effluent is discharged is overloaded with oxygen absorbing matter, the oxygen content of the water will be reduced to a level at which fish and some plant life cannot be supported. Secondly a class of bacteria which can live without oxygen will predominate in the sewage or in the waterway to which it is discharged. The bacteria associated with this condition produce hydrogen sulphide with its characteristic pungent smell. BOD is usually associated with a specific period and that normally taken is five days. This value, written BOD_5, is determined by incubating a one litre sample of sewage at $20°C$ diluted in sufficient well-oxygenated water. The amount of oxygen absorbed over the five day period is then measured.

Suspended solids are unsightly and over a period of time can give rise to silting problems. They are usually a sign of a malfunctioning sewage plant and when very high will be accompanied by a high BOD. Suspended solids are measured by filtering a sample through a pre-weighed asbestos pad which is then dried and re-weighed.

The e-coliform is a family of bacteria which live in the human intestine. They can be quantified easily in a laboratory test, the result of which is indicative of the amount of human waste present in a particular sewage sample. The result of this test is called the e-coli. count and is expressed per 100 ml.

Effluent quality standards

Legislation for the quality of effluent standards is far from settled and reference is best made to current National regulations and/or, when ratified, the 1973 IMCO Conference on Marine Pollution which sets out in Annexe IV the discharge standards a sewage treatment plant must be capable of achieving.

The general trend is towards approving particular designs of sewage plant. The United States Coast Guard recognises three distinct types of plant and then legislates on which type of plant is required for a particular circumstance. Known collectively as Marine Sanitation Devices (MSD) the types are:

Type 1. A flow-through device from which the effluent contains no visible floating solids and produces an e-coli count of less than 1000/100 millilitre.

Type 11. A flow-through device from which the effluent contains suspended solids of no more than 150 mg/litre and has an e-coli count of less than 200/100 millilitre.

Type 111. A zero discharge device i.e. holding tank or recirculation device.

A summary of the various effluent quality standards imposed around the world is shown in Table 4.3.

Recirculating holding system

Recirculating holding systems are not designed to produce an effluent suitable for discharge in a controlled area; they are designed

Table 4.3 IMCO and National effluent quality standards

Rules	Solids (mg/litre)	Faecal coliform (millilitre)	B.O.D.	Chlorine residual	Remarks
USCG Type I	No visible	1000/100	No requirements	No requirements	Allowed on new building deliveries until 31.1.80 and on existing vessels up to 31.1.78.
USCG Type II	150	200/100	No requirements	No requirements	Mandatory on all equipment ordered after 30.1.78 for existing vessels and after 31.3.80 for newbuildings.
USCG Type III			No discharge		Recirculation or on board retention.
IMCO	50	250/100	50 mg/litre	Not mandatory but low recommended	Whilst convention not yet ratified, regulations already being unilaterally adopted by some countries.
JAPAN	150	1000/100	50 mg/litre	No requirements	Vessels with complement of 100 plus at present in force.
CANADA (Great Lakes)	50	200/100	50 mg/litre arithmetic mean	Not less than 0.5 mg/litre or more than 1.0 mg/litre	Effective February 1982 for existing vessels. Effective Feb. 1980 for new vessels.
USSR (Black Sea)	—	—	No discharge		For vessels with 100 plus complement at present in force.
SWEDEN					For vessels with 100 plus complement at present in force.

In general Baltic states will accept M.S.D.'s approved by the regulating bodies of the country under whose flag the vessel sails only after examination of the test results and manufacturing specification of the various makers.

to retain a minimum amount of sanitary waste within a ship during the vessel's stay within the area. This can then be pumped out in a decontrolled area or to a shore receiving facility.

The retained liquor is minimised by discharging waste water from showers, baths and wash-hand basins directly overboard (not always permitted) and by using the liquid collected in the holding tanks as a flushing and transport medium. The parameters of this system are to produce a liquid for recycling which will be aesthetically acceptable and relatively harmless. The retained waste must also be acceptable — even after a long storage period — to a shore receiving facility.

Figure 4.17 shows diagrammatically the Elsan chemical recirculating system. It will be noticed that there are three chemical dosing tanks in this system. Odour and colour control chemicals are added to the first tank and in the second tank sterilising and breakdown chemicals are added.

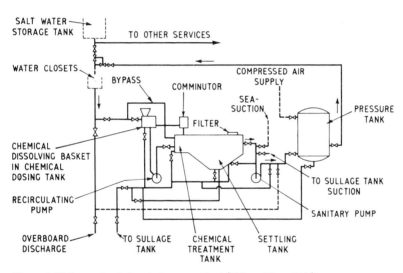

Figure 4.17 Example of chemical sewage plant (Wilson-Elsan Ltd.)

It is important to maintain the correct chemical dosage rates and these are determined by taking a daily sample and performing a simple chemical test. Failure to maintain the correct dosage may result in some chemical odour of the flush water and some darkening of its colour. By incorrect dosing it is possible for the liquor to develop a very high alkalinity which will cause some corrosion of pipes and tanks.

Biological sewage treatment

A number of biological sewage treatment plant types are in use at sea but nearly all work on what is called the extended aeration process. Basically this consists of oxygenating the liquor either by bubbling air through it or by agitating the surface. By so doing a family of bacteria is propagated which thrives on the oxygen content and digests the sewage to produce an innocuous sludge. The impression that bubbling air through the sewage serves to oxidise it thus reducing BOD is not strictly the case. It is the bacteria which reduce the BOD by converting the organic content of the sewage to a chemically and organically inert sludge.

In order to exist, the bacteria need air and nutrient. The nutrient is in the form of body and galley wastes. If the nutrient source is cut off, i.e. the plant shut down or by-passed for say an extended ocean passage, the bacteria die and the plant cannot function correctly until a new bacteria colony is generated. This process can taken from 7 to 14 days. Bacteria which live in the presence of oxygen are said to be aerobic. When oxygen is not present, the aerobic bacteria cannot live but a different family of bacteria is generated. These bacteria are said to be anaerobic. While they are equally capable of producing an inert sludge, in so doing hydrogen sulphide, carbon dioxide and methane are formed. Some sewage processes are designed to work anaerobically but these are not usually adopted for shipboard use.

Extended aeration plants used at sea are package plants consisting basically of three inter-connected tanks (Figure 4.18). The influent

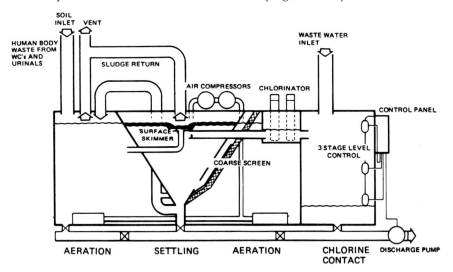

Figure 4.18 (a) Schematic diagram of Super Trident sewage treatment unit

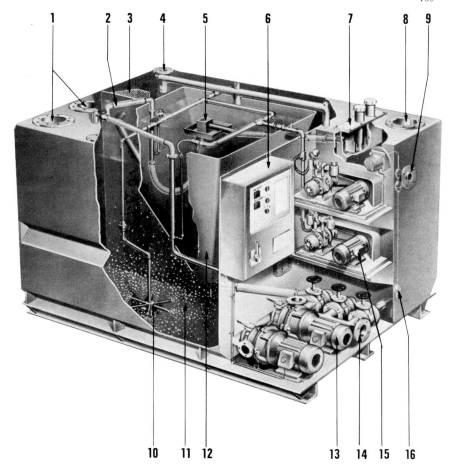

Figure 4.18 (b) Sectional view of Super Trident aeration sewage plant (Hamworthy Engineering Ltd)

1. Raw sewage inlet
2. Visual indication pipe for activated sludge return
3. Screen
4. Vent
5. Skimmer to remove floating debris
6. Control panel
7. Chlorinated for continuous chlorination of effluent

8. Waste water inlet
9. Emergency overflow
10. Air diffuser assemblies
11. Aeration tank
12. Settling tank
13. Discharge pump
14. Filling connection
15. Aeration compressors
16. Float switches

may be comminuted (i.e. passed through a device which consists of a rotating knife-edge drum which acts both as a filter and a cutter) or simply passed through a bar screen from where it passes into the first chamber. Air is supplied to this chamber via a diffuser which breaks the air up into fine bubbles. The air is forced through the diffuser by a compressor. After a while a biological sludge is formed and this is dispersed throughout the tank by the agitation caused by the rising air bubbles.

The liquid from the aeration tank passes to a settling tank where, under quiescent conditions, the activated sludge as it is known, settles and leaves a clear effluent. The activated sludge cannot be allowed to remain in the settling tank since there is no oxygen supplied to this area and in a very short time the collected sludge would become anaerobic and give off offensive odours. The sludge is therefore continuously recycled to the aeration tank where it mixes with the incoming waste to assist in the treatment process.

Over a period of time the quantity of sludge in an aeration tank increases due to the collection of inert residues resulting from the digestion process, this build up in sludge is measured in p.p.m. or mg/litre, the rate of increase being a function of the tank size. Most marine biological waste treatment plants are designed to be desludged at intervals of about three months. The desludging operation entails pumping out about three quarters of the aeration tank contents and refilling with clean water.

The clear effluent discharged from a settling tank must be disinfected to reduce the number of coliforms to an acceptable level. Disinfection is achieved by treating the clean effluent with a solution of calcium or solium hypochlorite, this is usually carried out in a tank or compartment on the end of the sewage treatment unit. The chlorinater shown in Figure 4.18 uses tablets of calcium hypochlorite retained in perforated plastic tubes around which the clean effluent flows, dissolving some of the tablet material as it does so. The treated effluent is then held in the collection tank for 60 minutes to enable the process of disinfection to be completed.

In some plants this disinfection is carried out by ultra-violet radiation.

5 Heat exchangers

Most coolers used on board ship transfer heat from a hot fluid to seawater. For the main propulsion engine the motor ship, the engine jacket water, lubricating oil and charge air must be cooled and generally also water or oil used in cooling the pistons. In a steam ship, apart from the heat yielded to the main condenser, the turbine and gearbox lubricating oils provide the principal sources of heat rejected to the circulating cooling water. Auxiliary prime movers require cooling and compressor intercoolers and aftercoolers have already been mentioned in another chapter.

Steam heated heat exchangers include heavy fuel oil heaters, boiler air pre-heaters, units to heat sea water for tank washing, evaporators, feed heaters and calorifiers.

Theory

In almost all heat exchangers, heat flows from the hot fluid to a cooler one through an intermediate heat-conductive wall, which takes up some intermediate temperature. The temperature profile across an element of wall surface may be considered as approximating to that depicted in Figure 5.1.

The temperature of the hot fluid falls through the boundary layer associated with that side of the wall from the bulk temperature t_h to the wall temperature t_{cw}. There is a small drop in temperature through the wall, due to its thermal resistance, from t_{hw} to t_{cw}. On the cold side of the wall, the fluid immediately in contact with the wall is at t_{cw} but its temperature falls through the boundary layer on that side to the bulk fluid temperature t_c.

Considering a rate of heat flow δQ through the element of wall surface area δA:

$$\delta Q = h_1(t_h - t_{hw})\delta A = (k/y)(t_{hw} - t_{cw})\delta A = h_2(t_{cw} - t_c)\,\delta A$$

171

where:

    h_1 = co-efficient of heat transfer on the hot fluid side;
    h_2 = co-efficient of heat transfer on the cold fluid side;
    k = thermal conductivity of the wall material;
    y = thickness of the wall.

If the overall co-efficient of heat transfer between the hot and cold fluid is defined as:

$$U = \frac{\delta Q}{(t_h - t_c)\,\delta A}$$

then $\dfrac{1}{U} = \dfrac{1}{h_1} + \dfrac{1}{h_2} + \dfrac{y}{k}$

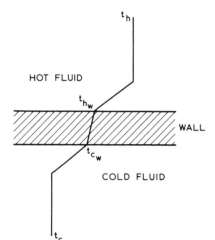

Figure 5.1 Temperature gradient between fluids
(Figures 5.1 to 5.11 are by courtesy of Serck Heat Transfer)

This is the basic equation governing the performance of a heat exchanger in which the heat transfer surface is completely clean. Additional terms may be added to the right hand side of the equation to represent the resistance to heat flow of films of dirt, scale, etc. The values of h_1 and h_2 are respectively determined by the fluids and flow conditions on the two sides of the wall surface. Under normal operating conditions, water flowing over a surface gives a relatively high co-efficient of heat transfer, as does condensing steam, whereas oil provides a considerably lower value. Air is also a poor heat transfer fluid and it is quite usual to modify the effect of this by adding extended surface (fins) on the side of the wall in contact with the air.

In a practical heat exchanger, the thermal performance is described by the equation.

$$Q = U \theta A$$

where:

Q = rate of heat transfer;

θ = logarithmic mean of the temperature differences at the inlet and outlet of the heat exchanger: this is a maximum if the fluids flow in opposite directions (counterflow);

A = surface area of heat transfer wall.

It is sometimes important to appreciate the effect of variation of cooling water flow through a heat exchanger. The graph in Figure 5.2

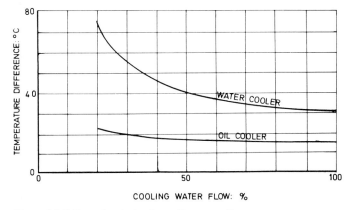

Figure 5.2 Effect of variation in cooling water flow

illustrates two typical instances, one a jacket water cooler and the other a lubricating oil cooler (both sea-water cooled), in which the difference in temperature between the hot fluid and the sea-water is plotted against sea-water flow, assuming constant hot fluid flow and rate of heat transfer.

TYPES OF HEAT EXCHANGER

Shell-and-tube

Most marine heat exchangers are of the shell-and-tube type, an example of which is shown in Figure 5.3. A tube bundle, or stack, is inserted inside a shell, whose branches are connected into the circulating system of the hot fluid. The stack comprises a number of tubes secured into a tubeplate at each end, and a series of baffles directs the flow of hot fluid back and forth across the tube bundle. At each end of the heat exchanger is a header, whose purpose is to conduct the other fluid (usually sea-water) through the tubes. These headers may be designed to give a single pass through the tubes or, as in Figure 5.3, two passes. Removable covers are normally provided on the headers, to facilitate access to the tubes for cleaning.

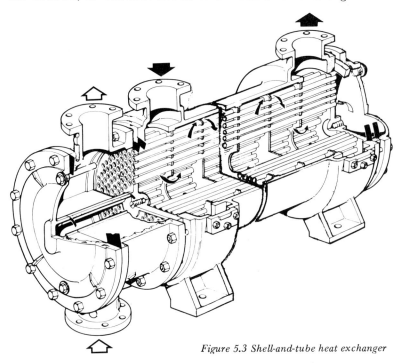

Figure 5.3 Shell-and-tube heat exchanger

At one end of the exchanger, gaskets are fitted between the tubeplate and both the shell and the header. At the other end, two separate elastomer seals lie either side of a 'safety leakage ring', as illustrated in Figure 5.4 so that, in the event of leakage past either seal, the two fluids cannot intermix. This arrangement also permits movement of the tube-plate to accommodate differential expansion between tube bundle and shell.

Depending on the size and duty of the cooler the cylindrical shell may be of fabricated steel, cast iron or occasionally aluminium bronze. The tube plates are usually of Naval brass and the tubes of aluminium brass; occasionally 70/30 copper-nickel tubes are used.

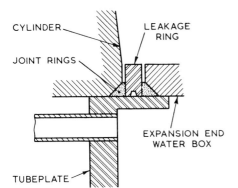

CYLINDER

LEAKAGE RING

JOINT RINGS

EXPANSION END WATER BOX

TUBEPLATE

Figure 5.4 Safety leakage ring

The usual method of securing the tubes to the tube-plates is to roll-expand them. The cooler headers may be of cast iron. In such instances the headers act as sacrificial anodes, wasting in preference to the aluminium-brass tubes. Unless soft iron or mild steel sacrificial anodes or impressed current cathodic protection is used to protect the cooler from corrosion such headers should not be painted internally.

To avoid impingement attack, care must be taken with the water velocity through the tubes. In the case of aluminium-brass the upper limit is about 2.5m/sec. While it is advisable to design to a lower velocity than this — to allow for poor flow control — it is equally bad practice to run tubes at less than 1m/sec since silting and the settling of marine organisms is more likely to occur at low velocities.

On some smaller engines, a variant of the shell-and-tube heat exchanger is used for oil cooling, in which fins are bonded to the tube to provide additional heat transfer surface in contact with the lubricating oil and so increase performance. This is illustrated in Figure 5.5.

Plate type

In more recent years, plate heat exchangers have frequently been employed for jacket water and oil cooling. A series of identical pressed plates, such as shown in Figure 5.6 each with an elastomer seal fitted around its periphery, is contained within a frame with clamping plates at each end. Four branches, attached to one end plate, align with the ports in the plates through which the two fluids pass. The seals between the plates are so arranged that one fluid flows in alternate passages between plates, and the second fluid in the intervening passages, usually in the opposite direction.

Figure 5.5 Oil cooler with fins

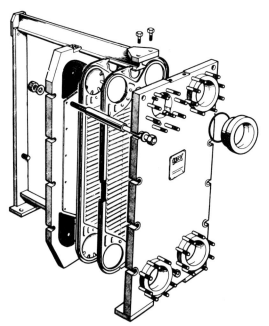

Figure 5.6 Plate type heat exchanger

(a)

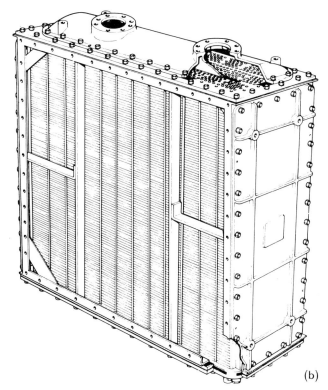

(b)

Figure 5.7 Charge air coolers

The plates are corrugated, principally to promote turbulence in the flow of both fluids, and so improve heat transfer, and dimples pressed in the plate surface provide support from one plate to the next to resist the load due to difference in pressure between the two fluids.

The plates are usually of titanium providing very high resistance to corrosion by sea water.

Charge air coolers

The great advantage to be gained, in terms of specific output of an engine, by cooling the charge air in a turbo-charged diesel engine is now well recognised. Because of the relatively poor heat transfer properties of air, fins are always used on the air-side of the heat transfer surface in a charge air cooler.

These coolers are normally rectangular in form as shown in Figures 5.7(a) and (b). In the former example, used on smaller engines, the whole cooler must be removed from the air trunking for cleaning and general maintenance. In the second example each header may be removed from the air trunking for cleaning and general maintenance. In the second example each header may be removed, with the cooler in situ, for cleaning through the tubes and removable sideplates give access for cleaning of the air-side surface. The cooler stack may be withdrawn, if required, for repair or replacement.

Electrical machine coolers

Some larger alternators and electric motors are designed for operation with closed-circuit air cooling. Heat must be extracted from the

Figure 5.8 Electric motor cooler

circulating air and, for this purpose, finned tube units are employed, inserted into the air ducting. An example is shown in Figure 5.8. Sea-water flows through the tubes, while the air passes over them, flowing between the fins. Spray baffles are usually fitted to these units so that, in the event of puncture of one of the tubes, sea-water will not flow directly into the alternator or motor.

Oil heaters

For burning heavy fuel oil in a boiler furnace, or a compression ignition engine, it is necessary to pre-heat it. This is done in shell-and-tube units either with plain tubes, or tubes with fins bonded to them (the oil flowing outside the tubes) and tubes fitted internally to promote turbulence in the oil flowing through them. The heating medium is condensing steam.

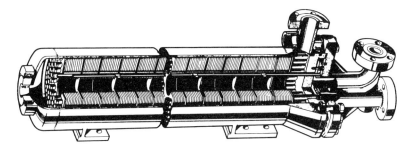

Figure 5.9 Heavy fuel oil heater

The tubestack of a finned tube design, which provides a very compact unit, is shown in Figure 5.9.

Air pre-heaters

The combustion air required for forced-draught boiler furnaces may be heated by the exhaust gas in either a rotary regenerative heat exchanger (so called Ljungstrom type) or in a conventional heat exchanger as part of the group of heat exchangers associated with steam raising. The air may alternatively be preheated by the use of steam bled from the turbine.

In the latter case, a finned-tube unit such as that illustrated in Figure 5.10 may be employed, the combustion air flowing over the banks of finned tubes and steam condensing inside the tubes. Air pre-heating improves the overall efficiency of the whole cycle.

Figure 5.10 Bled steam air preheater

Figure 5.11 Sea water heater

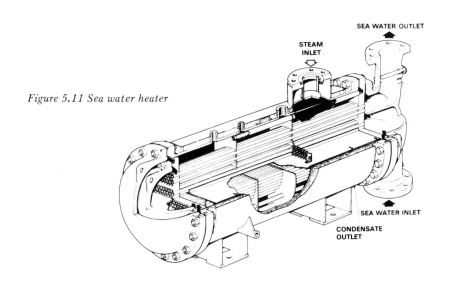

Sea-water heaters

These are usually of the shell-and-tube type construction, with a fully floating head as illustrated in Figure 5.11. The sea-water to be heated flows in several passes through the tubes, being heated by the condensation of steam in one part of the tube bundle and sub-cooling the condensate in another.

In the condensing section, baffles merely provide support for the tubes, while in the sub-cooling section they direct the flow of condensate. The two sections are separated by a longitudinal baffle, whose sides are in contact with a cylinder into which the tube bundle fits. A baffle is located beneath the steam inlet branch to prevent direct impingement of the steam on to the tubes.

CONTROL OF TEMPERATURE IN HEAT EXCHANGERS

There are three basic ways by which the temperature of the hot fluid being cooled in a heat exchanger may be controlled, when the cooling medium is sea-water:

(a) By by-passing a proportion of the hot fluid flow, the remainder being passed through the heat exchanger.

(b) By throttling the sea-water flow or, alternatively, by-passing a proportion of it.

(c) By controlling the temperature of the sea-water entering the heat exchanger — this is done in the sea-water system as a whole, by spilling part of the heated discharge back into the pump suction.

The last of these cannot provide a satisfactory degree of control by itself but is often used in conjunction with one of the other two.

Whilst automatic control equipment may be employed in any one of the three modes, only the second can normally be used for manual temperature control. As a general rule, the control valve on the sea-water side should be placed downstream of the heat exchanger, in order to avoid reduction of pressure within the heat exchanger itself leading to aeration of the water through cavitation. This is particularly important when the heat exchanger is mounted high in the sea-water system and especially if it is above the water line. Excessive reduction of sea-water flow through the heat exchanger may lead to the deposition of silt in horizontal tubes. Figure 5.2 indicates the extent to which the sea-water flow variation is necessary to maintain a fixed hot fluid tempearture when sea-water temperature alters.

The flow of hot fluid through a heat exchanger may be controlled by a valve directly actuated by a temperature sensor, but pneumatically operated valves are normal with all methods of temperature control. On steam heated heat exchangers, automatic temperature control equipment is usually fitted.

MAINTENANCE OF HEAT EXCHANGERS

The only attention that heat exchangers should require is to ensure that the heat transfer surfaces remain substantially clean and the flow passages generally clear of obstruction. Indication that undue fouling is occurring is given by a progressive increase in the temperature difference between the two fluids, over a period of time, usually accompanied by a noticeable rise in pressure loss at a given flow.

Fouling on the sea-water side is the most usual cause of deterioration in performance. The method of cleaning the sea-water side surfaces depends upon the type of heat exchanger. With shell-and-tube heat exchangers, the removal of the header covers or, in the case of the smaller heat exchangers, the headers themselves, will provide access to the tubes. Obstructions, dirt, scale etc., can then be removed, using the tools provided by the heat exchanger manufacturer. Flushing through with fresh water is recommended before a heat exchanger is returned to service. In some applications, such as piston oil cooling, progressive fouling may take place on the outside of the tubes. Most manufacturers recommend a chemical flushing process to remove this in situ, without dismantling the heat exchanger.

Plate heat exchangers may be cleaned by unclamping the stack of plates and mechanically cleaning the surface of each plate as recommended by the manufacturers. The plate seals may require replacement from time to time and here the manufacturers' instructions should be closely followed.

Corrosion by sea-water may occasionally cause perforation of heat transfer surfaces. This will cause leakage of one fluid into the other but this is not always easy to detect whilst the leakage is small, although substantial leaks may become evident through rapid loss of lubricating oil, jacket water etc.

Location of a perforation is a straightforward matter in the case of a tubular heat exchanger, whether this is of the shell-and-tube type or of other tubular construction. Having drained the heat exchanger of sea-water and removed the covers or headers to expose the tube ends, some flow of the liquid on the other side of the surface will be apparent, in the case of oil and water coolers, from any tubes which are perforated. To test for leaks in air coolers, drains coolers etc. each tube in turn can be plugged at each end and pressurised with air; inability to hold pressure indicates a leak.

To aid the detection of leaks in a large cooler such as a main condenser, in which it is difficult to get the tubes dry enough to witness any seepage, it is usual to add a special fluorescent dye to the shell side of the cooler. When an ultra-violet light is shone on to the tubes and tube plates any seepage is seen since the dye glows with a vivid green light. In plate heat exchangers, the only way to locate leaks is by visual inspection of the plate surfaces.

On docking for any protracted period, such as for repairs, refitting etc. it is advisable to drain the sea-water side of heat exchangers, clean and flush through with fresh water, after which the heat exchanger should be left drained, if possible until the ship re-enters service.

VENTING AND DRAINING

It is important that any heat exchanger through which sea-water flows should run full. In vertically-mounted single-pass heat exchangers of the shell-and-tube or plate types, venting will be automatic if the sea-water flow is upwards. This is also the case with heat exchangers mounted in the horizontal attitude, with single or multi-pass tube arrangements, provided that the sea-water inlet branch faces downwards and the outlet branch upwards. With these arrangements, the water will drain substantially completely out of the heat exchanger, when the remainder of the system is drained.

With other arrangements, a vent cock fitted at the highest point in the heat exchanger should be opened when first introducing sea-water into the heat exchanger and thereafter periodically to ensure full running. A drain plug at the lowest point should be provided.

CONDENSERS

A condenser is a vessel in which a vapour is deprived of its latent heat of vaporisation and so changed to its liquid state, usually by cooling at constant pressure.

In surface condensers, steam enters at an upper level, passes over tubes in which cold water circulates, falls as water to the bottom and is removed by a pump (or flows to a feed tank). The tube arrangement provides for condensation with minimum loss of heat: the cooling water normally circulates in two passes, entering at the bottom.

Construction

Main condensers and condenser design are outside the scope of this book; information can be found in the book 'Marine Steam Engines and Turbines' in the Marine Engineering Series published by Newnes-Butterworths.

Auxiliary condensers are usually built as a mild steel fusion-welded shell, the vertical cross-section decreasing in area from top to bottom. A water box, of cast iron or steel, is fitted at each end of the shell and, sandwiched between the flanges of the boxes and the shell are admiralty brass (70 Cu 29 Zn 1 Sn) tube plates. These are drilled, and when soft-packing is used, counter bored and tapped.

Tubes may be of Admiralty brass (not now common), cupronickel (70 Cu 30 Ni) or aluminium brass (76 Cu 22 Zn 2 Al) 16—20 mm. outside diameter. They may be expanded into the tube plates at both ends, expanded at the outlet end and fitted with soft packing at the other, or fitted with soft packing at both ends, in which case there is a screwed check ferrule at the outlet end. The soft packing takes the form of a short linen sleeve, encased in zinc foil, rammed home. The tubes are prevented from sagging by a number of mild steel tube plates. Access doors are provided in the water box doors for routine inspection and cleaning with one or more manholes in the shell bottom for the same purpose. To prevent gross wastage due to galvanic action of the cast iron, or steel, and dezincification of aluminium brass tubes where fitted, zinc or mild steel sacrificial anodes are fitted to the tube plates. Alternatively impressed current cathodic protection may be used.

The simplest method of degreasing the steam side of tubes is as follows. A vessel containing trichloroethylene is secured to a bottom manhole and is warmed gently. The trichloroethylene vaporises, rises among the tubes, condenses and falls into the vessel, bringing with it the grease and oil from them. This agent is toxic if inhaled and precautions must be taken.

Tube failure is a rare occurrence nowadays: it may occur from corrosion/stress cracking or dezincification of brass tubes, or by corrosion/erosion arising from entrained air in or excessive speed of, circulating water. When it occurs the defective tube may be fitted with a wooden plug or a capped ferrule until it can be renewed conveniently. Tube bores are cleaned by brushing out, by use of a compressed air lance, or by a 'devil' i.e. a small plug of nylon, shot through the tube by hydraulic or pneumatic pressure.

The salt water side of condensers out of use should either be drained or kept wholly full and circulated from time to time: stagnant cooling water may be destructive. Corrosion plates when fitted should be examined and scraped every two or three months.

DISTILLATION

A considerable amount of fresh water is consumed in a ship. The crew consumes on average about 70 litre/head/day and in a passenger ship consumption 'per capita' can be as high as 225 litre/day. In addition water will be consumed in any steam plant. In a steamship the consumption for the propulsion plant and hotel services can be as high as 50 tonnes/day.

Sufficient potable water may be taken on in port to meet crew and passenger requirements but the quality of this water will be too poor for use in water tube boilers and will require further treatment by distillation. It is common practice to take on only a minimal supply of potable water and make up the rest by distillation of sea water. Even in vessels which carry sufficient potable water for normal requirements it is a statutory requirement that such ships, when ocean-going, should carry distillation plant for emergency use.

The main object of distillation is to produce water essentially free of salts. Potable water should contain less than 500 mg/litre of suspended solids. Good quality boiler feed will contain less than 2.5 mg/litre.

Sea water, on the other hand, has a total dissolved solids (TDS) content in the range 30 000–42 000 mg/litre, depending on its origin but, for most cases, the TDS is taken as 32 000 mg/litre.

By bringing the water to its boiling point and drawing off the vapour the salts and other solids are left behind in the liquid, a proportion of which is discarded. The vapour produced is essentially solids-free although some solids are carried over, especially if the equipment is misused.

The equipment in which this process takes place is known as an evaporator of which there are two distinct types. One type boils the water at the saturation temperature corresponding to the pressure in the evaporator and is known as a boiling evaporator (Figure 5.12). The other type heats the water in one compartment before it is released into a second compartment in which the pressure is substantially lower, causing some of the water to 'flash' into a vapour. This type is known as a flash evaporator. Thus in a boiling evaporator the water is maintained continuously at its saturation temperature – in other words latent heat is added – while in the flash evaporator it is sensible heat that is supplied.

The submerged tube type of boiling evaporator was extensively used for many years. These evaporators were bulky and heavy and, in the absence of constant, careful attention they primed readily, their thermal performance was poor and they required frequent de-scaling, usually by physical effort.

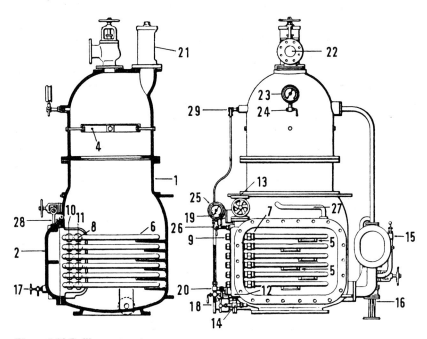

Figure 5.12 Boiling evaporator

1. *Shell*
2. *Main door for withdrawing coils*
3. *Hand cleaning door*
4. *Baffle plate or deflector*
5. *Support for coils*
6. *Evaporating tube coils*
7. *Inlet steam copulings for coils*
8. *Drain outlet couplings for coils*
9. *Coupling nuts for (7) and (8)*
10. *Inlet steam header*
11. *Drain header*
12. *Drain collecting pocket*
13. *Inlet valve for steam coils*
14. *Valve for drain from coils to hot-well*
15. *Feed check valve*
16. *Brine ejector*
17. *Salinometer valve*

18. *Cock for blowing off to sea*
19. *Top cock for water gauge*
20. *Bottom cock for water gauge*
21. *Safety valve*
22. *Outlet valve for generated steam*
23. *Compound gauge for generated steam in shell*
24. *Cock for compound gauge*
25. *Pressure gauge for inlet steam to coils*
26. *Cock for pressure gauge*
27. *Swing crane bar for door*
28. *Eyebolt for supporting door on crane bar*
29. *Connection from top of water gauge to steam space in evaporator shell*

Evaporator-distillers

The many advantages of evaporation at sub-atmospheric pressures namely, the improved heat transfer between the heating steam and the salt feed water, the greatly reduced formation of (a much softer) scale, the facility for using otherwise waste heat and the increased output per unit of weight and bulk have brought about the development of a number of compact, simply-controlled evaporator-distiller units. These enable even the largest passenger ships to produce economically all their requirements of fresh water, with an attendant freedom of use.

Basically, these units incorporate an evaporating section beneath a condensing or distilling section in a common vessel of appropriate shape. A controlled flow of filtered feed, taken preferably from a salt circulating outlet, enters the evaporating section and ascends through a battery of vertical tubes, surrounded by steam or hot water, vaporising as it goes, to the condensing section through a labyrinth or screen (generally called a demister) which ensures that no droplets of salt water enter the condenser with the vapour. The vapour, directed by suitably placed baffles, passes over the condenser tubes and falls as water to an outlet duct, from which it is removed by a distillate pump, via a salinometer.

If acceptably pure, i.e. not having a salinity over say 3 mg/litre, the distillate goes to the appropriate tanks; if not, a solenoid-operated valve, energised by the salinometer, either dumps it to bilge or alternatively, the distillate pump is stopped (by electric relay from the salinometer) and the impure distillate is returned to the evaporator for re-processing.

The evaporator-distiller vessel may be of cupro-nickel or other non-corrodible material or again, more usually, of mild steel lined with a protective coating. The heating element tubes may be circular in section, of aluminium brass, expanded top and bottom into RNB tube-plates, alternatively, they may be thin-walled ellipses in section (to facilitate scale-shedding by cold-water shocking) of monel metal, welded into similar tube-plates. The distiller tubes are U-form, made of aluminium brass and expanded into RNB tube-plates.

The heating medium may be live steam or preferably and more commonly, exhaust or bled steam or again, in motor ships, hot fresh water, taken from the cooling main between the engine and the f.w. coolers. The distilling unit may be circulated by salt water or, depending upon the heat recovery attainable in the boiler feed system, by main or auxiliary condensate. Figures 5.13 and 5.14 show the advantage of the latter arrangement. Sub-atmospheric pressure is produced

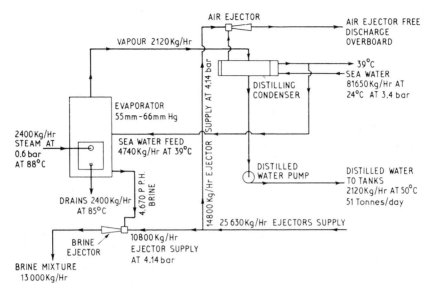

Figure 5.13 Heat flow diagram — sea water cooled evaporator (Caird & Rayner Ltd.)

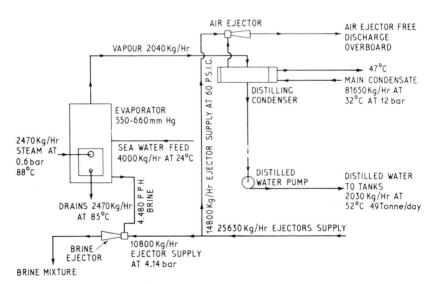

Figure 5.14 Heat flow diagram — condensate cooled evaporator (Caird & Rayner Ltd.)

in the units by water-operated air ejectors; the brine density in the evaporating section is maintained at 0.062 g/ml.

Flash evaporators

The temperature at which water boils is related to its pressure, e.g., 100°C at atmospheric pressure. This principle is employed in 'flash' evaporators, i.e. heated sea water if fed into a vessel maintained at sub-atmospheric pressure, flashes into steam, is condensed by contact with tubes circulated with the salt feed and is removed by a distillate pump. Baffles suitably placed and demisters, similar to those already described, prevent carry-over of saline droplets; the arrangements for continuous monitoring for purity of the distillate are similar to those described above.

If two or more vessels in series are maintained at progressively lower absolute pressure, the process can be repeated, the incoming salt feed absorbing the latent heat of the steam in each stage, with a resultant gain in economy of heat and fuel. This is known as 'Cascade' evaporation, a term which is self-explanatory. Figure 5.15 shows a two stage flash evaporator distiller. The flash chambers are maintained at very low absolute pressure by ejectors, steam or water operated, the salt feed is heated initially by the condensing vapour in the flash chambers, secondly in its passage through the ejector condenser (when steam-operated ejectors are used) and is raised to its final temperature in a heater supplied with l.p. exhaust steam. Brine density is maintained, as in the case of the evaporator-distillers described previously, by an excess of feed over evaporation and the removal of the excess by a pump: some re-circulation of brine may be provided for, in certain circumstances.

It should be noted that when distillate is used for drinking, it may require subsequent treatment, see pages 17 and 45.

FEED WATER HEATERS

Attention has been drawn in Chapter 1 to the need for the greatest practicable, economic recovery of the latent heat in steam. Feed heaters play an important part in this process; they may be either surface or direct contact, now known usually as de-aerators.

Surface feed heaters

These are shell and tube heat exchangers built with materials and scantlings appropriate to their working temperatures and pressures.

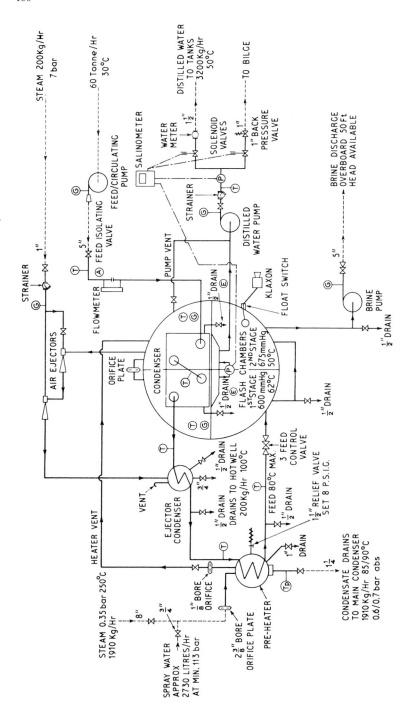

Figure 5.15 Flow diagram — cascade evaporator (Caird & Rayner Ltd.)

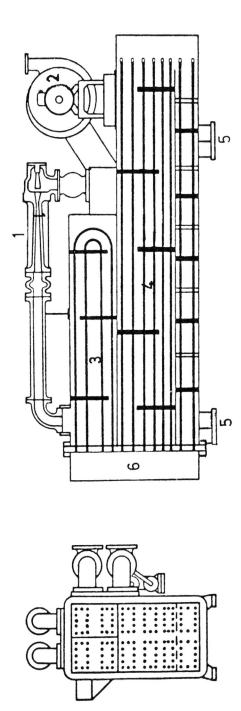

Figure 5.16 This module combines four closed feed system components and is claimed to offer a 14% reduction in weight and a 19% reduction in volume.

1. Steam jet air ejector
2. Electrically-driven vapour extractor
3. Gland steam condenser

4. L.P. heater/drains cooler
5. Supporting feet
6. Water heater

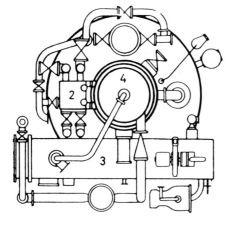

1. Extraction pumps
2. Steam jet air ejectors
3. L.P. heater and drain cooler
4. De-aerator
5. Flash chamber
6. De-aerated water storage tank

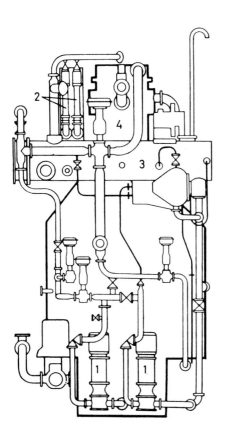

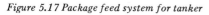

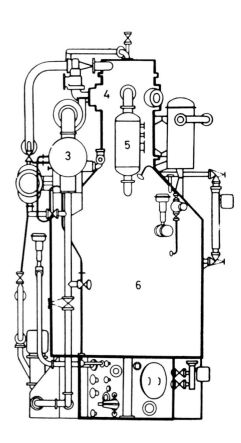

Figure 5.17 Package feed system for tanker

It can be shown that the minimum economic terminal difference (i.e. the temperature difference between the heating fluid inlet and the heated fluid outlet) is about 6°C. If, however, the heating fluid (steam) is super-heated by at least 110°C above saturation temperature and the heated fluid (feed water) leaves the heater at this saturation temperature or near to it, this terminal difference will be very small or zero. This can be achieved by using the exit section of the feed heater as a desuperheater.

By grouping a number of surface feed heaters in one module, an economy in space and pipe connections can often be achieved and there is a movement in that direction in practice.

Many parts of the feed system are now installed as packaged units or modules. Figure 5.16 shows a package arrangement which combines no less than four different units, viz. steam jet air ejectors, glands steam condenser, the de-aerator vent condenser and the l.p. heater drains cooler. Such packaging obviously leads to compactness, centralisation and reduction of branches to be connected by the shipbuilder. Other 'packaged' feed systems incorporate complete feed systems and Figure 5.17 shows an arrangement of a packaged feed system which may be found on a 67 000 tonne tanker.

DE-AERATORS

In Chapter 1, brief mention is made of the need for clean, neutral boiler feed, free from dissolved gases and of the consequent use of efficient de-aerators. Figure 5.18 shows one of several which liberate the dissolved gases from the feed and provide a measure of feed heating simultaneously. This type of de-aerator has a great range of capacity and given a temperature rise of at least 20°C, an oxygen content of 0.2 cc/litre can be reduced to 0.005 cc/litre, when working between one-half full load and full load in a closed feed system.

Normally, the de-aerator is mounted directly on a storage tank, into which the de-aerated water falls, to be withdrawn through a bottom connection by a pump or by gravity. The tank usually has a capacity sufficient for 10 minutes' running supply of water, but this is not necessarily the case.

The feed water enters the de-aerator head and so that its surface area may be increased to the maximum possible, it is divided into sprays of minute droplets by being forced through the spray nozzles into the shell; here it meets the heating steam and is brought rapidly to its saturation temperature. Most of the dissolved gases are released

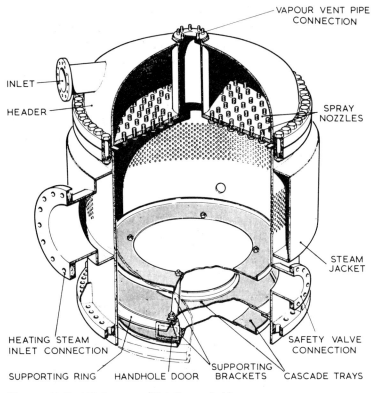

Figure 5.18 Typical de-aerator (Weir Pumps Ltd.)

and with some vapour, rise to the vapour release opening. The
header may be divided and provided with two feed inlet connections,
so that the efficiency of de-aeration may be maintained at low rates
of flow, by reducing the number of nozzles in use.

Cascade trays

Three cascade trays are set one above the other in the lower part
of the shell. The upper and lower of these trays have a raised lip on
the outer periphery, have the central opening blanked and have a
series of perforations arranged in rings towards the raised lip. The
middle tray has a central opening with a raised lip and is perforated
similarly. The falling spray collects on the upper tray and is again
broken up as it passes through the perforations to the middle tray,
where the process is repeated, to be repeated again as it passes
through the lower tray to the tank below. The combination of spray,
heating and cascade ensures the liberation of all but a minute

fraction of the gases in solution or suspension. The final water temperature depends upon the pressure of the controlled steam supply.

It will be apparent that de-aerators of this type must be installed at such a height in the engine room that the pressure head at the extraction or feed pump suction is greater than that corresponding to the water temperature.

The maintenance required is the control of corrosion, the cleaning of the nozzles, the renewal of those showing a sign of erosion (which will impair seriously the efficiency), the overhaul of fittings and the maintenance of the safety valve.

Devaporisers

If the deaerator cannot be vented to atmosphere or to a gland condenser satisfactorily, a devaporiser (Figure 5.19) is connected to the vapour outlet condensing the vapour vented with the non-condensable

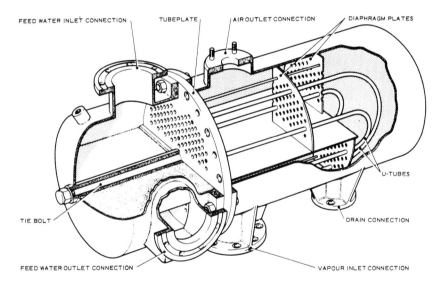

Figure 5.19 Sectional view of devaporiser (Weir Pumps Ltd.)

gases and cooling these gases before they are discharged. In the process the feed water is raised slightly in temperature.

In design and construction devaporisers are similar to other small heat exchangers working at moderate pressures (see pages 173–174). Leaving the devaporiser the feed water enters the deaerator header.

SALT-WATER LEAKS AND DETECTION

It is imperative that any inadvertent introduction of salt water to the boiler feed system, from tube or tube plate leakage in heat exchangers, mal-operation of the distillation plant or any other reason, is detected and rectified as soon as possible.

Samples will be drawn from the boilers and from the boiler feed at regular intervals (at least once a day) and tested with a hydrometer for any change in density which would indicate a change in dissolved solids and by chemical titration to show the chloride content, but the first indication of salt water ingress will be given by an electric salinometer. This device continuously monitors the salinity of the condensate and motivates an audible alarm if the water condition deteriorates below a pre-determined value. In some cases the salinometer also initiates the action of a dump or diversion valve. Such an instrument is capable of detecting 0.28 mg/litre of chlorine in water.

The electric salinometer

Pure distilled water may be considered a non-conductor of electricity. The addition of impurities such as salts in solution increases the conductivity of the water, and this can be measured. Since the conductivity of the water is, for low concentrations, related to the impurity content, a conductivity meter can be used to monitor the salinity of the water. The instrument can be calibrated in units of conductivity (micromhos) or directly in salinity units (older instruments in grains/gall., newer instruments in p.p.m or mg/litre) and it is on this basis that electric salinometers operate. A typical instrument is depicted in Figure 5.20 and its associated probe in Figure 5.21.

The probe type electrode cell shown in Figure 5.21 is fitted direct into the water system (pipeline) co-axially through a retractable valve which permits it to be withdrawn for examination and cleaning. The cell cannot be removed while the valve is open and consists of two stainless steel concentric electrodes having a temperature compensator located within the hollow inner electrode. It operates within the following limits:

Water pressure up to 10.5 bar

Water temperature $15^\circ - 110^\circ C$

The incoming a.c. mains from control switch S2 and fuses FS feed transformer T; a pilot lamp SL1 on the 24 V secondary winding indicates the circuit is live.

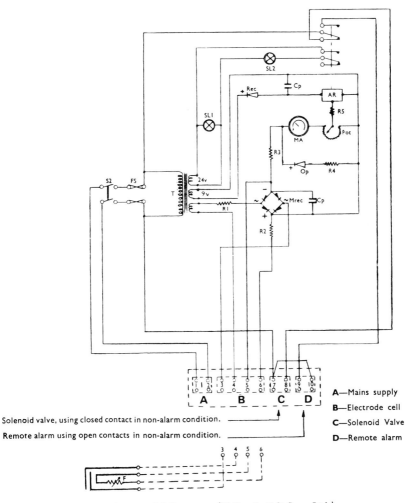

Figure 5.20 Schematic diagram of Salinometer (W Crockatt & Sons Ltd.)

The indicating circuit comprises an applied voltage across the electrode cell and the indicator. The indicator shows the salinity by measuring the current which at a preset value actuates the alarm circuit warning relay. The transformer cell tapped voltage is applied across a series circuit comprising the bridge rectifier M_{rec} the current limiting resistor R1 and the electrode cell.

The current from rectifier M_{rec} divides into two paths, one through the temperature compensator F via resistor R2 and the other through the alarm relay potentiometer (Pot) indicator MA and resistor R3, the two paths joining in a common return to the low potential side of the rectifier.

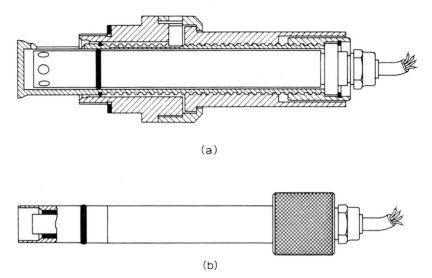

(a)

(b)

Figure 5.21 Electrode cell for Salinometer probe (W Crockatt & Sons Ltd.)
(a) valve open and electrode inserted. (b) electrode cell withdrawn.

The indicator is protected from overload by a semi-conductor in shunt across the indicator and potentiometer. When the water temperature is at the lower limit of the compensated range the total resistance of the compensator is in circuit and the two paths are as described above. As the temperature of the water rises, the resistance of the compensator device drops progressively, the electrical path through the compensator now has a lower resistance than the other and a large proportion of the cell current. The compensator therefore ensures that the alteration in the balance of the resistances of the two paths corresponds to the increased water conductivity due to the rise in temperature and a correct reading is thus obtained over the compensated range.

The alarm setting is adjustable and the contacts of warning relay AR close to light a lamp or sound a horn when salinity exceeds an acceptable level.

The salinometer is also arranged to control a solenoid operated valve which dumps unacceptable feed water to bilge and the complete installation, salinometer and valve reset automatically when the alarm condition clears.

6 Auxiliary engines

Auxiliary engines are used on board for a number of purposes. This chapter deals with diesel engines and steam turbines used for driving electrical generators and large pumps.

DIESEL ENGINES

The engine power range installed at sea for electric generation extends from as little as 30 kW to about 1.5 MW and engines are available with powers far in excess of this. The prime mover is generally direct-coupled to its generator, frequently with a fly-wheel interposed between the two. Fly-wheels of larger engines invariably have barring teeth to enable the engine to be turned by hand and they usually bear timing marks.

The generator may produce a.c. or d.c. although generally, the latter is not found in ships built after the mid-1960's. The engines may be medium or high speed; medium speed infers a rotating speed of between 300—1000 rev/min and high speed 1000 rev/min upwards although it would be more relevant to talk in terms of piston speeds; 'medium' inferring a speed in the range 6.5—9.5 m/sec and 'high' a mean piston speed of over 9.5 m/sec.

Medium speed engines have been used in merchant ships for many years, in both Vee and in-line configurations. In higher powered engines brake mean effective pressures of 10 to 20 bar are common.

High speed engines are relative newcomers to merchant ships, although they have been used extensively in fighting navies where their high power/weight ratio gives them a marked advantage. This class of engine is found throughout the power range of ships' generator sets; the engines at the lower end of the range are frequently modified automotive diesels such as Foden and Gardner with b.m.e.p.'s in the range of 10 bar. Most auxiliary engines used at sea are four stroke machines (one of the exceptions is the Foden) and in ships built from about 1965 onwards, are mostly turbo-charged.

Programmed control of 'duty' engines, with automatic starting and stopping of stand-by engines as demand rises and falls, automatic engine change-over if a fault develops on the running machine, automatic load sharing and synchronisation are now commonplace. As a result, these virtually unattended installations require high dependability which demands intimate knowledge of the machines and strict attention to the maintenance schedule.

MEDIUM SPEED DIESEL ENGINES

Over recent years many owners have elected to install auxiliary engines capable of burning the low grade residual fuels bunkered for main engines.

Typical of medium speed engines in common use capable of running on heavy fuel is the Allen type A12. The S12-D and higher speed S12–F versions are in-line engines having three, four, six and eight cylinders and four, six, eight and nine cylinders respectively; a Vee engine of similar cylinder size, designated VS12-F is also produced having twelve or sixteen cylinders. A special low-profile version having three or four cylinders is also found in service where head room is restricted.

Construction

In all the in-line S12-D and S12-F engines (Figure 6.1) the engine structure is in two basic parts, flanged and bolted together, namely a deep section bedplate of cast iron and a cast iron A-frame of mono-bloc construction suitably ribbed to form a rigid structure. The bed-plate carries thin wall, steel-backed, white-metal or aluminium-tin lined main bearings. An additional bearing is incorporated to carry the combined loads of the flywheel and part of the weight of the generator. Access doors are provided at the front and back. Those on the back of the engine are fitted with crankcase explosion relief valves.

In this style of construction, which is common to many medium speed engines it is necessary to lift the A-frame if the crankshaft is to be removed. Some designs incorporate a C-frame, permitting side removal e.g. the Allen S30 and the Smit-Bolnes two-stroke.

The one-piece alloy steel crankshaft is slab-forged, oil-hardened and tempered. A solid half coupling forged integrally carries the fly-wheel to which the generator is coupled. The main coupling bolts pass through the crankshaft half coupling, the flywheel and the

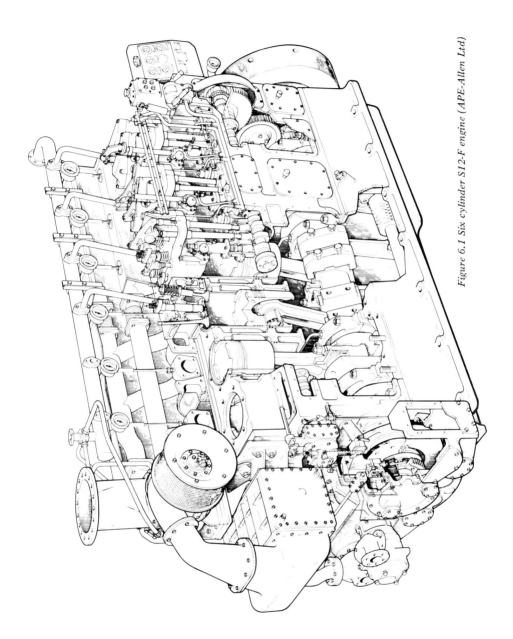

Figure 6.1 Six cylinder S12-F engine (APE-Allen Ltd)

Figure 6.2 Secondary balancing gear (APE-Allen Limited)

generator half coupling; two additional bolts are incorporated to
retain the flywheel on the crankshaft when the generator is un-
coupled. Additional machines may be driven from the free end of
the crankshaft, e.g. bilge pump or compressor, through a clutch.
Axial location of the crankshaft is maintained by renewable thrust
rings. Drilled passages in the crankshaft feed oil from the main to
the connecting rod bearings. In the S12-F engine these oil passages
are arranged to provide a continuous supply of oil to the cooled

pistons. Where necessary, balance weights are bolted to the crank-webs; in four-cylinder engines having cranks at 180°, secondary balancing gear is required (Figure 6.2).

The connecting rods are H-section forgings, bored to carry oil to the gudgeon pin bush, which is a light interference fit in the rod. Crankpin bearings are thin-walled steel, lined with aluminium-tin, split horizontally and secured by four fitted bolts. Connecting rods of differing designs will be found in other engines, e.g. steel stampings, crankpin bearing housings split diagonally and having serrated joints for location or again, split horizontally and having the bearing cap extended to fit precisely over the connecting rod foot, giving a very secure assembly.

The fully-floating gudgeon pin of the Allen S12 is steel, case-hardened and ground, retained in the aluminium alloy piston by circlips. The piston has an integrally cast alloy iron carrier for the top piston ring, two additional pressure and one slotted oil scraper rings, all above the gudgeon pin and a toroidal crown. Pistons for the S12-F are one-piece aluminium-alloy castings incorporating oil cooling cavities. The oil provides intensive coding of the piston particularly in the region of the ring belt. Wet-type cylinder liners of close-grained cast iron flanged at the top for support, are used in the range. The flange is pulled down by the cylinder head bolts on to a synthetic compound ring gasket located in a spigot on the engine frame. To permit vertical expansion of the liner it is free to move at its lower end, a seal being effected by two synthetic compound O-rings carried in grooves in the liner wall.

Camshaft and cylinder head

The camshaft, which runs in its own oil bath, is built up in sections, one for each cylinder and the sections, flanged at each end, are connected by fitted bolts. The cams are pre-formed and their individual settings cannot be altered. To allow accurate phasing of crankshaft and camshaft during initial set up, elongated holes are provided between the camshaft drive wheel and the camshaft. Adjustable packing pieces inserted into the elongated holes ensure that correct timing is maintained.

Lubrication of the camshaft bearings is by a forced feed system, an oilway bored through the full length of the camshaft conducting the oil to the bearings. On the S12-D the camshaft is driven by a roller chain while on the S12-F a spur gear train is used, in both cases at the flywheel end of the engine. An extension of the camshaft at the driving end drives both the hydraulic governor and the tacho-meter via a flexible coupling.

The cylinder heads are of alloy cast iron and separately detachable. Each head carries a single inlet valve, a single exhaust valve and a central fuel injector. The valve gear is totally enclosed and is lubricated from the engine oil system.

Combustion chambers are of the open type designed to ensure efficient combustion. The centrally placed fuel injector is situated outside the valve covers so that fuel oil contamination of the lubricating oil is avoided. This also enables the injectors to be easily withdrawn for servicing without disturbing the valve covers. Each valve has two springs and is fitted with a rotator. The valves seat in the cylinder heads on renewable inserts of iron alloy.

While the S12-D has one inlet and one exhaust valve, the S12-F, because of its higher running speed, is fitted with heads having two inlet and two exhaust valves. The valve pairs are parallel and operated by rocking levers and guided bridges.

Fuel pump

Separate camshaft-actuated helix-type feed pumps are employed for each cylinder. These deliver fuel to the injectors which are set to lift at a pressure of 176 kg/cm^2 on the S12-D and 211 kg/cm^2 on the S12-F. Each fuel pump delivery volume is controlled by a rack which alters the cut-off or spill point. The racks are linked through a control shaft to the engine governor which thus regulates the end of the fuel delivery period to the cylinders and hence the quantity of fuel delivered according to the power required. Fuel injection commences at approximately $15°$ before top-dead-centre and takes place over a period of about $35°$ of crank angle. Combustion should be completed within this period, this is important. Typical valve timing diagrams are shown in Figure 6.3 and 6.4. The normally aspirated diagram is shown for information only — the engine is now only supplied in the turbo-charged version. On turbo-charged engines the turbo-blower is mounted at the free end of the engine and a filter silencer is fitted on the air intake. An intercooler is interposed between the blower and the cylinders. It comprises a matrix of finned tubes through which sea or raw water may be passed. The cooling water circuit also serves the lub. oil and jacket water coolers.

Water pump

One or two engine driven cooling water pumps may be fitted. The materials used in their construction are stainless steel for the impeller shaft, gun metal for the impeller and cast iron or gun metal for the volute casing.

205

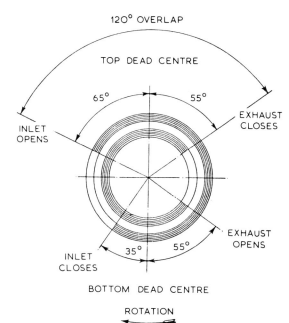

Figure 6.3 Valve timing diagram for turbo-charged four-stroke engine (APE-Allen Limited)

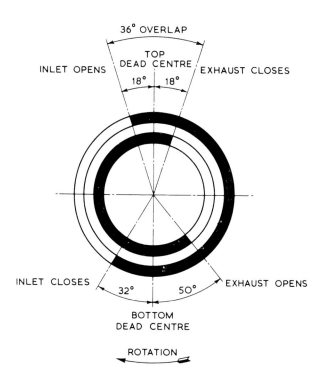

Figure 6.4 Valve timing diagram for normally-aspirated four-stroke engine (APE-Allen Limited)

Hydraulic governor

When used for electric power generation the engine is fitted with a hydraulic governor. This incorporates a centrifugal speed sensing device controlling a suitably damped oil-operated servo-cylinder through a pilot valve. The governor is fitted with speed droop and load limit controls. Motor operated speeder gear to enable remote alteration of engine speed is frequently fitted.

A typical example of a hydraulic governor is the Woodward. This has adjustable droop for parallel operation and it is fitted with a speed-adjusting control. If required it can be fitted with a synchronising motor to provide remote control from a switchboard. Figure 6.5 shows the dial control governor without its auxiliary equipment, but it is adequate for our purpose.

A gear pump driven from the engine supplies oil under pressure to accumulator pistons under which is a by-pass to regulate

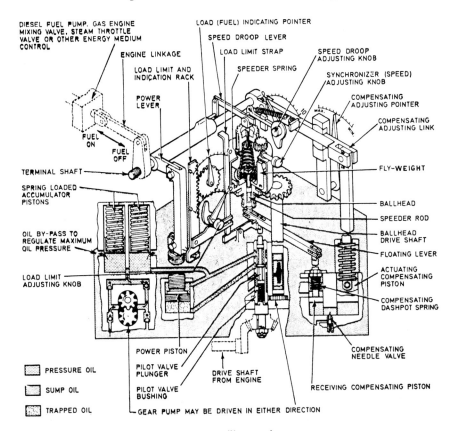

Figure 6.5 Dial control governor without auxiliary equipment

maximum pressure. One branch supplies oil which acts on top of the power piston, the pressure always tending to turn the terminal shaft to shut off fuel, while the other branch supplies oil to the pilot valve which is operated by the linkage from the flyweights above.

Should the speed of the engine decrease due to increased load the flyweights will move towards their centre of rotation and lower the position of the pilot valve plunger, admitting oil pressure from the top port to the bottom port. The pressure is now equal on top and bottom of the power piston. Since, however, the area on the bottom is much greater than the top the net resultant force causes the piston to move upwards. This operates the power lever turning the terminal shaft which varies the cut-off point in the fuel pump through linkage, but as the power piston moves up, the actuating compensating piston moves down. Oil under this piston is now forced through the needle valve to the receiving compensating piston, raising the outer end of the floating lever and the pilot valve plunger until it has returned to its normal position. This stops further movement of the terminal shaft so that the fuel control is now set in a position corresponding to the increased fuel required to run the engine at normal speed under the increased load. Figure 6.5 shows the speed-adjusting mechanism.

HIGH SPEED DIESEL ENGINES

A typical example of a high speed engine used for generator, pump and compressor drive, is the Ruston Paxman RPH range (Figure 6.6). It is manufactured in 6, 8, and 12 cylinder versions and has the suffix Z for normally aspirated marine machines, XZ for pressure charged machines and CZ for pressure-charged and intercooled machines. The engines are Vee-type four stroke machines with a bore of 178 mm and a stroke of 197 mm and provide outputs in the range 121-580 kW with shaft speeds from 900-1500 rev/min. At 1500 rev/min the mean piston speed is 9.8 m/sec. The normally aspirated version has a compression ratio of 17.25:1 and the pressure-charged engine a compression ratio of 15.5:1.

Construction

A cast iron crankcase carries an underslung crankshaft in thin shell bearings of steel, lined with aluminium alloy. The bearings are located in housings and the bearing caps have deep fitting faces to

208

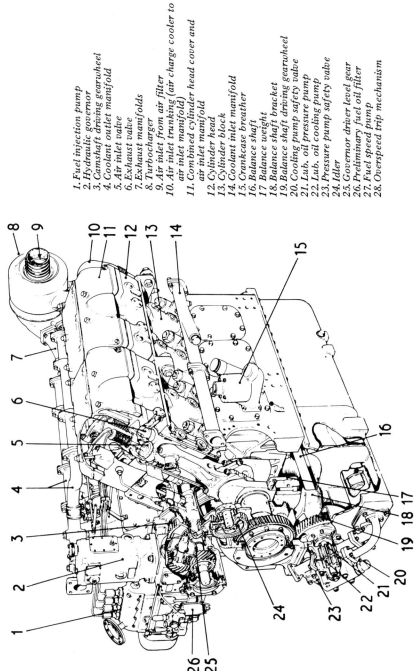

1. Fuel injection pump
2. Hydraulic governor
3. Camshaft driving gearwheel
4. Coolant outlet manifold
5. Air inlet valve
6. Exhaust valve
7. Exhaust manifolds
8. Turbocharger
9. Air inlet from air filter
10. Air inlet trunking (air charge cooler to air inlet manifold)
11. Combined cylinder head cover and air inlet manifold
12. Cylinder head
13. Cylinder block
14. Coolant inlet manifold
15. Crankcase breather
16. Balance shaft
17. Balance weight
18. Balance shaft bracket
19. Balance shaft driving gearwheel
20. Cooling pump safety valve
21. Lub. oil pressure pump
22. Lub. oil cooling pump
23. Pressure pump safety valve
24. Idler
25. Governor driver level gear
26. Preliminary fuel oil filter
27. Fuel speed pump
28. Overspeed trip mechanism

Figure 6.6 Sectional arrangement of Ruston Paxman RPH diesel engine

accommodate side-thrust. Machined faces on the top of the crankcase carry the cylinder blocks, these being arranged in a 60° included angle, and dry type renewable cylinder liners are fitted.

The forged steel crankshaft has ground journals and crank-pins and is drilled for lubrication of the crankpin bearing; it is flanged at the driving end to receive coupling bolts and a smaller flange at the free end is provided for auxiliary power take off. The main bearings comprise thin steel shells lined with aluminium tin alloy. Longitudinal location of the shaft is achieved by independent thrust rings in the crankcase at one bearing. The engine crankshaft is counter-balanced and a secondary balancing system is gear-driven from the free end of the crankshaft. Sometimes a torsional vibration damper is also fitted at this end of the engine.

Fork and blade connecting rods drilled for piston cooling are used in the engine. They are machined from steel stampings. A bottom end block attached to the forked rod houses a steel-backed aluminium tin thin-shell crankpin bearing. The blade rod bottom end runs on the fork rod block, the outer surface of the block being chrome plated. A copper-lead lined, lead-flashed bearing shell is

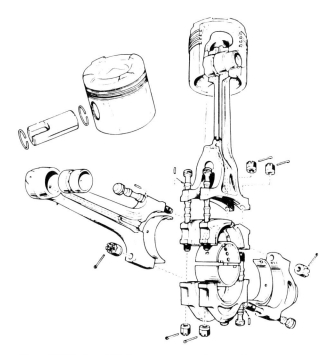

Figure 6.7 Fork and blade connecting rod assembly (Ruston Paxman Diesels Ltd.)

fitted in the blade rod bottom end. The top end bushes are of gunmetal and run on a fully floating gudgeon pin; the pin is retained in the piston by circlips (Figure 6.7).

Aluminium alloy pistons having an austenitic steel insert for the chrome-faced top pressure ring are used in the engine and each piston is fitted with three pressure rings and one oil scraper ring. The RPHZ engines with speeds of more than 1250 rev/min and all RPHXZ and RPHCZ engines feature oil-cooled pistons. This cooling takes the form of an oil-jet directed from the top of the connecting rod on to the underside of the piston crown. The pistons and connecting rods can be removed through the crankcase doors.

Cylinder heads and valve gear

Cast iron heads with renewable hardened valve seat inserts cover 1, 2 or 3 bores, depending on the cylinder configuration. Cast aluminium covers on the heads fully enclose the hardened inlet and exhaust valves (one of each per cylinder) and the valve gear is pressure-lubricated. On the 6- and 8-cylinder engines a one-piece steel camshaft, hardened and ground on all working surfaces, is housed centrally within the crankcase and driven by a gear-train from the free end of the crankshaft. The gear teeth are helical and are spray-lubricated. The camshaft operates push rods via roller type followers, all totally enclosed within the cylinder blocks. On the 12-cylinder engines, the camshaft is located in a separate housing and is chain driven from the crankshaft.

The engine fuel injection pumps are mounted in one unit at the free end of the engine and gear-driven from the camshaft. The pumps are of the Bosch type but, unlike the medium speed engine pumps, these are grouped together under the control of one rack. The rack is activated by a hydraulic governor according to the loading of the engine. The fuel pumps can be separately adjusted manually to power balance the engine.

Pressure charged engines are fitted with exhaust gas driven air-cooled turbo-blowers; the turbo-blowers are lubricated from the engine system. Where fitted, intercoolers may be water or air cooled.

CATERPILLAR ENGINES

Another make of four-stroke high speed engine in common use at sea is the Caterpillar engine. This company manufactures a family of ranges many of which are used as auxiliary engines or as main propulsion units on smaller ships. For electrical power generation, units ranging from the 3300 range, producing 50 kW in the four cylinder version, up to the 399 range which in V16 form produces 860 kW are in common use at sea.

The 3304 and 3306 engines have a bore of 121 mm and a stroke of 152 mm and are produced as in-line engines only. The 399 engines have a bore of 159 mm and a stroke of 203 mm; The most powerful in the 399 range is the D 399 V 16 TC which is a turbo-charged 16-cylinder vee engine producing 860 kW at 1200 rev/min.

All the Caterpillar marine diesels feature one-piece aluminium alloy pistons, slightly elliptical and tapered to allow for heat expansion and provided with integrally cast nickel-iron inserts in way of the compression ring grooves. In all models the pistons are fitted with only three rings. The upper ring is a pure compression ring and the lower ring is a spring-loaded oil control ring. Between these two is a specially formed 'twist' ring which acts as a compression ring on the up-stroke but cants slightly on the down-stroke to act as a second oil scraper. By using only three piston rings Caterpillar claim to reduce ring friction. The inlet and exhaust valves are fitted with rotators and the valve seats are renewable.

Steel backed aluminium alloy bearings are used on main and crankpin journals and underslung crankshafts supported in dismountable bearings are used throughout the range.

Fuel system

The 399 range features helix-type fuel pumps driven off a separate camshaft. Their principle of operation can be followed from the description given subsequently under 'diesel engine fuel pumps' (see page 215) although the pump shown in Figure 6.11 is similar in principle but different in detail.

In the smaller 3300 range engine a sleeve-metred fuel pump, unique to Caterpillar is used and is worthy of further comment here.

A sleeve-metred fuel injection pump is shown in Figure 6.9a and cross-sections of the plunger are shown in Figure 6.9b. The cams and the plungers are completely immersed in diesel oil supplied under pressure. The diesel oil can enter the space above the plunger whenever either the fuel inlet port 5 or the fuel outlet port 9 is uncovered. The cam 12 moves the plunger 7 through a constant stroke. The fuel outlet port 9 in the plunger is uncovered earlier or later in the stroke by sleeve 8 controlled by sleeve control lever 10 which is positioned by the engine governor. This controls the spill point of the pump.

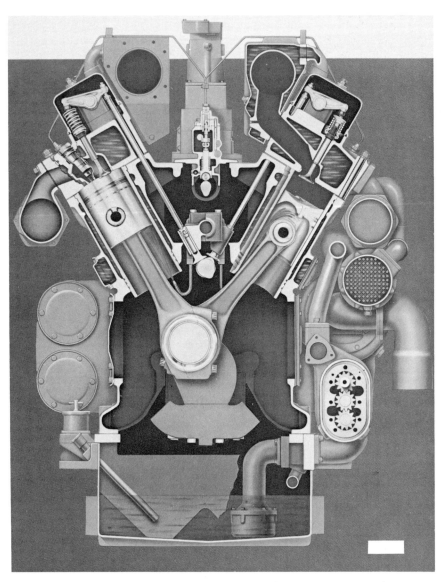

Figure 6.8 Cross-section of Caterpillar Type D398 engine (Caterpillar Tractor Co.)

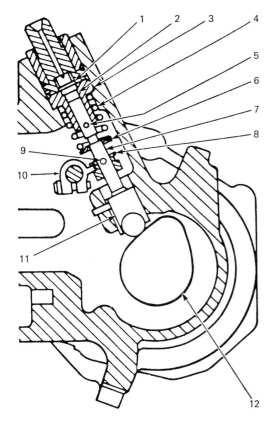

Figure 6.9 (a) Fuel injection pump operation
1. Reverse flow check valve
2. Chamber
3. Barrel
4. Spring
5. Fuel inlet (fill port)
6. Retainer
7. Plunger
8. Sleeve
9. Fuel outlet (spill port)
10. Sleeve control lever
11. Lifter
12. Camshaft
    (Caterpillar Tractor Co.)

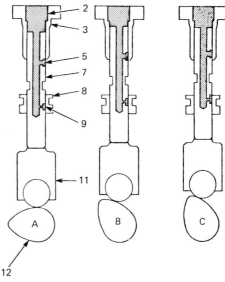

Figure 6.9 (b) Fuel injection pump operation
2. Chamber
3. Barrel
5. Fuel inlet (fill port)
7. Plunger
8. Sleeve
9. Fuel outlet (spill port)
11. Lifter
12. Camshaft
A Before injection
B Start of injection
C End of injection
   (Caterpillar Tractor Co.)

Cooling systems

Whilst a variety of cooling systems may be adopted for marine auxiliary engines the most commonly used is the simple closed circuit system (Figure 6.10). Sea-water is passed through the inter-cooler, the oil cooler and then the jacket water cooler in series flow. Engine driven fresh water circulating pumps are normally fitted but the sea water pump may be either an independent unit or engine driven in tandem with the fresh water pump. The cooling systems may be arranged so that in an emergency sea water can be circulated through the engine jackets, after removal of certain blanks installed in the pipework.

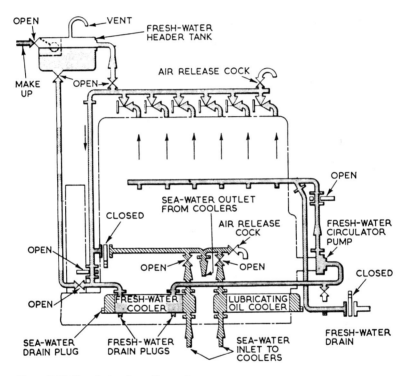

Figure 6.10 Closed circuit cooling system

In ships with diesel main propulsion engines cross-connections between the main and auxiliary engine jacket water systems are sometimes found. This enables the main engine to be kept warm in port from the heat in the auxiliary engine jacket water. To enable the auxiliary engine to be run in dry dock it is also customary to arrange a connection from a convenient double bottom F.W. tank.

DIESEL ENGINE FUEL PUMPS

By far the most common type of fuel pump used on auxiliary diesel engines is the Bosch type, a jerk pump with a helical groove on the plunger which controls the delivery cut-off point. These may be arranged singly along the camshaft, one at each cylinder position, as described for the Allen engine, or ganged together as described for

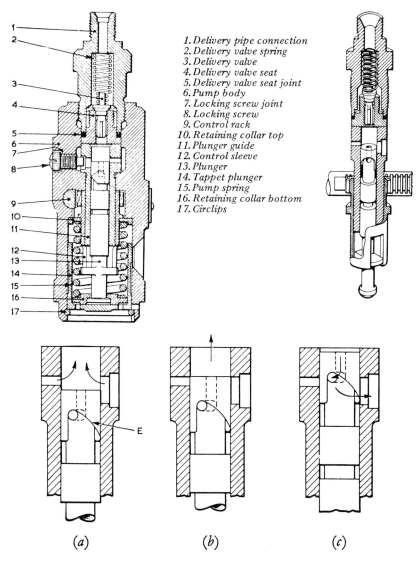

1. Delivery pipe connection
2. Delivery valve spring
3. Delivery valve
4. Delivery valve seat
5. Delivery valve seat joint
6. Pump body
7. Locking screw joint
8. Locking screw
9. Control rack
10. Retaining collar top
11. Plunger guide
12. Control sleeve
13. Plunger
14. Tappet plunger
15. Pump spring
16. Retaining collar bottom
17. Circlips

(a)   (b)   (c)

Figure 6.11 Sectional views of fuel pump assembly

the Ruston-Paxman engine, but have the same operating principle. Each pump contains one pumping element which may be divided into two main components, i.e. pump plunger and guide; delivery valve and seating.

It should be noted that plungers and guides are not interchangeable — they should be treated as combined units or elements.

Operation

The operation of the pump element is shown diagrammatically in Figure 6.11. At *a* the plunger E is at the lower limit of its travel and fuel can enter the barrel by the two ports from the surrounding suction chamber. As the plunger rises, the fuel displaced returns through the ports until the plunger reaches position *b*, where the top of the plunger has closed both ports. The fuel above the plunger is then trapped, and its only outlet is via the delivery valve mounted on top of the pump barrel.

The pressure exerted by the rising plunger causes fuel to lift the valve and to enter the pipe which connects the pump to the injector. As the pipe is already full, the extra fuel which is being pumped in causes a rise in the pressure throughout the line and lifts the nozzle valve of the injector. This permits fuel to be forced into the engine combustion chamber in the form of a fine spray.

At *c* the lower edge of the control helix has uncovered the barrel port, thus allowing the fuel to be by-passed from the barrel to the suction chamber by way of the drilling in the plunger. This allows the delivery valve to shut under the action of its spring, and with the consequent collapse of pressure in the pipeline, the injection valve also shuts.

The plunger stroke is always constant, but the effective part of it during which pumping takes place is variable. By means of the helical edge, which itself can be rotated within the guide, it is possible to vary the point in the stroke at which cut-off occurs. The plunger is rotated axially by the toothed quadrant (Figure 6.11), which is clamped to the sleeve, having two rectangular slots engaging the lugs of the plunger at its lower end. The toothed quadrant meshes with the rack on the control rod, which is externally connected by suitable linkage to the governor and hand control.

Injector

The injector assembly (Figure 6.12) consists of two main components, the nozzle valve and nozzle holder.

The nozzle valve takes the form of a plunger lapped into the nozzle body within which it works freely, having at its inner end a stem upon which a valve face is formed. The outer end has an extension engaging with the valve spindle.

Fuel is fed to the mouth of the nozzle through small holes, drilled vertically in the nozzle body, which terminate in an annular gallery just above the valve seating. The nozzle valve is raised from its seating in the nozzle body by the pressure of fuel from the pump. Thus, the fuel in the gallery is forced by the upward movement of the plunger in the pump, through the holes in the nozzle to form a spray in the engine combustion chamber.

Note that the dismantling, cleaning and testing of injector assemblies calls for the use of special tools; unless, therefore, the operator is in possession of this equipment the complete injector assembly should be sent to the nearest authorised repairer when investigation proves that the injector requires servicing.

Injectors should be taken out for examination at regular intervals. It is not easy to state just how long the intervals should be, owing to the widely different conditions under which engines operate, but under normal conditions an inspection every three months should suffice.

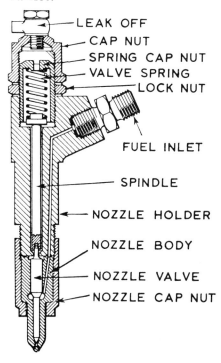

LEAK OFF
CAP NUT
SPRING CAP NUT
VALVE SPRING
LOCK NUT

FUEL INLET

SPINDLE

NOZZLE HOLDER

NOZZLE BODY

NOZZLE VALVE
NOZZLE CAP NUT

Figure 6.12 Injector assembly

After the engine has been in service for a considerable period it may be found that starting has become difficult owing to a fall in the injection pressure. This can be corrected by an adjustment to the injector spring compression, but such adjustment should be carried out only with the injector removed from the engine and coupled to a testing outfit capable of indicating release pressures.

For cleaning the fuel injectors, a special set of tools is available and contains the following main items: brass wire brush, nozzle body groove scraper, probing tool, nozzle body seat cleaner, nozzle body, dome cavity cleaner. Cleaning is carried out as follows. Hold injector in vice with nozzle pointing upwards. Remove nozzle cap with spanner. Examine nozzle for carbon and note if the valve lifts out freely. Immerse in paraffin or other approved cleansing agent to soften the carbon. Clear nozzle holes with a probing tool of correct size.

Tracing faults

The symptoms of faulty pumps and the possible cause are given below.

Difficult starting. Fuel system not primed through to the nozzles.
Starting-air pressure too low.
Water in the fuel.
Air intake filter choked.
Poor compression due to any of the following causes:
    Exhaust valve seats in bad condition.
    Exhaust valve spindle sticking.
    Starting-air valve sticking.
    Piston rings stuck or liners badly worn.

Engine running Sticking fuel-pump control racks or linkage.
unsteadily Fuel-injection-pump delivery valves sticking.
or hunting. Overload or governor fault.

Smoky exhaust. Check fuel pump timing.
Check maximum firing pressure and exhaust temperature.
Injectors may be sticking or sprayer holes choked or enlarged.
Injector spring screwed down too tightly.
Overload or unsuitable fuel oil.

STEAM TURBINES

Auxiliary steam turbines are prime movers for pumps, electric generators and fans. Power outputs vary from about 200 kW for pump and fan applications to about 1.5 MW for generator sets. They are always single cylinder engines and may be horizontal or vertical, sometimes with overhung wheels; both condensing and back pressure turbines are used designed for steam conditions ranging from about 6 bar to about 62 bar at 510°C.

Construction

For power generation, turbines are horizontal axial flow machines built similarly to the propulsion turbines described in 'Marine Steam Engines and Turbines' in the Marine Engineering series published by Newnes-Butterworths. They are of the impulse reaction type and may be condensing, exhausting either to an integral condenser — invariably underslung — or to a separate condenser. This may be a central auxiliary or the ship's main condenser (usually with cross-connection to provide operational flexibility in port). A typical auxiliary condensing turbine is shown in Figure 6.13. Alternatively the turbine may be a back-pressure unit in which the exhaust is used as a source of low pressure steam for other services. The casings, split horizontally and supporting the rotors in plain journal bearings are cast mild steel or, for temperatures exceeding 460°C ½% Molybdenum steel, with cast or fabricated mild steel for parts not subject to high temperatures. Solid gashed rotors of chrome-molybdenum alloy steel are usual though some may be encountered having rotor spindles of this alloy, with shrunk and keyed bucket wheels. Blades may be stainless iron, stainless steel or Monel Metal, with shrouded tips, fitted into the rotors in a number of root forms.

Depending on steam conditions and power the turbine will have a two row velocity compounded stage followed by a suitable number, probably five or more, single row pressure compounded stages, each separated by a cast steel nozzle. Steam enters the turbine at the free end via a cast steel nozzle box and flows towards the drive end which is connected to the pinion of the reduction gearing by a fine tooth or other flexible coupling designed to accommodate longitudinal expansion of the rotor. Typical rotating speed of the rotor is about 6500 rev/min, geared down to suit an alternator speed of 1800 rev/min.

The diaphragms separating each stage are split horizontally and fitted in grooves machined in the casing, to which they are securely

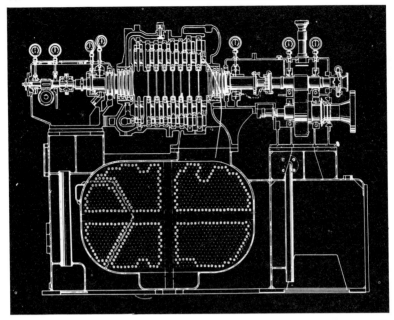

Figure 6.13 Sectional arrangement of condensing turbine (Peter Brotherhood Ltd.)

fixed so as not to be disturbed when the top half casing is lifted. The diaphragms may be of steel or cast iron depending on the stage pressure.

Interstage leakage, where the rotor shaft passes through the diaphragm, is minimised by labyrinth glands of a suitable non-ferrous alloy such as nickel-bronze. Labyrinth packing may also be used for the turbine shaft/casing glands which are steam-packed. In some turbines contact seals utilising spring-loaded carbon segments as the sealing media are used instead of the labyrinth gland (Figure 6.14). A typical labyrinth gland arrangement is shown in Figure 6.15. The low pressure labyrinth is divided into three separate groups so as to form two pockets. The inner pocket serves as an introduction annulus for the gland sealing steam; this flows inwards into the turbine and some escapes through the centre labyrinth into the outer pocket. The supply of sealing steam is regulated to keep the pressure in the outer pocket just above atmospheric. Surplus steam in the outer pocket is usually led to a gland steam condenser. The gland at the high pressure end of the turbine is subject to a considerable pressure range from sub-atmospheric at low load to considerably above atmospheric at full load and is therefore arranged with three pockets. Gland steam is supplied to the centre pocket. The innermost pocket is connected to a lower pressure stage further down the turbine enabling the

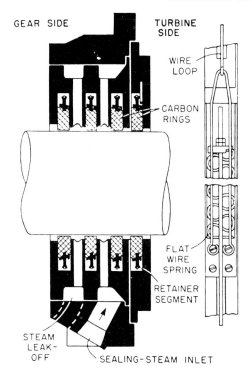

GEAR SIDE

TURBINE SIDE

WIRE LOOP

CARBON RINGS

FLAT WIRE SPRING

RETAINER SEGMENT

STEAM LEAK-OFF

SEALING-STEAM INLET

Figure 6.14 Example of carbon ring shaft seal

leakage steam to rejoin the main stream and do further work while the outermost pocket, connected to the gland condenser, prevents excessive leakage to atmosphere.

The labyrinth packings at both ends of the turbine and in the diaphragms are retained by T-heads on the outer peripheries which slot into matching grooves. Each gland segment is held in position by a leaf spring. The retaining lips of the T-head prevent inward movement and the arrangement permits temporary outward displacement of the segments. Rotating of the segments is prevented by stop plates or pegs fitted at the horizontal joint.

Although there is little residual end thrust on the rotor it is necessary to arrange a thrust bearing on the rotor shaft and it is normal to make this integral with the high-pressure end journal bearing. Sometimes the thrust is of multi-collar design but is more frequently a Michell-type tilting pad bearing.

Governing

Unlike propulsion turbines, generator turbines work at constant speed and must be governed accordingly. Classification Society Rules

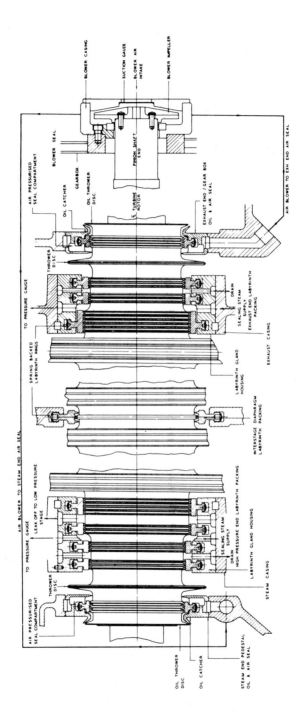

Figure 6.15 Arrangement of steam labyrinth glands and air sealing system (Peter Brotherhood Ltd.)

require that there must be only a 10% momentary and a 6% permanent variation in speed when full load is suddenly taken off or put on. On an a.c. installation it is required that the permanent speed variations of machines intended for parallel operation must be equal within a tolerance of ± 0.5%. In addition to the constant speed governor an overspeed governor or emergency trip is also fitted.

Speed-governing system

Speed governing systems consist of three main elements:-

1. A speed sensing device, usually a centrifugal flyweight type governor driven through worm and bevel gearing from the turbine shaft.
2. A linkage system from the governor to the steam and throttle valve; on larger turbines this is an oil operated relay consisting of a pressure balanced pilot valve controlling a supply of high pressure oil to a power piston.
3. A double-beat balanced steam throttle valve which regulates the amount of steam passing to the turbine nozzles, according to the speed and electrical load.

To ensure stability, i.e. freedom from wandering or hunting of the speed, the system is designed to give a small decrease in speed with increase in load. The usual amount of this decrease, called the 'speed droop' of the governor, is 3% between no load and full load. If the full load is suddenly removed the speed will at first increase to a value of 7-10% above normal. This is called the governor 'fly-up' or momentary variation. The speed will then decrease to a value of 3% above normal (the droop value). Similarly if the full load is suddenly applied a momentary fall in speed of 4-7% below normal will occur. This is called governor 'dip'.

Figure 6.16 is a simplified schematic arrangement of a typical speed governing system from which the sequence of events during load changes may be more easily followed. In the diagram the throttle valve is operated via lever Y.

Overspeed trip

Overspeed occurs when the load is suddenly thrown off and while this is normally rectified by the speed governor, an emergency trip is always fitted, a usual type is illustrated in Fgiure 6.17.

An unbalanced steel valve 3, located in the pinion shaft extension,

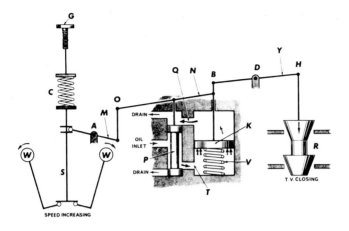

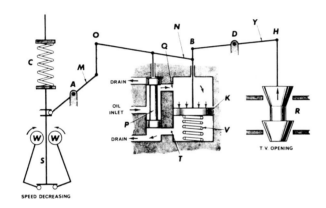

Figure 6.16 Schematic arrangement of speed governing system

a, D. Fulcrums   Q.   Port
C.   Adjusting spring   R.   Throttle valve
G.   Hand-operated wheel   S.   Sleeve
K.   Piston   T   Port
M,N. Levers   V.   Spring
O.   Fulcrum   W.   Weights
P.   Pilot valve
(Peter Brotherhood Ltd.)

is held onto the valve seat by a helical spring 2, while the speed of
pinion shaft remains below tripping speed. If the speed increases 10—
15% above the turbine rated speed, centrifugal force exerted by the
trip valve, at this higher speed, overcomes the spring force and moves
away rapidly from the valve seat. This allows lubricating oil, which is
fed into the centre of the shaft extension via an orifice plate, to
escape; thus dropping the oil system pressure downstream from the

orifice to zero. This causes the low pressure oil trip to operate and drain oil from the relay cylinder, the relay cylinder spring thus raising the relay piston and closing the throttle valve which cuts off the steam supply to the inlet of the turbine. It is vital to maintain the trip gear in good working order and this can be greatly aided by testing at regular intervals. This should be done whenever the generator is to be loaded.

In addition to an overspeed trip it is customary to fit a low pressure oil trip to steam turbines and frequently a back pressure trip (Figure 6.18) is fitted.

Back pressure trip

The purpose of the back-pressure trip is to protect the turbine and exhaust system in the event of over-pressure due to loss of condenser cooling water, extraction pump failure or accidental closure

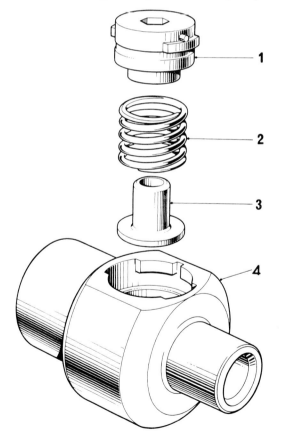

1. Cap
2. Emergency valve spring
3. Emergency valve
4. Casing
Figure 6.17 Overspeed trip gear
(Peter Brotherhood Ltd.)

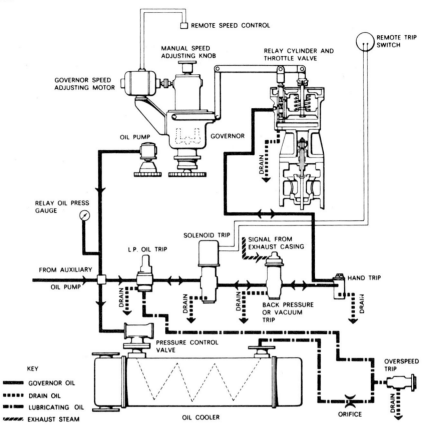

Figure 6.18 Governor control oil system (Peter Brotherhood Ltd.)

of the exhaust valve. The trip consists of a spring loaded bellows, connected to a hydraulic spool valve interposed in the governing oil circuit between the main oil pump and the governor relay.

When the turbine is running the valve is held upwards by the spring. In this position high pressure oil can pass freely across the upper ports of the valve to actuate the governor relay. If the back pressure increases to a predetermined level the load on the bellows unit is sufficient to overcome the adjusting spring and allow the operating spindle to move downwards and push the ball valve off its seat. In so doing, oil at relay pressure, is admitted through the drilled passages in the trip body to the piston valve, depressing the valve against its spring. The valve will, simultaneously, cut off and drain the high pressure supply to the governor relay. The throttle valve is consequently closed by the relay pistons under the action of the spring V (Figure 6.16) and the turbine is stopped.

BACK PRESSURE TURBINES

Many ships now utilise an auxiliary steam turbine as a primary pressure reducing stage before passing the steam to other auxiliaries demanding steam at a substantially lower pressure than that available. Such an arrangement is shown in Figure 6.19 and the heat balance of a scheme such as this is far more favourable than using a pressure reducing valve.

These back pressure turbines have most features of a condensing turbine but no condenser; the most important difference is in the governing. Designed to work against back pressures in the range 1—3.5 bar and with much lower available heat drops than with a condensing turbine, governing by the simple opening or closing of the throttle

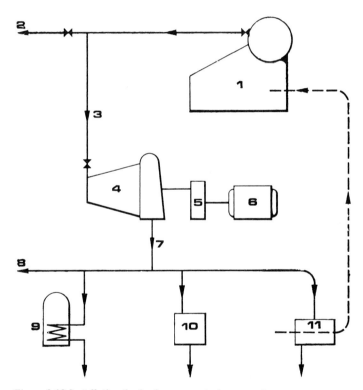

Figure 6.19 Installation for back-pressure turbo-generator
1. *Main high pressure boilers*
2. *H.P. steam to main propulsion turbines*
3. *H.P. steam to back presure turbo-generator*
4. *Auxiliary turbine*
5. *Reduction gear*
   (Peter Brotherhood Ltd.)
6. *Generator*
7. *Auxiliary turbine exhaust*
8. *Steam to ship's services*
9. *Evaporator*
10. *De-aerator*
11. *Steam/air heater*

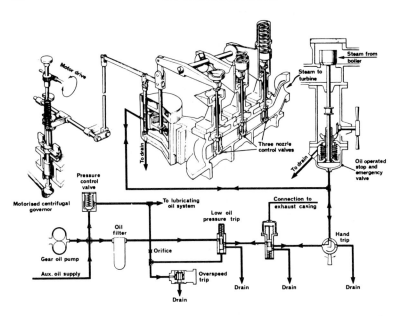

Figure 6.20 Back-pressure governor and control system (Peter Brotherhood Ltd.)

valve is inadequate. Instead the governor, as shown in Figure 6.20 sequentially controls the opening of a number of nozzle control valves. The control system is arranged for a straight line regualtion (load/speed) with a speed droop of about 4% between full load and no load. The turbine governor also incorporates speed droop control adjustment with a range of approximately 2.5% to 5.5% to enable load sharing between a group of generators to be readily adjusted.

VERTICAL STEAM TURBINES

Vertical steam turbines are extensively used for cargo, ballast and other pump drives. Like the horizontal machines used for power generation they can be condensing or back-pressure units. They are however invariably single stage machines having an overhung wheel as shown in Figure 6.21. The steam casing of this turbine is a single steel casting bolted to the top of the exhaust casing. The nozzles are fitted and seal welded into the underside of the steam casing inlet belt forming a ring which provides an uninterrupted arc of admission. The exhaust casing is split in the vertical plane, allowing removal of the front half for rotor inspection without disturbing the steam or exhaust piping.

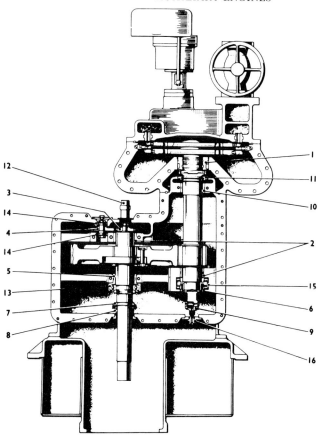

Figure 6.21 Sectional arrangement of vertical turbine

1. Labyrinth gland housing
2. Pinion and gear shaft bearing
3. Spur gear (govnr. and oil pump drives)
4. Idle gear (govnr. and oil pump drives)
5. Gear wheel shaft bearing
6. Thrust bearing oil seal
7. Oil thrower gear shaft
8. Oil seal gear shaft
(Peter Brotherhood Ltd.)

9. Overspeed trip unit
10. Pinion shaft upper bearing
11. Pinion shaft upper oil seal
12. Tachometer generator
13. Gear shaft location bearing
14. Idler shaft bearings
15. Pinion thrust bearing
16. Trip oil inlet fitting

The rotor shaft is bolted to a head flange on the pinion shaft of the single reduction gearing. A thrust bearing located below the pinion supports the weight of the rotor and absorbs any vertical thrust. This bearing is usually of the Michel multi-pad type.

7 Propellers and propulsion

The most common form of propeller today is the fixed pitch, the blades being cast integral with the boss (Figure 7.1), although there are still some ships in service in which the blades are cast separately and secured to the propeller boss by studs and nuts.

Controllable-pitch propellers

As its name implies, it is possible to alter, at will, the pitch of this type of propeller to suit the prevailing resistance conditions. This change in pitch is effected by rotating the blade about its vertical axis, this movement usually being carried out by hydraulic or mechanical means. The most obvious application is for the double-duty vessel, such as the tug or trawler where the operating conditions when towing or running free are entirely different. Since it is usually possible to reverse the pitch completely, this type of propeller may be used with a uni-directional engine to give full ahead or astern thrust.

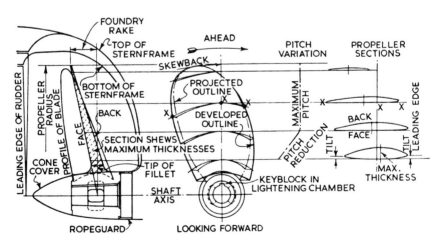

Figure 7.1 Typical arrangement of solid propeller

Many propellers of this type are now in operation for power ratings varying from 500 to over 30 000 hp on a wide range of vessels. One of the most widely used c.p. propellers is the KaMeWa, an hydraulically operated Swedish propeller first introduced in 1937. In this unit the blade pitch is altered by a servomotor housed within the hub body. This consists of one or two hydraulic cylinders with piston and rod. The piston moves in response to the difference in oil pressure on its ends. The oil flow to and from the servomotor is controlled by a slide valve in the piston rod; the slide valve is mechanically operated by a hollow rod which passes through a hole bored in the propeller shaft and is connected to operating levers located in an oil distribution box.

Figure 7.2 shows a single piston servomotor. If the slide valve is moved aft the valve ports are so aligned that oil under pressure flows along the hollow valve rod to the forward end of the piston, causing the piston to move aft, in sympathy with the valve rod, until the ports are again in a neutral position. Conversely, if the valve is moved forward, the piston will move in a forward direction.

When the piston moves the whole piston rod assembly, including the sliding shoe in the crosshead, moves with it. A cranking pin on each propeller blade locates in each of the sliding shoes so that any movement of the servomotor piston causes exactly the same pitch change in each of the propeller blades (Figure 7.3).

Oil enters and leaves the hub mechanism via an oil distribution box mounted inside the ship — either at the end of the gearbox or on a section of an intermediate shaft. Oil pressure — about 40 bar max. in the single piston hub and about 130 bar max. in the two piston unit — is maintained by either a positive displacement pump driven from the propeller shaft or by separate motor-driven pumps. An inlet valve on the oil distribution box regulates the pressure in the high pressure chamber from where the oil passes to the hub mechanism via the hollow rod in the propeller shaft. Oil passes from the hub mechanism to the low pressure chamber of the distribution box along the outside of the valve rod. An outlet valve maintains a slight back pressure on the oil-filled hub when the vessel is underway. When in port this pressure is maintained by the static head of an oil tank mounted above the ship's waterline and connected to the oil distribution box.

The forward end of the valve rod connects to a key which is moved fore or aft by a sliding ring within the oil distribution box. A servomotor mounted externally to the box is used to move the sliding sleeve through a fork mechanism. In the event of a failure in the servomotor an auxiliary servomotor, located near the oil distribution box, can be used to shift the propeller blades to the full ahead

232

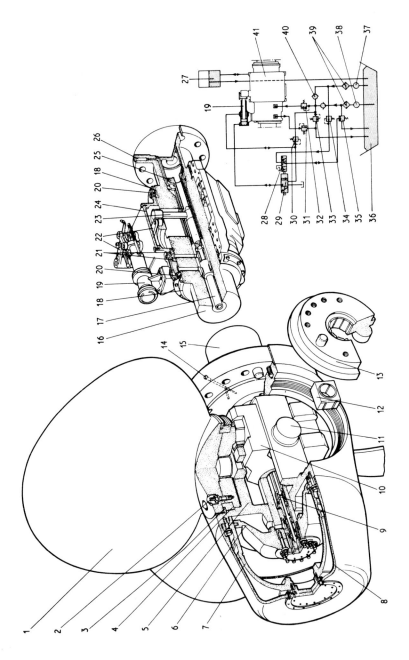

Figure 7.2 Single piston servomotor
(KaMeWa)

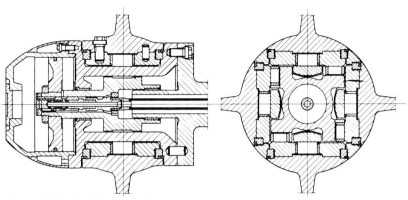

Figure 7.3 Detail of KaMeWa S1 propeller hub

Key to Figure 7.2

1. *Blade with flange*
2. *Blade stud with nut and cover*
3. *Blade sealing ring*
4. *Bearing ring*
5. *Hub body*
6. *Servometer piston*
7. *Hub cylinder*
8. *Hub cone*
9. *Main regulating valve assembly*
10. *Piston rod with cross head*
11. *Centre post (integrated with hub body)*
12. *Sliding shoe with hole for crank pin*
13. *Crank pin ring*
14. *Safety valve for the low pressure part of the propeller hub*
15. *Propeller shaft with flange*
16. *Intermediate shaft*
17. *Valve rod*
18. *End cover*
19. *Pitch control auxiliary servomotor assembly*
20. *Low pressure seal assembly*
21. *High pressure seal assembly*
22. *Yoke lever*
23. *Valve rod key*
24. *Oil distribution box casing*
25. *Stand-by servo*
26. *Non-return and safety valve for stand-by servo*
27. *Oil tank*
   e.g. Oil tank for static over-pressure in propeller hub
28. *Regulating valve for unloading pump*
29. *Regulating valve for auxiliary servo-motor*
30. *Reducing valve (auxiliary servo-motor*
31. *Back pressure maintaining valve*
32. *Sequence valve*
33. *Safety valve*
34. *Reducing valve (unloading)*
35. *Unloading valve*
36. *Main oil tank*
37. *Main pump*
38. *Unloaded pump*
39. *Main filter*
40. *Check valve*
41. *Oil distribution box*

pitch position by causing the valve rod to mechanically push the crosshead into the correct position.

The propeller pitch and engine speed can be remotely controlled from a single lever known as a Combinator. Any number of Combinators may be installed in a ship. The lever controls a number of cam-operated transmitters. These may be electrical or pneumatic devices.

In some installations an electronic system may be incorporated which keeps the engine load at a preset value irrespective of variations in external conditions. In this system the control unit operates a hydraulic correction servo in the feedback of the main propeller control servo system.

Design considerations

In the space available it is impossible to go into detailed considerations of propeller design. The following notes are, therefore, only
intended to show in outline just what factors have to be considered.

Definition of terms

Diameter. The diameter of a propeller is the diameter of the circle
described by its tips.

Pitch. When a point rotates about an axis and, at the same time,
moves parallel to that axis, it describes a helix. The distance the
point moves measured parallel to the axis in one revolution is the
pitch.

Similarly, if a line at right angles to the axis rotates about that axis
and, at the same time, moves parallel to the axis, it will generate a
helicoidal surface.

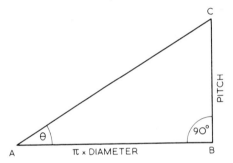

Figure 7.4 Pitch angle

The face of a screw propeller is part of such a helicoidal surface.
If the pitch of a propeller is constant throughout the length and
breadth of the blade, it is said to be of uniform pitch. If the pitch
varies radially or circumferentially, the propeller is said to be of non-
uniform pitch.

Pitch angle. The pitch angle is best described by Figure 7.4 where
$AB = \pi \times$ dia. of the helix, BC = pitch and θ = pitch angle.

It should be noted that, in the case of a constant-pitch propeller,
although BC remains unaltered, AB will increase from the boss
towards the tips, and consequently θ, the pitch angle, will decrease.

Pitch ratio. The pitch ratio of a propeller is the mean pitch
divided by the diameter. When the ratio is unity, the propeller is
described as a square propeller.

Projected area. This is the sum of the area bounded by the outline
of the blades when projected onto a plane perpendicular to the axis
of the screw.

Developed area. The developed area (or expanded area as it is sometimes called) is the total surface area of the driving forces of the blades. This may be expressed inclusive or exclusive of the boss area.

Disc area ratio (d.a.r). This is the developed area divided by the area of the circle described by the tips.

Direction of rotation. When viewed from astern, if the propeller revolves in a clockwise direction when going ahead, it is known as right-handed. If in an anti-clockwise direction, it is left-handed.

Wake

In moving forward, a ship imparts forward motion to the water at the stern; this is known as the wake. Thus the flow into the propeller disc, although not uniform, has a mean speed less than that of the speed of the ship. It will be seen, therefore, that the speed of advance of the propeller relative to the water in which it is working will be less than the observed speed of the vessel. This difference in speed, expressed as a percentage of the ship's speed, is known as the wake fraction. From the foregoing it will be evident that it is of the utmost importance to determine accurately the value of the wake when deciding the correct pitch of the propeller.

Wake varies considerably with the form of the vessel, but, generally speaking, the bluffer the ship the higher the wake value. The two standard systems of notation for wake are the Taylor system and the Froude system. The former expresses wake as follows:

$$\frac{V_S - V_A}{V_S} = \text{wake fraction,}$$

where V_A = speed of advance propeller;

V_S = speed of ship

The Froude system relates wake to the speed of advance and is expressed as follows:

$$\frac{V_S - V_A}{V_A} = \text{wake fraction.}$$

Diameter and pitch determination

In settling the leading dimensions of the propeller the designer has at his disposal the results of the many series of experiments which have been conducted at the various testing tanks. Among the best known of these series are those conducted by Taylor and by Professor Troost (*N.E.C. Trans.,* Dec. 1950).

236

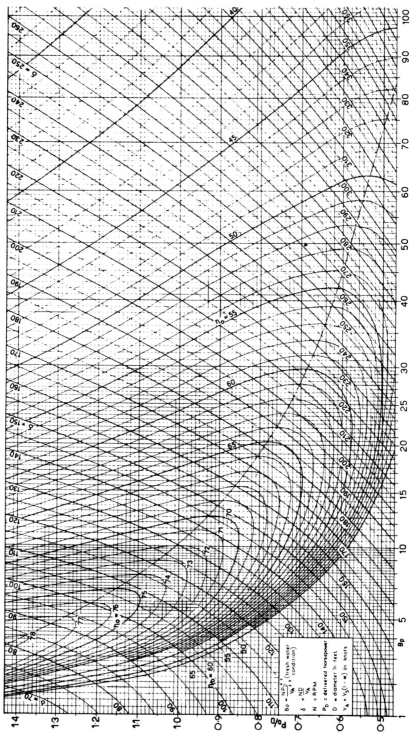

Figure 7.5 Bp-δ diagram (British Ship Research Association)

Results of experiments on model propellers are commonly presented on a chart known as a $B_p-\delta$ diagram. This type of diagram was originally introduced by Taylor and families of such charts are now in use for the design of merchant ship propellers. A typical $B_p-\delta$ diagram is shown in Figure 7.5. B_p represents a value known as Taylor's propeller coefficient and is represented on the diagram along the horizontal axis. δ represents a value known as Taylor's advance coefficient and is represented by the series of contours running almost diagonally across the chart. The series of loops on the chart are lines representing the propeller efficiency in open water η_o. A point of maximum efficiency exists for each value of B_p and these are indicated on the chart by the chain-dotted curve running from top left to bottom right. The vertical axis shows the pitch ratio.

A full account of the use of the diagram is beyond the scope of this book, but further reading is listed in the bibliography at the end of this chapter. It can be said, however, that from such diagrams and certain operating parameters of the vessel to which the propeller is to be fitted the following can be determined:

1. Optimum diameter.
2. Pitch ratio.
3. Open water efficiency.

From these and other factors at his disposal the designer can determine:

1. s.h.p. required to achieve a specific trial speed, given the vessel deadweight, rev/min and ship resistance data
2. Trial speed given the deadweight, rev/min and ship resistance data
3. s.h.p. and optimum rev/min given the deadweight, maximum propeller diameter, trial speed required and ship resistance data.

When using these diagrams it should be remembered that the model propeller had a certain combination of disc-area ratio, thickness and section shape, and any variation from this combination in the full-size propeller will have to be allowed for.

It will be found that it is not always possible to accommodate the best propeller from the point of view of diameter in the stern frame, and certain losses in efficiency may have to be borne through having to fit a propeller of reduced diameter. Co-operation between the builders and propeller designer at an early stage can often do much to overcome this difficulty.

Blade number

The impression gained from a study of a family of $B_p-\delta$ propeller diagrams mentioned above will probably be that for given conditions three-bladed propellers are better from considerations of efficiency than four-bladed ones. This may be so for models tested in the ideal conditions of the tank, and even for full-size propellers under good trial conditions, but for average service performance four-bladed propellers can be shown in almost every instance to give the best results.

This question must, however, be examined, particularly in the case of diesel machinery, with regard to the incidence of the blade number on the firing sequence of the engine. For example, a four-bladed propeller working with a four-cylinder engine might easily set up undesirable vibrations, and consequently, regardless of any other considerations of performance a three-bladed propeller should be chosen.

Some super tankers are now fitted with six-bladed propellers, up to 7.9 m diameter.

Area

The outline of most modern propellers is based on an elliptical form. The developed area is settled, bearing in mind an allowable loading per unit area and also with a view to giving adequate width in relation to thickness in order to avoid a flow breakdown with resulting cavitation and erosion. This latter consideration is very evident in the case of high-power fast-running propellers, as fitted to destroyers and cross-channel vessels, which are usually of high disc-area ratio with wide blades, giving a cloverleaf appearance.

Thickness

The determination of the thicknesses of the propeller blades is a complex problem. The blade is essentially a cantilever upon which are acting various forces.

The normal procedure is to calculate these forces, torque, thrust and centrifugal, and then resolve them into a bending moment about the tip of the fillet. Given the mechanical properties of the material, a limiting stress is decided upon and the thicknesses so arranged that the stresses will not be exceeded at the full output of the machinery.

Boss

The dimensions of the boss are settled in relation to the shafting, but care should also be taken to make its outline fair in with that of the ropeguard which is normally fitted and with the stern-frame boss. To obtain a smooth flow, cone covers or cone nuts are also arranged on most propellers.

Cavitation

Cavitation is normally encountered only in high-speed heavily loaded propellers of the cross-channel, high speed container ship or destroyer type. It is caused by the breakdown in flow over the back of the blade referred to above, due to excessive lift or suction. Small bubbles of aqueous vapour and air are formed, which collapse violently allowing the water to hammer the back of the blade. Under particularly onerous conditions, for example, the full-speed trial of a destroyer, it is possible for this to erode holes of measurable depth in a very short time.

Similar erosion can occur on the face at the leading edge; this is due to a bad angle of flow, and can be avoided by tilting the leading edge.

Back cavitation may be reduced by using finer blade sections, but blade thickness will always be determined by the tensile strength of the blade material.

Slip

A note on the question of slip is appropriate here, although the following remarks are of a practical nature and do not relate to design. The apparent slip, which the engineer calculates from speed (V_S), revolutions and pitch, is a comparison of the logged speed of the ship with the theoretical speed derived from pitch and revolutions of the screw, and should not be confused with the real slip, which is related to the speed of advance (V_A). The actual value of the figure for apparent slip (S_a) is determined from the expression

$$S_a = \frac{PN - V}{PN} = 1 - \left(\frac{V}{PN}\right)$$

where:
S_a = apparent slip;
P = pitch;
N = revolutions per minute;
V = observed speed of ship.

It is only of practical use for the ship concerned or when compared with an exact sister vessel having an identical propeller.

The following is a simple illustration of why apparent slip should not be used to compare performance of vessels with dissimilar propellers, yet having the same hull and machinery characteristics.

A propeller of 5.65 m diameter, having a mean pitch of 4.88 m, turning at 72 rev/min, at a speed of 11 knots, will give

$$S_a = 1 - \left(\frac{11.00 \times 30.89}{4.88 \times 72} \right)$$
$$= 1 - 0.967$$

i.e. S_a = 3.3 per cent.

Now a propeller of 5.5 m diameter, but having a mean pitch of 5.04 m would absorb the same power at the same revolutions at the above with almost the same efficiency, but the apparent slip would be quite different, e.g.:

$$S_a = 1 - \left(\frac{11.00 \times 30.89}{5.04 \times 72} \right)$$
$$= 1 - 0.936$$

i.e. S_a = 6.4 per cent.
(30.89 converts knots to m/min.)

Finally, before leaving the question of slip, mention must be made of negative slip. It is a growing tendency for modern propellers to give negative apparent slip and, quite often, engineers believe there is something wrong with the propeller on this account. If one considers for a moment that we are dealing with apparent slip only and that the real slip will always be positive, much of the mystery will disappear. It arises largely from improved ship forms, especially in the after body.

MATERIALS

Propeller material must be resistant to sea water corrosion and cavitation erosion; it should be capable of withstanding severe shock loading and be suitable for casting into the complex shape of a propeller, which in the case of a VLCC can weigh as much as 80 tonnes. At the same time the material should be repairable and capable of good surface finish. To allow thin blade sections — important from the point of view of increased efficiency and reduced cavitation — the material should be of high tensile strength.

For many years nearly all large propellers were of high tensile brass (often called manganese bronze) of one formula or another. This is a copper-zinc alloy containing manganese, aluminium, iron, tin and sometimes nickel, with tensile strengths in the range 40–50 MN/m^2 depending on actual make-up. The high tensile brasses have now been largely superseded by aluminium bronzes. These are given proprietary names but generally contain about 8–10% aluminium together with manganese (10–12%), iron (2–3%), nickel (2–5%) and the balance mainly copper. These metals have greater tensile strength, about 65 MN/m^2 and corrosion resistance and lower specific gravity than the high tensile brasses. There are also some stainless steel propellers in service.

MAINTENANCE

At every dry docking, propellers should be carefully examined and the slightest defect at once rectified. Similarly, any unusual vibration or noise whilst at sea should be investigated immediately and, if the cause cannot be found inboard, then the first opportunity must be taken of an under-water inspection. The expert examination and repair of propellers is a service provided by the leading manufacturers. Wherever possible, use should be made of this service because, first, it is not always easy to detect slight damage and, secondly, inexpert work can quite easily cause irreparable harm.

Cracks formed at or near blade edges, no matter how small, are potentially dangerous. If a crack should be found close to the boss (within 0.45 of the propeller radius) the makers should be consulted as any repair in this region involving heating can leave very high residual stresses which can only be removed by annealing the complete propeller. Small cracks can be temporarily stopped from spreading by drilling a hole of at least 10 mm in diameter at the extremity of the crack. The hole should then be plugged to avoid any risk of cavitation.

Damage

Damage may vary from a slight deformation, as would arise from a glancing blow on a submerged object, to severe bending or breaking away of portions of a blade, which may result from a heavy impact, such as striking a dock wall or barge. Such damage causes a disturbance of the flow of water across the blade, which may result in loss of efficiency, vibration and erosion of the metal surface. Distortion can be straightened and breakages repaired by burning or welding.

A further result of an impact may be the formation of cracks at the blade edge. Alternatively, internal stresses induced by sudden cooling after local heating may cause cracking. The effective section is thus weakened, causing overloading which, in turn, results in further growth of the crack, with the possibility of ultimate fracture.

Corrosion

Corrosion is a chemical or electro-chemical attack on the metal surface which may be further increased if the sea-water is polluted. It causes pitting and dezincification. If severely attacked the blade's surface breaks down and becomes rough due to the partial removal of the soft corrosion product by the scouring action of the water flow across the blade, with resulting loss of efficiency and in extreme cases loss of effective section thickness. As such an attack is usually widespread, a smooth surface can be restored only by heavy grinding.

Erosion

Erosion is a mechanical attack on the metal surface which may be due to a disturbance in the flow of the water over the blade arising from damage, in which case it tends to be localized in the form of comets. Alternatively, it may be caused by cavitation or incorrect design of the blade as mentioned previously. Such an attack produces cavities, often of considerable depth, which can be repaired by welding.

METHODS OF REPAIR

Repairs which can be undertaken with the propeller on the shaft are limited to the fairing of minor edge damage and to the elimination of light corrosion and erosion by surface grinding. In no circumstances should localised heating, particularly in the form of welding, be applied in this position, as adequate stress-relieving treatment cannot be undertaken.

It is always preferable for the propeller to be returned to the manufacturers' works, where the precise extent of the damage is first determined. It is checked for pitch, any variation or bends being marked, and the edges are ground, polished and etched to assist in the detection of cracks. Brief details of the various repair processes are given below.

Straightening

The bent portion of the blade is slowly heated over a large open propane gas or coke brazier to a low red heat. The distortion is then slowly pushed back to shape by means of heavy weights placed directly on the blade. After very slow cooling, the surface is again checked for pitch and, if necessary, the process repeated.

Burning

This process consists essentially of cutting out the damaged or cracked portion of the blade, enclosing the cavity thus produced with a dried sand mould and pouring molten metal directly on to the blade edge until the whole of it is melted down and the cavity filled.

If a large portion of the tip of a blade is missing (not usually exceeding one-third of the whole), a precast section is often manufactured and then burnt on to the blade stump. Since the burn metal used is of the same composition as the propeller, abnormal corrosion (such as may be experienced with welding) does not arise in the vicinity. metallurgical examination of such burns is made, but stress-relieving treatment is not normally necessary, provided that pre-heating and subsequent slow cooling have been employed. This process provides a very satisfactory permanent repair, particularly as the mechanical properties in the region of a burn are at least equal to those elsewhere in the propeller.

Welding

The repair of small cracks, torn edges and erosion holes is usually carried out by means of oxy-acetylene or electric-arc welding, the latter being the more satisfactory. A coated aluminium-bronze or phosphor-bronze electrode is used as it is not possible to arc-weld with manganese bronze. The affected area is first pre-heated and the actual welding carried out with great care. Afterwards the weld is annealed in order to maintain the correct micro-structure of the alloy, and to relieve the stresses induced in the metal. Failure to do this will often result in cracking under corroding conditions.

The repaired propeller is then polished, carefully checked dimensionally and statically balanced. It is finally subjected to minute dimensional and metallurgical examination.

PROPELLER MOUNTING

Traditionally, the fixed pitch propeller has been fitted to the tail-shaft with key and taper as shown in Figure 7.6, c.p. propellers are normally fitted to a flanged tailshaft since the operating mechanism is housed in the propeller boss. The conventional key and taper has however been under critical review due to a high incidence of fretting or fatigue cracks near the forward end of the keyway in high-powered single screw ships.

The fixed pitch propeller could, of course, be flange-mounted but this means that the tailshaft cannot then be drawn into the vessel for inspection and results in a larger boss diameter for the propeller. a more fundamental approach is to dispense with the key and rely entirely on a good interference fit. Two principal methods of doing this have been developed.

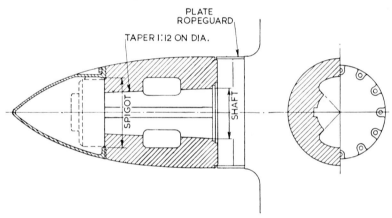

Figure 7.6 Typical arrangement of solid propeller boss

The first, and perhaps the simplest, method, is the oil injection system, commonly associated with the name of SKF. In applying this system to marine propellers, SMM have had the full co-operation of SKF (UK) who have considerable experience in this field. Oil is injected between the shaft taper and the bore of the propeller by means of high pressure pumps, and this is assisted by a system of small axial and circumferential grooves machined in the propeller bore, with the result that the coefficient of friction during oil injection fitting is reduced to about 0.015.

A hydraulic ring jack is arranged between the shaft nut and the aft face of the propeller boss, and with this it is a simple matter to push the propeller up the shaft taper by the required amount, over-coming the friction force and the axial component of the radial pressure. When the oil injection pressure is released, the oil is forced back from between the shaft/bore surfaces leaving an interference fit with a coefficient of friction of at least 0.12.

Lloyds require that the degree of interference be such that the frictional force at the interface can transmit 2.7 times the nominal torque when the ambient temperature is 35°C. Lloyds also require that at 0°C the stress at the propeller bore, as given by the Von Mises stress criterion, shall not exceed 60% of the 0.2% proof stress of the propeller material as measured on a test bar. These requirements are usually easily met and the whole process is relatively quick, simple and effective.

When it is required to remove the propeller, the process is equally simple and even quicker; thus obviating the need for using any form of heating. In the past oxy-propane torches have been used inadvisedly for boss heating in order to remove propellers from the shaft and this has led to a number of premature propeller failures due to stress-corrosion cracking. The other method is the Stone-Pilgrim keyless system in the development of which SMM have co-operated closely with P and O Research Ltd. and which is associated with the name of T.W. Bunyan. In the latest development of this method, a cast-iron sleeve is bonded into the propeller boss with a special form of Araldite under pressure. The propeller assembly is then ready to be force-fitted to the tail-shaft by means of a Pilgrim nut with the shaft and bore surfaces dry and degreased.

STERN BEARINGS

The tailshaft is supported in a bearing of one of a number of designs. The traditional tailshaft or stern bearing was water-lubricated and consists of a number of lignum vitae staves located in longitudinal grooves in a gun-metal bush. Lignum vitae is a hard wood with quite good wear characteristics and is compatible with water. The staves in the lower part of the bearing are cut so that the grain runs radially in the bearing. This gives it longer life. As an alternative to wood both Tufnol and reinforced rubber are used as stave material (Figure 7.7).

The gun-metal bushes in which the staves are mounted are pressed into a cast iron stern-tube let into the stern-frame and after bulkhead, to which it is secured by a flange and studs. A shoulder on the after end is pulled against the stern-frame by a locked nut screwed onto the stern-tube. Two staved bushes are fitted in each stern-tube. The aftermost bush has a length four times the diameter of the shaft and

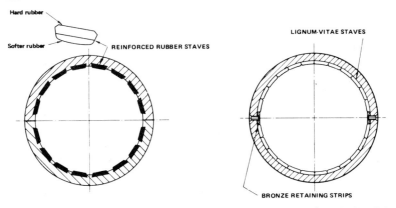

Figure 7.7 (left) Rubber stave bearing (right) Lignum vitae bearing (Glacier Metal Co.)

is the main bearing unit. The forward bush is relatively short and acts mainly as a guide; it is sometimes omitted. The centre, unbushed portion of the stern-tube is connected to a sea-water service line which, together with ingress of water between the shaft and after bush, provides lubrication. A packed gland seals the forward end of the bearing and is adjusted to permit a slight trickle of water along the shaft and into the tunnel well where it can be regularly removed by the bilge pump. A continuous gun-metal liner is shrunk to the shaft for the whole length between the propeller and a point beyond the stuffing box gland. A rubber ring is the usual seal between propeller and aft end of liner. *It is essential that the rubber has freedom to flow when compressed.* Bearing clearances are liberal both to accommodate the swelling which occurs when the staves are immersed in water and to permit the essential flow of water through

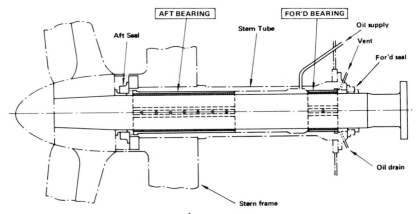

Figure 7.8 Two-bush sterntube system (Glacier Metal Co.)

the bearing. The permissible wear, measured by inserting a wedge between the shaft and bearing from the outside, is in the region of 9–12 mm on large diameter shafts.

A large number of ships with water-lubricated bearings of this type continue to be built but oil-lubricated whitemetal bearings are almost exclusively used in vessels of over 100 000 dwt and are being used increasingly in smaller vessels. In oil lubricated bearings the shaft is not sleeved.

Some whitemetal bushes are arranged in much the same way as the water-lubricated bearings, two bushes being pressed into a stern-tube which itself is pressed into the stern-frame (Figure 7.8). In some instances a steel tube welded to the after bulkhead and stern-frame is used.

Because of alignment difficulties a single bush whitemetal bearing (Figure 7.9) is often used. In this case the stern-tube is not necessary and the bearing is fitted directly into the sternframe. The aftermost intermediate shaft bearing is located further aft and more or less

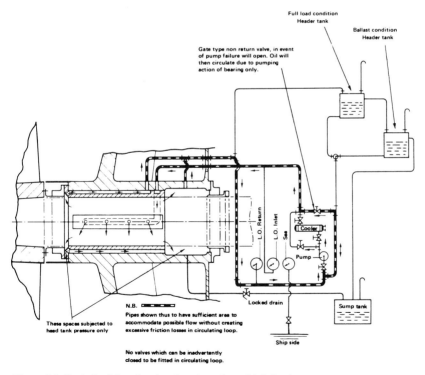

Figure 7.9 Single-bush bearing showing also a forced lubrication system (Glacier Metal Co.)

248

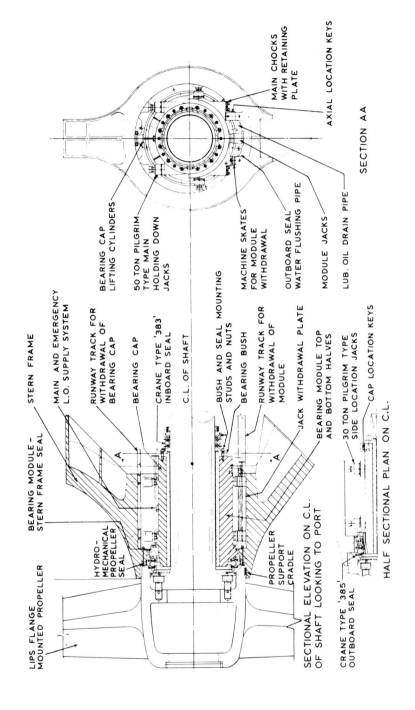

LIPS FLANGE
MOUNTED PROPELLER

BEARING MODULE -
STERN FRAME SEAL

STERN FRAME

HYDRO-
MECHANICAL
PROPELLER
SEAL

MAIN AND EMERGENCY
L.O. SUPPLY SYSTEM

RUNWAY TRACK FOR
WITHDRAWAL OF
BEARING CAP

BEARING CAP
LIFTING CYLINDERS

BEARING CAP

50 TON PILGRIM
TYPE MAIN
HOLDING DOWN
JACKS

CRANE TYPE '383'
INBOARD SEAL

C.L. OF SHAFT

BUSH AND SEAL MOUNTING
STUDS AND NUTS

BEARING BUSH

MACHINE SKATES
FOR MODULE
WITHDRAWAL

RUNWAY TRACK FOR
WITHDRAWAL OF
MODULE

OUTBOARD SEAL
WATER FLUSHING PIPE

JACK WITHDRAWAL PLATE

MODULE JACKS

BEARING MODULE TOP
AND BOTTOM HALVES

LUB. OIL DRAIN PIPE

PROPELLER
SUPPORT
CRADLE

30 TON PILGRIM TYPE
SIDE LOCATION JACKS

CAP LOCATION KEYS

MAIN CHOCKS
WITH RETAINING
PLATE

AXIAL LOCATION KEYS

SECTION AA

CRANE TYPE '385'
OUTBOARD SEAL

SECTIONAL ELEVATION ON C.L.
OF SHAFT LOOKING TO PORT

HALF SECTIONAL PLAN ON C.L.

Figure 7.10 General arrangement of Ross-Turnbull Mark IV bearing (Ross Turnbull Ltd.)

fulfils the function that the forward sterntube bush performed in the earlier arrangement. In oil lubricated bearings, classification societies will generally accept a bearing length of only twice the shaft diameter, providing that the specific loading does not exceed 0.6 N/mm^2. The bush is normally grey or nodular cast iron, centrifugally lined with whitemetal or occasionally, whitemetal-lined gun-metal or bronze.

A typical analysis of whitemetal would be 3% copper, 7.5% antimony and the remainder tin. Whitemetal thicknesses vary according to the classification society. Lloyds Register recommends 3.8 mm for a shaft 300 mm dia. to 7.4 mm for a 900 mm dia. shaft. The corresponding bearing clearances are 0.51–0.63 mm and 1.53–1.89 mm before the bush is pressed in. There is some close-up of the bore on installation.

Ross-Turnbull and Glacier-Herbert split bearings

Due to the problems associated with examining stern bearings — the tailshaft must be drawn out of the bush, entailing dry-docking the vessel — attention has been given in the last few years to developing a bearing which is split along the horizontal axis. By providing a suitable outboard seal the two halves of the bearing can then be drawn into the ship exposing the running part of the shaft and the whitemetal bush. The two most widely adopted types are the Ross-Turnbull bearing, Figure 7.10, and the Glacier-Herbert bearing, Figure 7.11.

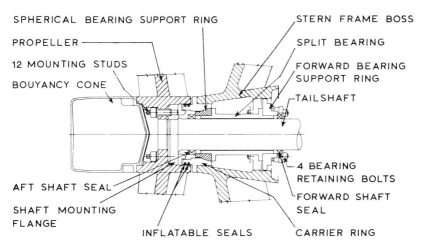

Figure 7.11 Solid propeller with hollow cylindrical boss and internal flanged mounting

Figure 7.10 shows the latest design of the Ross-Turnbull bearing and it can be seen that the bottom half of the bearing is chocked on to two horizontal fore and aft machined surfaces within the stern-frame. The bearing is held in position vertically by two 50-tonne Pilgrim-type jacks, the chock thickness determining the bearing height. These jacks also hold the two halves of the bearing together. Lateral positioning of the bearing is by two 30 tonne Pilgrim-type jacks arranged on each side of the bearing. A running track is arranged above the bearing to allow easy transport of the top half. Roller race skids are provided below the bearing to provide easy transport of the bottom half. When removing the bearing bottom half, a jack is first placed underneath it to lift it free of its chocks. The chocks are then removed and the bottom half then rests on the roller skids. The weight of the propeller and shaft is taken by a support cradle built into the stern-frame.

In the Glacier-Herbert system the two completely symmetrical bearing halves are flanged along the horizontal centre line and held together by bolts. The after end of the bearing carries a spherical support ring to which is bolted the outboard seal housing. The forward end is supported by a circular diaphragm which is bolted to a bulkhead provided in the stern frame casting. This diaphragm also acts as a carrier for the forward seal.

A series of axial bolts, fitted with Belleville washer packs to ensure virtually constant loading of these bolts and those securing the spherical seating ring, hold the diaphragm firmly in position. This arrangement permits final alignment of the bearing. Chocks are used to hold the bearing positively in its final position. The arrangement is such that it allows for the differential expansion of the bearing and its housing without detracting from the rigidity of support at the forward end of the bearing.

Sealing arrangements

There are basically three sealing arrangements used for preventing uncontrolled ingress of sea water through, or loss of oil from, the stern bearings. These are:

1. Simple stuffing boxes filled with proprietary packing material, usually rove cotton impregnated with tallow or graphite as a lubricant. In the case of high duty packing the material may be whitemetal clad.
2. Lip seals in which a number of flexible membranes are held in contact with the shaft preventing the passage of fluid along the shaft.

3. Radial face seals in which a wear-resistant face fitted radially around the shaft is in contact with similar faces fitted to the after bulkhead and to the after end of the stern-tube. A spring system is necessary to keep the two faces in contact.

Figure 7.12 shows an example of a lip seal. Known as the Simplex-Compact seal it consists of a number of rings of special cross-section (Figure 7.13) which are held in contact with a renewable sleeve fitted to the shaft. The rings are made of an elastomer such as Perbunan or Viton and are scarfed rings so that they can be renewed by passing them around the shaft and vulcanising the ends in situ. Figure 7.12a shows a forward seal having two rings and Figure 7.12b shows an after seal having three rings. The rings are identical in each case. Both forward and after seals are built up from three basic assemblies, namely the flange, intermediate, and cover rings and these parts can be used for either seal. It will be noticed that the garter spring holding the sealing ring aginst the shaft is located aft of the ring anchoring bulb in the case of both forward sealing rings. In the case of the after seal the two outboard sealing rings have their garter springs located aft of the ring anchoring bulbs while the inboard ring has its garter spring located inboard of the anchoring bulb.

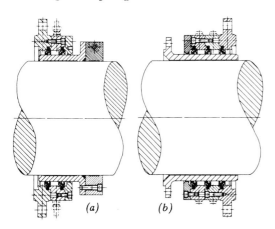

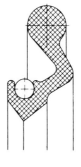

Figure 7.13 Simplex seal ring section

Figure 7.12 Simplex shaft seal

In some instances four or more sealing rings are installed. These are arranged so that one ring does not normally run on the shaft liner. In the event of leakage from the working seals the tailshaft can be shifted slightly to bring the reserve ring into play.

An example of a radial face seal is the Crane seal shown in Figure 7.15. One of the principal features of the design and construction of

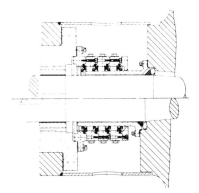

Figure 7.14 Simplex multiple ring seal

this type of seal is the split construction of all component parts. This facilitates installation, and subsequent inspection and maintenance.

The function of sealing against leakage around the shaft is effected by sustaining perfect mating contact between the opposing faces of the seal's seat which rotates with the shaft, and of the main seal unit which is stationary and clear of the shaft.

This mating contact of the seal faces, which are hydraulically balanced, is sustained by spring pressure and by the method of flexibly mounting the face of the main seal unit, the flexible member consisting of a tough, but supple, reinforced bellows. Thus the main seal unit is able to accommodate the effects of hull deflection and the amplitude of vibrations which may be encountered under all conditions, including those which are consequential to damage to a ship's propeller, which may occur, for example, when vessels operate in northern waters, where ice is present.

The bellows member is clear of the shaft, and its flexibility therefore cannot be impaired, as may happen when a flexible member is mounted on the shaft and hardens, seizes or becomes obstructed by a build-up of solids. The mechanical design principles also ensure continued sealing under fluctuating pressure conditions, i.e. changing draught. Spray leakage, frequently encountered with packed glands, is eliminated, together with the associated corrosion around the hull.

An emergency sealing device can be incorporated in the design. The device, when inflated with air or liquid, forms a tight temporary seal around the shaft, enabling repairs to the seal or replacements to be made when the ship is afloat, without the shaft being drawn, or dry-docking being necessary.

Shaft bearings

The intermediate shafting between the tailshaft and main engine, gearbox, or thrustblock may be supported in plain, tilting pad or

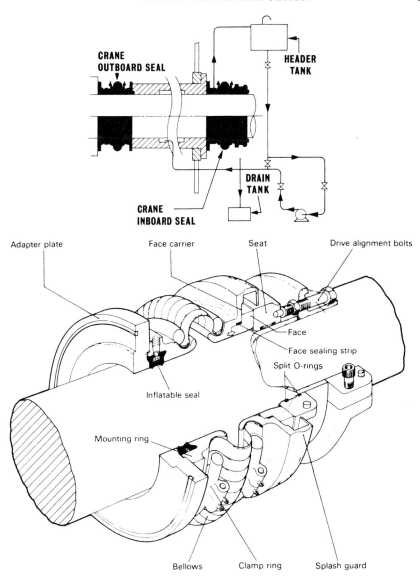

Figure 7.15 Example of a radial face seal (Deep Sea Seals Ltd.)

roller bearings. The two former types usually have individual oil sumps, the oil being circulated by a collar and scraper device, although in some high performance ships a force-lubricated system is fitted. The individual oil sumps usually have cooling water coils, or a simple cooling water chamber, fitted. Cooling water is provided

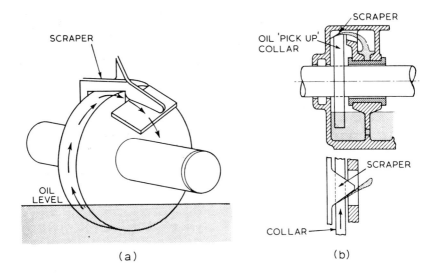

Figue 7.16 Bearing disc and scraper oil transport (Michell Bearings Ltd.)

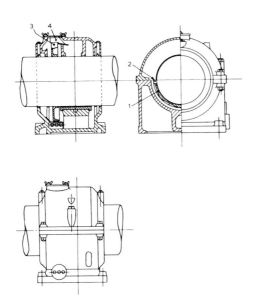

Figure 7.17 Intermediate shaft bearing of the tilting pad type
  1. Pivoted journal pad   *2. Bearing stop pins*
  3. Oil thrower ring   *4. Oil scraper*
(Michell Bearings Ltd.)

from a service main connected to a convenient sea water pump. The cooling water may pass directly overboard or be led to a bilge although the latter practice is now seldom found.

In many plain and tilting pad bearings only a bottom bearing half is provided, the top acting purely as a cover. The aftermost bearing is, however, always a full bearing.

Roller bearings, where fitted, are grease-lubricated.

BIBLIOGRAPHY

1. Hannan, T.E. *Strategy of propeller design,* Thomas Reed Publications Ltd.
2. Wright, B.D.W. *Propeller data and their applications,* The N.S.M.B. Standard Series. British Ship Research Association (not in general circulation)
3. Sinclair, L. and Emerson, A. 'The design and development of propellers for high powered merchant vessels', *Trans. I. Mar. E.,* **80**, 5 (May 1968)
4. Bille, T. 'Experiences with controllable pitch propellers', *Trans. I. Mar. E.,* **80**, 8 (August 1970)
5. Crombie, G. and Clay, C.F. Design feature of and operating experience with Turnbull split stern bearings, *Trans. I. Mar. E.,* **84**, 11 (January 1972)
6. Herbert, C.W. and Hill, A. Sterngear design for maximum reliability – the Glacier-Herbert system, *Trans. I. Mar. E.,* **84**, 11 (January 1972)
7. Rose, A. 'Hydrostatic stern gear' *N.E. Coast I. of Engineers and Shipbuilders* (January 1974)
8. *Sterntube bearings.* The Glacier Metal Co. Ltd.

8 Steering gears

The general requirements of a steering gear, based on the regulations of SOLAS 74 and the major classification societies may be summarised as:

(1) Ships must have a main and an auxiliary steering gear, arranged so that the failure of one does not render the other inoperative. An auxiliary steering gear need not be fitted however when the main steering gear has two or more identical power units and is arranged such that after a single failure in its piping system or one of its power units, steering capability can be maintained. To meet this latter alternative the steering gear has to comply with the operating conditions of paragraph 2 – in the case of passenger ships while any one of the power units is out of operation. In the case of large tankers, chemical tankers and gas carriers the provision of two or more identical power units for the main steering gear is mandatory.

(2) The main steering gear must be able to steer the ship at maximum ahead service speed and be capable at this speed, and at the ship's deepest service draught, of putting the rudder from 35° on one side to 30° on the other side in not more than 28 secs. (The apparent anomaly in the degree of movement is to allow for difficulty in judging when the final position is reached due to feedback from the hunting gear). Where the rudder stock, excluding ice strengthening allowance, is required to be over 120 mm diameter at the tiller, the steering gear has to be power operated.

(3) The auxiliary steering gear must be capable of being brought speedily into operation and be able to put the rudder over from 15° on one side to 15° on the other side in not more than 60 seconds with the ship at its deepest service draught and running ahead at the greater of one half of the maximum service speed or 7 knots. Where the rudder stock (excluding ice-strengthening allowance) is over 230 mm diameter at the tiller then the gear has to be power operated.

(4) It must be possible to bring into operation main and auxiliary steering gear power units from the navigating bridge. A power

failure to any one of the steering gear power units or to its control system must result in an audible and visual alarm on the navigating bridge and the power units must be arranged to restart automatically when power is restored.

(5) Steering gear control must be provided both on the bridge and in the steering gear room for the main steering gear and, where the main steering gear comprised two or more identical power units there must be two independent control systems both operable from the bridge (this does not mean that two steering wheels are required). When a hydraulic telemotor is used for the control system a second independent system need *not* be fitted except in the case of a tanker, chemical carrier or gas carrier of 10 000 gt and over. Auxiliary steering gear control must be arranged in the steering gear room and where the auxiliary gear is power operated, control must also be arranged from the bridge and be independent of the main steering gear control system. It must be possible, from within the steering gear room, to disconnect any control system operable from the bridge from the steering gear it serves. It must be possible to bring the system into operation from the bridge.

(6) Hydraulic power systems must be provided with arrangements to maintain the cleanliness of the hydraulic fluid. A low level alarm must be fitted on each hydraulic fluid reservoir to give an early audible and visual indication on the bridge and in the engine room of any hydraulic fluid leakage. Power operated steering gears require a storage tank arranged so that the hydraulic systems can be readily re-charged from a position within the steering gear compartment. The tank must be of sufficient capacity to recharge at least one power actuating system.

(7) Where the rudder stock is required to be over 230 mm diameter at the tiller (excluding ice strengthening) an alternative power supply capable of providing power for paragraph 3 above is to be provided automatically within 45 seconds. This must supply the power unit, its control system and the rudder angle indicator and can be provided either from the ships emergency power supply or an independent source of power, located within the steering compartment and dedicated for this purpose. Its capacity shall be at least 30 minutes for ships of 10 000 gt and over and 10 minutes for other ships.

LARGE TANKERS AND GAS CARRIERS

Tankers, chemical carriers and gas carriers of 10 000 gt or over require two or more identical power units and the steering gear must be arranged so that loss of steering capability due to a single failure in one of the power actuating systems of the main steering gear (excluding tiller etc.), or seizure of the rudder actuators, must be regained in not more than 45 seconds. The main steering gear must comprise either two independent and separate power actuating systems each capable of producing the performance stipulated in paragraph 2 above or two identical power actuating systems which, during normal operation, will together produce this performance and be so arranged that the loss of hydraulic fluid from one system can be detected and the defective system automatically isolated to permit operation of the remaining system. Non-hydraulic steering gears must achieve similar standards.

In the case of tankers of less than 100 000 tonnes deadweight some relaxation of this single failure criterion of the actuators may be permitted. Additional requirements are stipulated for the electrical circuitry of electric and electrohydraulic steering gear but these requirements are not covered here.

While the above summary gives a guide to the general requirements desired for steering gears in the light of the 'Amoco Cadiz' incident, the degree to which ships built before 1984 need to comply with these regulations vary according to size and type and to some extent by National regulations. Those readers requiring explicit knowledge of the SOLAS 74 regulations, as amended in November 1981 are well advised to study the original document.

In the following pages several types of steering gear are described. Not all fully comply with the 1981 SOLAS amendments. Units such as these will still be found at sea for several years, however, except in the case of large tankers.

An historical note

In 1866, James Macfarlane Gray fitted a successful steam steering gear of his own design in the 'Great Eastern'. This is probably the first on record.

Steam engines driving right-and-left handed screws, shafts, drums, gypsy wheels, etc. and hydraulic gears followed, many of which were ingenious. Andrew Betts Brown of Edinburgh played an important part in their development. As means of remote control were lacking, the engines were placed close to the wheel and their effort taken aft to the rudder crosshead or tiller by systems of chains, rods and levers, a practice which survived until the thirties.

In 1888, Mr. Brown brought out his patent hydraulic telemotor, described later, which is still in use today, almost unchanged. The

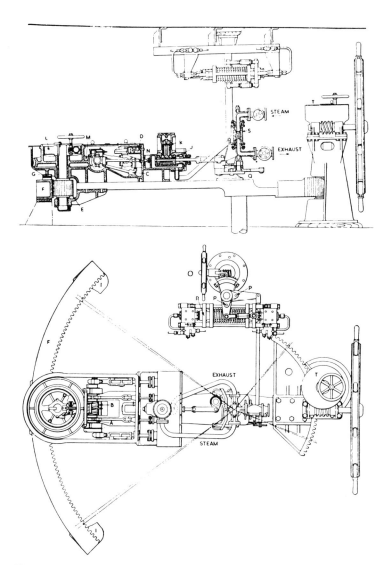

Figure 8.1 Brown's patent steam tiller steering gear

A. Crankshaft
B. Worm (cast on A)
C. Lubricating pumps
D. Oil tank
E. Pinion and shaft
F. Rack
G. Slipper
I. Stops
J. Control valve
K. Economic valve

L. Brake
M. Wormwheel
N. Distance piece
O. Local control steering
    wheel
P. Portable pin (telemotor
    to local control)
Q. Floating lever
R. Telemotor
S. Steam trunnion

(Brown Bros Co Ltd)

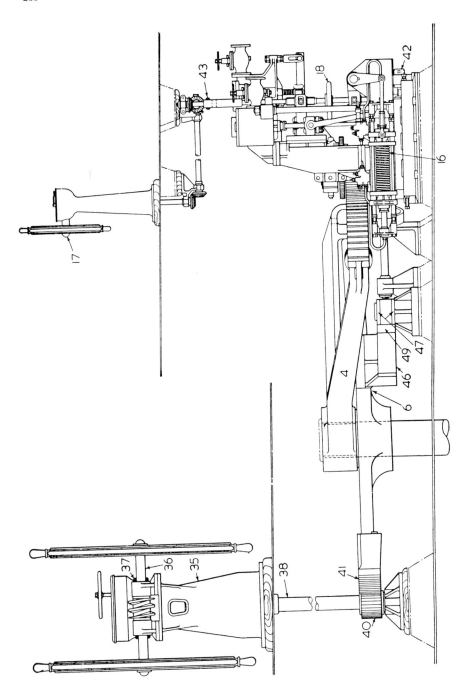

261

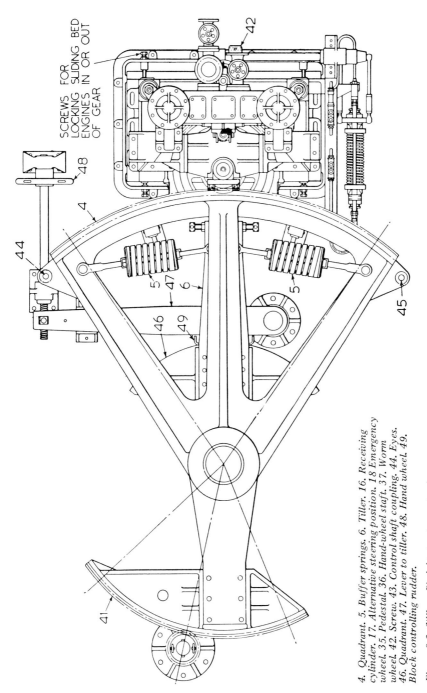

SCREWS FOR LOCKING SLIDING BED ENGINES IN OR OUT OF GEAR

4. Quadrant. 5. Buffer springs. 6. Tiller. 16. Receiving cylinder. 17. Alternative steering position. 18 Emergency wheel. 35. Pedestal. 36. Hand-wheel shaft. 37. Worm wheel. 42. Screw. 43. Control shaft coupling. 44. Eyes. 46. Quadrant. 47. Lever to tiller. 48. Hand wheel. 49. Block controlling rudder.

Figure 8.2. 'Wilson Pirrie' type steam steering gear

telemotor, reliable and sensitive, made it possible to place the
steering gear above the rudder and set in train a series of improvements and new developments notably, in 1890, the steam tiller
(Figure 8.1), also Mr. Brown's invention and later, the Wilson-Pirrie
gear (Figure 8.2). The extremely large torque efforts (upwards of

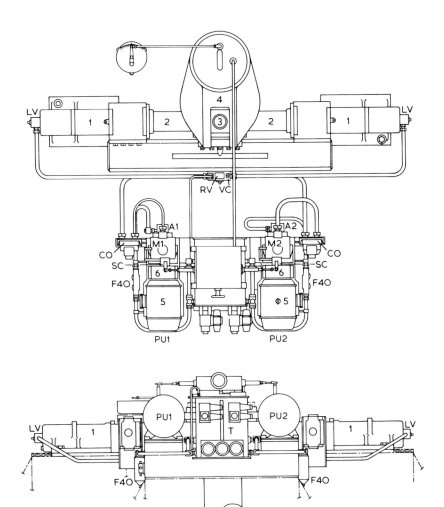

Figure 8.3 Two-ram electro-hydraulic steering gear.
1. *Cylinders*   5. *Motors*
2. *Rams*   *M1*
3. *Cod piece*   *M2 Variable delivery pumps*
4. *Tiller*   *LV Locking valve*

Key continued on facing page

1800 kNm) which were demanded as size and speed increased, thus became available.

The economic valve (again Mr. Brown), improved greatly the thermal efficiency of these gears. In both cases, two-cylinder simple engines with cranks at 90°, revolved in directions determined by a

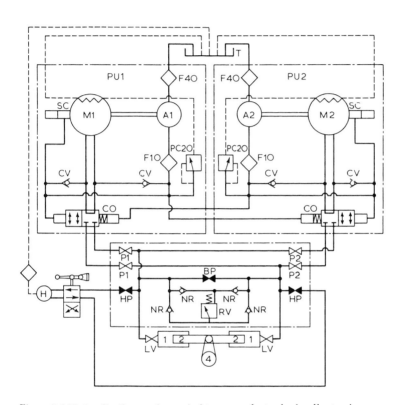

Figure 8.4 Hydraulic diagram for typical two-ram electro-hydraulic steering gear

PU1, PU2,	Power units
A1, A2,	Auxiliary pumps
T	Reservoir
F10	Filters
SC	Servo-controls
CO	Changeover valves
PC20	Pressure limiting valves
CV	Check valves
P1, P2	Isolating valves
LV	Locking valves
BP	By-pass valve
RV	Relief valve
HP	Hand-pump shut-off valves
WP	Non-return valves

sliding control valve, moved by the telemotor. Hunting gear of the floating-lever type brought the engine to rest when the rudder had moved as desired and the economic valve cut off steam.

In 1912, the first electro-hydraulic steering gear appeared, a two-ram type of 275 kNm torque; a 15 kW motor, driving a swash-plate variable delivery pump working at 70 bar, put the rudder hard-over to hard-over in 27 seconds. The Rapson Slide was introduced in 1919, giving a mechanical advantage equal to the secant[2] of the rudder angle and therefore, maximum torque at maximum rudder angle.

From these beginnings sprang the four-ram electro- (or steam-) hydraulic gear, with its inherent safeguards against failure or mishap and its relative ease of manufacture and maintenance.

The hydraulic cylinders may be double-, or more usually, single-acting, working at pressures as high as 275 bar. They may be fixed and their effort applied to the tiller by the Rapson Slide or variants of it, e.g. sliding cod-pieces in open jaws (see Figures 8.3 and 8.7) or they may be connected directly to the tiller, oscillating on trunnions to preserve alignment.

Cylinders, rams (or piston rods), crossheads and guides are, nowadays, of cast or forged steel; the continuously-running variable-delivery pumps may have axial or radial cylinders, the direction and quantity of flow being varied by swash-plate or floating ring respectively. These are described later in this chapter. Control may be by hydraulic telemotor (see Figure 8.25) or by an electric control system see Figure 8.6.

Many direct-acting electric and rotary-vane gears are in service. The rotary-vane gear is of course, electro-hydraulic but is defined separately for the sake of clarity. A section on the hydraulic tele-motor is provided at the end of this chapter with a special note on the floating lever.

Two-ram hydraulic steering gear

Figure 8.3 shows an arrangement of a two-ram gear; such gears may have a torque capacity of 120–650 kNm. Figure 8.4 shows the hydraulic system diagrammatically.

The steering cylinders are cast steel; the rams are one-piece steel forgings, with integral pins to transmit their effort to cod-pieces free to slide between the jaws of the fork-end tiller, a steel forging, suitably machined; they are ground to slide in the gun-metal neck bushes and hydraulic seals in the cylinders. The tiller may be of single-pin type, working in a ball crosshead (the Rapson Slide) as in the four-ram gear, see Figure 8.7.

The duplicate power units, PU1 and PU2, each have a continuously-running motor (or steam engine) driving, through a flexible coupling, a variable delivery axial-cylinder pump (page 303) and an auxiliary pump, A1, A2; the latter draw filtered oil from the reservoir T and discharge through a 10 micron filter F10 to lubricate the main pumps, to supply oil at constant pressure to the servo-controls SC and to the automatic change-over valves CO; also to maintain a flow of cool oil and to make up any loss in the main system. When the main pumps are at 'no-stroke', the auxiliary pumps discharge to the reservoir via a pressure-limiting valve PC20, set at 20 bar, and to the pump casings; when the main pumps are 'on-stroke', they discharge to the main pump suction.

Either or both pump(s) may be brought into operation at any position of the gear at any time by starting the motor(s). The servo-operated automatic change-over valves (Figure 8.5), are held in the bypass condition by the spring while the associated pump(s) are at rest; when a pump is started, the auxiliary pump pressure builds up, overcomes the spring, closes the bypass and connects the main pump to the hydraulic system. Thus, the main pump starts in the unloaded condition, it cannot be motored when idle by cross pressure flow and load is held off until the electric starting currents are dissipated; when a pump is stopped, the spring returns the valve to the bypass condition. The spring end of the valve is connected to the constant-pressure line and, to obviate hydraulic locking, the spring chamber has a bleed line. From the automatic change-over valves CO (Figure 8.6) the main pump discharge passes to the pump isolating valves, P1 and P2, and to the cylinders, through the locking valves LV. These valves are incorporated in a group valve chest so arranged as to provide cross-connections with the bypass valve BP, relief valve, RV, and the emergency hand pump shut-off valves, HP, with appropriate non-return valves NR.

In open water it is usual to have one power unit in use. If quicker response is required, two units may be run simultaneously, doubling the flow of oil and, almost, the speed of operation.

Normally, the gear is controlled from the bridge by a manually operated hydraulic or electric remote control system or by a gyro-pilot, but a local control handwheel (a 'trick' wheel) is provided and a steering position, wheel, standard, etc. is commonly fitted aft. Electric remote systems are commonly used because of their inherent simplicity and reliability and they are adaptable to the control of

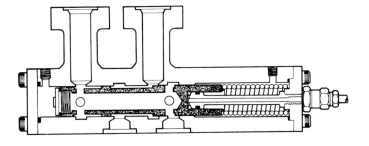

VALVE IN STRAIGHT THROUGH CONDITION

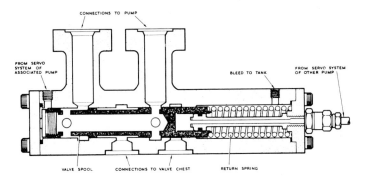

VALVE IN BY-PASS CONDITION

Figure 8.5 Automatic change-over valve
(Brown Bros & Co Ltd)

any type of steering gear. Figure 8.6 shows the control box of such a system.

The construction is light as the forces concerned are small; the control and hunting gears are combined as a self-contained unit, self-lubricating, in an oil-tight case. Pump output and therefore rudder movement, is controlled by a floating lever, one end of which is moved by the control motor (or telemotor), the other end by the movement of the tiller. A rod attached to its mid-point and to the pump control levers puts the pump(s) on stroke in response to movement of the floating lever by the control motor or telemotor. As the tiller moves, the cut-off linkage acts to counteract the initial movement (i.e. hunts) and brings the gear to rest by restoring the pump control levers to the no-stroke position. Spring links, suitably disposed, obviate over-stressing of the mechanism.

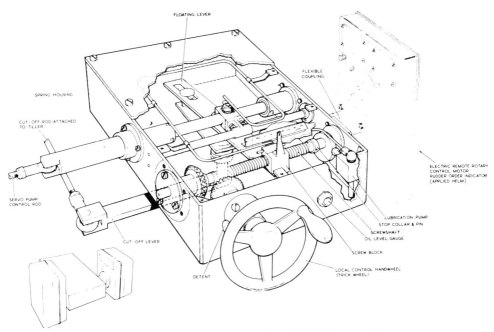

Figure 8.6 Control box. Electrical remote input and local mechanical input version (Brown Bros & Co Ltd)

The end of the floating lever connected to the remote control mechanism is attached to a block which moves along a screw shaft when the latter is rotated by the control motor; the stroke is restricted by stop collars. This shaft carries a mitre wheel engaging with the local control arrangement.

Four-ram hydraulic steering gear

Figure 8.7 shows a four-ram gear; this type may have torque capacities of 250 to 10 000 kNm. Figure 8.8 shows the hydraulic circuit.

The gear incorporates the Rapson Slide and the arrangement of stop and bypass valves in the chest **VC** enables the gear to be operated on all four or on any two *adjacent* cylinders (but *not* on two diagonally disposed cylinders), the inactive cylinders being isolated from the pumps by valves and the bypass valves connecting them opened to permit free flow of idle fluid. Either or both duplicate independent power units may be employed with any usable combination of cylinders. It will be seen that the torque available from two cylinders is only one-half of that from four, even

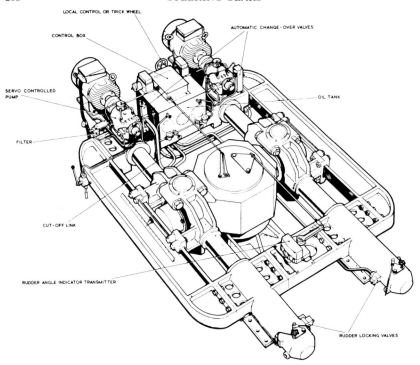

Figure 8.7 Four-ram electro-hydraulic steering gear with electric control (Brown Bros & Co Ltd)

when both power units are working, though the *speed* of operation will be increased if both are used.

The mechanical arrangement of the control gear and the basic hydraulic system, in all but their layout, are identical with the two-ram gear already described. The valve chest however, must cater for four cylinders in all useful combinations; this demands four cylinder isolating valves, C_1-C_4, and four bypass valves, B_1-B_4. The emergency hand pump arrangement, its directional control valve, the main system relief and the locking valves remain unchanged, as do the remote, local and emergency control arrangements.

Referring to Figure 8.8, normally, the pump and the four cylinder isolating valves, P_1, P_2, C_1-C_4, and the rudder locking valves LV are *open;* the bypass valves B_1-B_4 and the emergency hand pump isolating valves HP are *closed.* Either or both power unit(s) may then be brought into action or shut down by starting or stopping the associated motor(s).

To change from four-ram to two-ram working, it is only necessary to make two cylinders inoperative by closing their isolating valves,

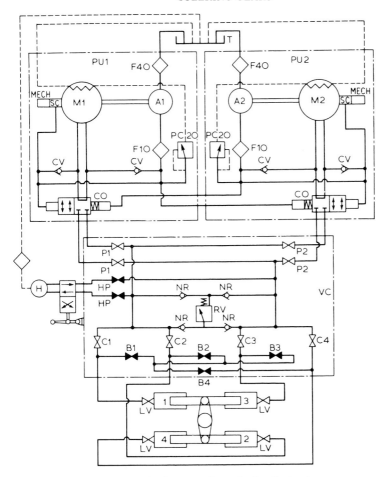

Figure 8.8 Hydraulic circuit for four-ram electro-hydraulic steering gear
B1-B4. *By-pass valves* *For other key letters see Figure 8.4*
C1-C4. *Cylinder isolating valves* *(Brown Bros & Co Ltd)*

C_1-C_4, and opening the bypass valves between them. For example, to steer on cylinders 1 and 3, valves C_2 and C_4 are *closed* and B_2 and B_4 *opened* so that cylinders numbers 2 and 4 are isolated from the main hydraulic system and the oil in them is free to flow from one to the other. The cylinder isolating valves and the bypass valves were shown as separate items in Figure 8.8 but, each pair may be combined as a double seating valve so that, as any cylinder is isolated from the main hydraulic system it is automatically opened to a bypass manifold and thence to another inoperative cylinder.

Figure 8.9 illustrates a four ram gear modified to comply with the SOLAS (1974) amendments of 1981 relating to tankers, chemical

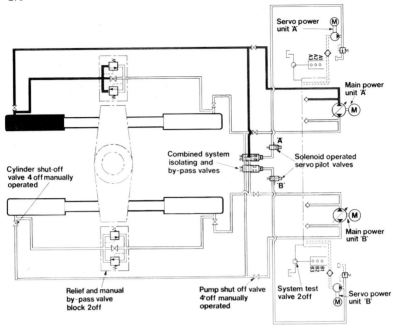

Figure 8.9 The Hastie-Brown split system, shown here in split operation, is arranged to give two ram operation automatically in the event of loss of fluid from one system.

tankers or gas carriers of 10 000 gt and upwards (Chapter II–1 Regulation 29, Paragraph 16). Two main power and servo-power units draw from a two-compartment tank fitted with oil level switches arranged at three levels. Level 1 gives an initial alarm following loss of oil from either system. In normal operation one or both power units provides hydraulic power to all four rams. Continued loss of oil initiates one or both of the level 2 switches. These energise their respective solenoid operated servo valves, causing the combined isolating and by-pass valves to operate, splitting the system such that each power unit supplies two rams only. At the same time if one power unit is stopped it is automatically started.

Further loss of oil, and the system on which it is occurring will operate one of the level 3 switches. This will close down the power units on the faulty side. Steering then continues, uninterrupted, but at half the designed maximum torque on the sound system. The defective system is out of action and isolated.

Four-ram gear with axial-piston pumps

One variant of the four-ram steering gear uses a servo-controlled swash plate axial cylinder pump, Figure 8.21, capable of working at

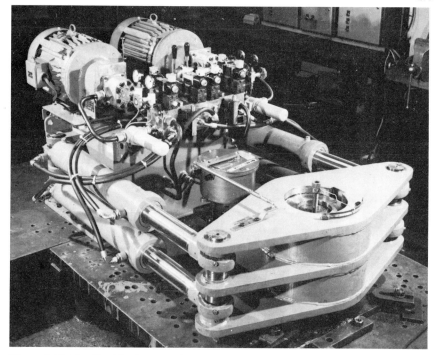

Figure 8.10 The Donkin four-ram hydraulic steering gear designed to meet the fail-safe criteria of IMO Resolution A.210 (V11) et al. (Donkin & Co. Ltd)

210 bar. Each pump is complete with its own torque motor, servo-valve, cut-off mechanism, shut-off valve and oil cooler. These pumps are brought into operation as described earlier and an idle pump is prevented from motoring.

The rotating assembly of the pump, which consists of cylinder, nine pistons, valve plate, slippers, slipper plate and retaining ring, is manufactured from EN8 steel, which is finally machined, heat treated and then hardened for long wear by Tuftriding. The nine pistons are fitted with return springs. The casing and covers are of nodular cast iron. The main valve block is EN8 steel and houses five check valves, main pump relief valve, boost and servo relief valves, boost gear pump and servo gear pump. Piping from the valve block supplies the servo pistons via the servo valve. The main drive is through a splined shaft to the cylinder body of the pump. The pump swash plate is actuated by two servo cylinders which receive oil at the desired pressure through a directional servo-valve. The servo-valve is displaced initially by the torque motor acting on the input signal demand and is returned to the neutral position by the hunting linkage connected to the swash plate. The hunting action is achieved

272

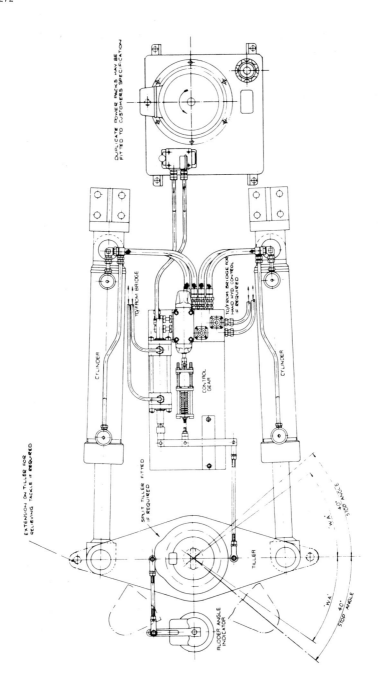

Figure 8.11 Electro-hydraulic steering gear (hydraulic control). General arrangement of actuator type steering gears (*Brown Bros & Co Ltd*)

through the application of a simple lever system connecting input displacement servo-valve and the hunting action from the pump swash plate angle. This allows for a very fast response.

SMALL HAND AND POWER GEARS

A simpler variant of the electro-hydraulic gear, for small ships requiring rudder torques below say, 150 kNm is shown in Figure 8.11. The hydraulic circuit is shown diagrammatically in Figure 8.12.

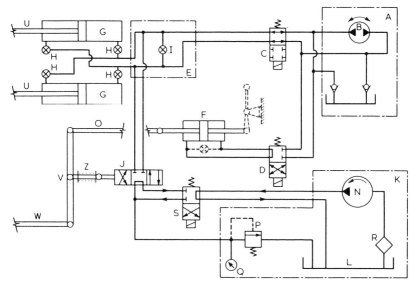

Figure 8.12 Hydraulic circuit diagram for hand and power steering. For key to letters, see text

The rams U in the double-acting steering cylinders G, which are free to oscillate on chocked trunnions, are linked directly to the tiller. F is a double-acting control cylinder, linked to the floating lever V by rod O; J is a directional control valve linked to the mid point of the floating lever by spring link Z; W is the cut-off link from tiller to floating lever. H is a locking valve and I a by-pass valve.

Valves C, D and S are solenoid controlled. When unenergised, C is open and D is closed to through flow but acts as a bypass between the ends of control cylinder F, solenoid S is closed. When the solenoids are energised, C is closed, D and S are open. J is only operative when steering by power.

When steering by hand, C, D and S are unenergised. A pump B

in the steering pedestal, coupled to the wheel, delivers fluid under pressure direct to the steering cylinders G, so moving the tiller in the sense and to the extent appropriate to the movement of the wheel. There is no 'hunting' action.

To change to power steering, the power pump is started and C, D and S are energised, i.e. C is closed, D ceases to be a bypass and it connects the steering pedestal pump to the control cylinder F. S, now open, allows fluid to pass from the power pump to J which now comes under the influence of the floating lever. Steering wheel movement now moves the piston in F, the floating lever pivots on its attachment to W and J opens to allow the power pump to discharge to the appropriate ends of the steering cylinders G. As the tiller moves, a 'hunting' movement occurs, the cut-off link W acting on the floating lever which, pivoting on its attachment to O, closed J and brings the gear to rest with the rudder at the angle required.

Rudder angle indicators etc. mechanically linked or electric are fitted as required. For local control, the ends of the control cylinder are made common by opening a bypass valve and the control piston is moved by a hand lever (shown dotted) or e.g. wheel, rack and pinion. It will be seen that in the off-loaded condition, the pump discharge circulates through J.

Emergency steering is by relieving tackles, fitted when the rudder is locked by closure of the valves H.

Figure 8.13 Hydraulic steering gear

If hand steering only is required, the gear is reduced to tiller, cylinders and rams, locking and bypass valves, rudder indicator and steering pedestal with pump. A simple form of this gear for torques below 11 kNm is shown in Figure 8.13.

Rotary vane gears

These may be regarded as equivalent to a two-ram gear, with torque capacities of some 3000 kNm. Figure 8.14 illustrates the principle.

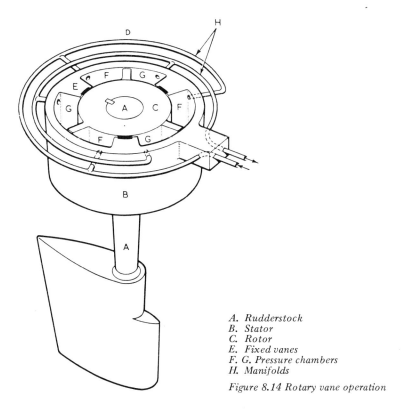

A. Rudderstock
B. Stator
C. Rotor
E. Fixed vanes
F. G. Pressure chambers
H. Manifolds

Figure 8.14 Rotary vane operation

The rotor C is fitted and keyed to a tapered rudder-stock; the stator B is secured to the ship's structure. Fixed vanes, secured equidistantly in the stator bore and rotating vanes secured equidistantly in the rotor, form two sets of alternative pressure chambers in the annular space between rotor and stator; they are interconnected by a manifold. Fluid supplied at pressure to one set of these chambers will rotate C clockwise and the rudder will turn to port, or to starboard if the alternative set is put under pressure.

Three of each type of vane, rotating and fixed, are usual and permit a total angular rudder movement of 70°; two of each, 130°. The vanes act as rudder stops. Figure 8.15 shows the details of a typical unit.

The rotary vanes D and the fixed vanes E are made in SG iron and are securely fixed to the cast-steel rotor C and the stator B respectively, by high-tensile steel dowel pins and cap screws. Keys are

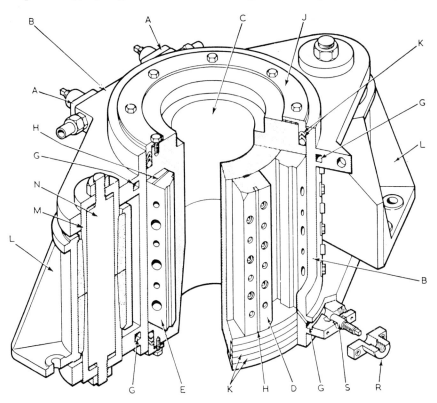

Figure 8.15 Rotary vane unit

A.	Stop valves	J.	Gland
B.	Stator	K.	Gland packing
C.	Rotor	L.	Anchor bracket
D.	Rotary vanes	M.	Span bush
E.	Fixed vanes	N.	Anchor bolt
G.	Upper (starboard) and lower	R.	Stop valve casting
	(port) manifolds	S.	Stop valve spindle
H.	Steel sealing strips		

also fitted over the whole length of each rotary vane in order to give them a mechanical strength at least equal to that of the rudderstock so that they are entirely suitable to act as rudder stops. Steel sealing strips H backed by synthetic rubber are fitted in grooves along the working faces of the fixed and rotary vanes, thus ensuring a high volumetric efficiency, 96–98% even at the relief valve pressure of 100 bar or over. Rotation of B is prevented by means of two anchor brackets, and two anchor pins. The anchor brackets are securely bolted to the ship. Vertical clearance is arranged between the inside of the stator flanges and the top and bottom of the anchor brackets to allow for vertical movement of the rudderstock. This clearance varies with each size of the rotary vane unit, but is approximately 38 mm total and it is necessary that the rudder carrier should be capable of restricting the vertical movements of the rudderstock to less than this amount.

The method of control for these gears and also the hydraulic supply system is as described for electro-hydraulic gears.

VARIABLE-DELIVERY PUMPS

The steering torque required in a 90 000 tons d.w. ship at 18 knots may exceed 2000 kNm. Three types of variable-delivery pumps are in common use and are described below. These work on the principle of varying the amount of oil displaced by altering the stroke of the pump pistons in either radial or axial cylinders, by floating ring, swash-plate or slipper pad.

The radial cylinder pump

The pump (Figure 8.16) consists of case A, to which are attached two covers, the shaft cover B and the pipe connection cover C. This latter cover carries the D tube (or central valve), which has ports E and F forming the connections between the cylinders and branches G and H. The cylinder body J is driven by shaft K, and revolves on the D tube, being supported at either end by ball bearings T.

The pistons L are fitted in a row of radial cylinders, and through the outer end of each piston is a gudgeon pin M, which attaches the slippers N to the piston. The slippers are free to oscillate on their gudgeon pins and fit into tracks in the circular floating ring O. This ring is free to rotate, being mounted on ball bearings P, which are housed in guide blocks R. The latter bear on tracks formed on the covers B and C and are controlled in their movement by spindles S, which pass through the pump case A.

The maximum pump stroke is restricted by the guideblock ends coming in contact with the casing. Further restriction of the pump stroke is effected externally.

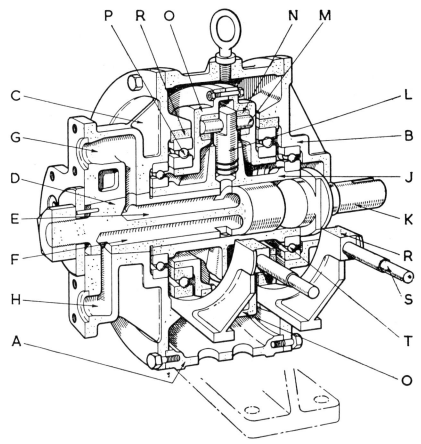

Figure 8.16 Variable delivery pump. See text for key to letters.
(John Hastie & Co Ltd)

Figure 8.17 shows sections through the D tube, cylinder body, pistons and slippers at right angles to the axis. XY is the line along which stroke variations take place. The arrow indicates direction of rotation.

With the floating ring central, i.e. concentric with the D tube, the slippers move round in a circle concentric with the D tube, and consequently no pumping action takes place. With the floating ring moved to the left, the slippers rotate in a path eccentric to the D tube and cylinders, consequently the pistons, as they pass above the line XY, recede from the D tube and draw oil through the ports E, whilst the pistons below XY approach the D tube and discharge oil through ports F.

With the floating ring moved to the right the reverse action takes

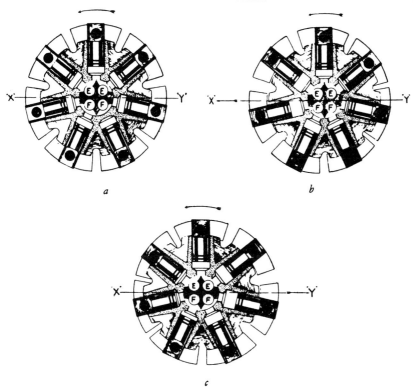

Figure 8.17 Operation of variable delivery pump
(a) Floating ring central   *(c) Floating ring moved to right*
(b) Floating ring moved to left

place, the lower pistons moving outwards drawing oil though ports F and the upper pistons moving into the cylinders and discharging oil through ports E.

The direction of flow depends on the location of the floating ring to left or right of the centre. The floating ring can be moved to any intermediate position between the central and maximum positions, the quantity of oil discharged varying according to the amount of displacement of the floating ring from its mid-position.

Non-reverse locking gear

When two pumping units are fitted and only one is running, the idle pump might be driven in the reverse direction by fluid under pressure from the running pump, if non-reverse locking gear were not fitted. This gear is integral with the flexible coupling connecting motor and pump. It consists of a number of steel pawls so mounted on the

motor coupling that, when pumping units are running, they fly
outward due to centrifugal force and remain clear of the stationary
steel ratchet secured to the motor supporting structure.

The limit of this outward movement is reached when the pawls
contact the surrounding casing, which revolves with the coupling.

When the pumps stop, the pawls return to their normal, inward
position and engage the ratchet teeth, so providing a positive lock
against reverse rotation. This action is automatic and permits instant
selection and commissioning of either unit without needing to use
the pump isolating valves, which are normally open — and only
closed in an emergency.

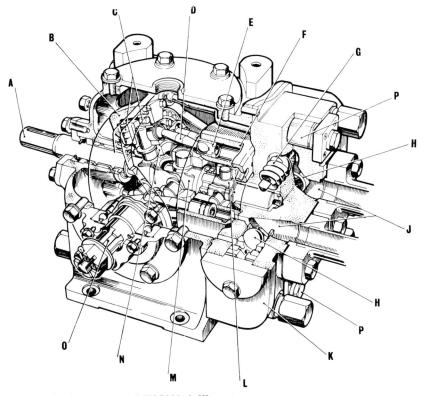

Figure 8.18 Arrangement of 'VSG' Mark III pump

A. *Input shaft*
B. *Tilting box*
C. *Roller bearings*
D. *Connecting rod*
E. *Piston*
F. *Cylinder barrel*
G. *Relief valve*
H. *Replenishing valve*
J. *Ports*
K. *Valve plate*
L. *Barrel joint*
M. *Universal joint*
N. *Socket ring*
O. *Control trunnion*
P. *Control cylinder*

(Vickers Ltd)

The swash-plate axial-cylinder pump

A circular cylinder barrel is bored and splined centrally to suit the input shaft with which it revolves. Several cylinder bores are machined in the cylinder barrel, concentric and parallel with the shaft, one end of each terminating in a port opening into that end face of the barrel which bears against the stationary valve plate, maintained in contact by spring pressure, compensating automatically for wear. Ports in the valve plate match those in the barrel and are connected by external pipes to the steering cylinders, through a valve chest. In the current design, the cylinder barrel is driven by the input shaft through a universal joint and the valve plate contact springs are supplemented by hydraulic pressure in operation. Each cylinder contains a piston, connected by a double ball-ended rod to a socket ring driven by the input shaft through another universal joint and rotating on roller thrust bearings (in some cases on Michel pads) within a tilt box. This is carried on trunnions and can be tilted on either side of the vertical by an external control, e.g. a telemotor. Figure 8.18 shows a cut-away section of the pump.

When the tilt box is vertical, the socket ring, cylinder barrel and pistons all revolve in the same plane and the pistons have no stroke. As the box is tilted, and with it the socket ring, stroke is given to the pistons at each half-revolution, the length of stroke determined by the angle of tilt.

The slipper pad axial-cylinder pump

This is a later development of the pump described above, suitable for the higher pressures demanded for steering gears and fin stabilisers. The socket ring and connecting rods are replaced by slipper pads in the tilt box, the spherical ends of the pistons being carried in the pads. Inclination is given to the tilt box by a control piston, using pressure from the auxiliary pump. Figure 8.19 shows a cut-away section.

Another variant is the 'Sunstrand' pump. Figure 8.20, in which a reversible swashplate, vertically disposed and given the desired angular rotation by an integral servo-piston, is used to vary the quantity and direction of the hydraulic fluid.

A still later variant is that shown in Figure 8.21. The control system is shown in Figure 8.22. As will be seen, a torque motor, receiving the appropriate signal from the bridge through an amplifier unit actuates the floating lever, putting the pumps on stroke in response. The hunting action of the floating lever is no longer required as the normal control of the steering gear from the bridge is

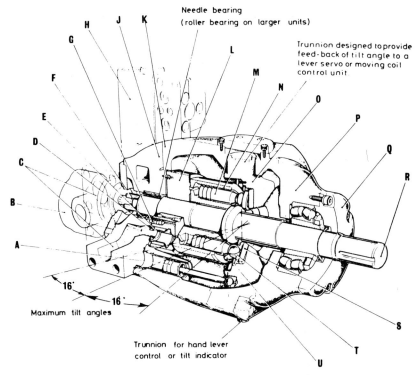

Needle bearing
(roller bearing on larger units)

Trunnion designed to provide
feed-back of tilt angle to a
lever servo or moving coil
control unit

16°

16°

Maximum tilt angles

Trunnion for hand lever
control or tilt indicator

Figure 8.19 Arrangement of 'VSG' slipper pad

A. Control piston	*H. Valve block*	*P. Casing, end section*
B. Auxiliary pump	*J. Casing, centre*	*Q. End cover*
C. Ports	*K. Casing, centre*	*R. Shaft*
D. Floating seal	*L. Floating film face ring*	*S. Thrust plate*
E. Floating piston	*M. Working piston*	*T. Slipper pad*
F. Splined coupling	*N. Cylinder barrel*	*U. Retaining plate*
G. Compensating piston	*O. Tilt box*	

(Vickers Ltd)

by electric signal. The signal is directed to the torque motor which operates the servo valve which in turn controls the pump. When the steering gear has attained the required rudder angle, the electric feed back unit connected direct to the rudder-stock cancels the input signal to the control amplifier, and the steering gear is held at that angle until another rudder movement is required.

This form of control eliminates the need for mechanical linkage and hunting gear on the steering gear.

ELECTRO-MECHANICAL STEERING GEAR

The Reid patent electric steering gear is used in relatively small craft, where rudder torques of less than 120 kNm are appropriate. It may

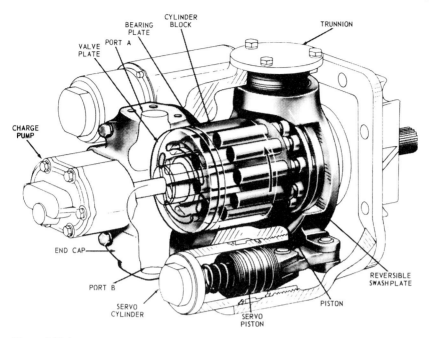

Figure 8.20 Sunstrand pump

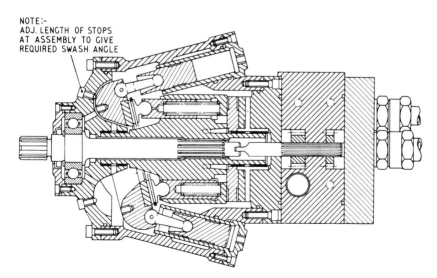

Figure 8.21 Hastie axial-cylinder pump (John Hastie and Co Ltd)

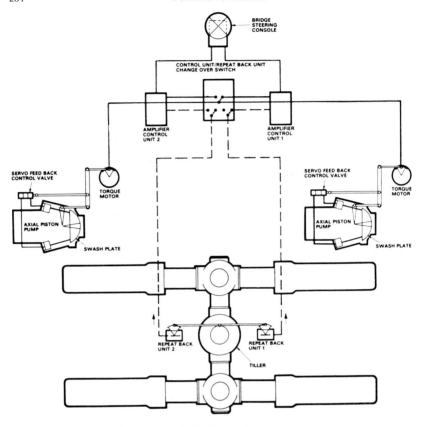

Figure 8.22 The control system for the Hastie pump
(John Hastie & Co Ltd)

be controlled by hydraulic telemotor or shafting. Figure 8.23 shows
a sectional arrangement.

The motor runs continuously at constant speed in one direction
and is connected directly to a bevel pinion A which meshing with
bevel wheels B and C gives opposite alternative rotation to the power
shaft D, in this way. In a steel case between bevels B and C, and
mounted on power shaft D, is a double series of friction plates
separated by a web on the power shaft. The friction plates are
alternately carried in ribs in a steel casting and ribs on the power
shaft. The power shaft is solid, and at the extremity E, telemotor
links are mounted, coupled direct to the telemotor through con-
necting-rod M, which is connected to telemotor links R by ball and
socket joint at the telemotor connection and ball thrust bearings on
end of control shaft E.

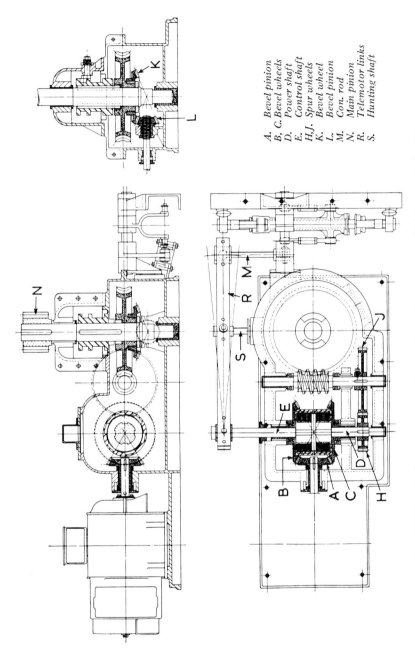

A.   Bevel pinion
B, C. Bevel wheels
D.   Power shaft
E.   Control shaft
H.J. Spur wheels
K.   Bevel wheel
L.   Bevel pinion
M.   Con. rod
N.   Main pinion
R.   Telemotor links
S.   Hunting shaft

*Figure 8.23 Sectional arrangement of direct-acting steering gear controlled from bridge by Telemotor
(Thomas Reid & Sons (Paisley) Ltd)*

When the steering-wheel on the bridge is moved, the telemotor rams on the receiver, alongside the gear, move in end direction in direct synchronism with the steering-wheel, and the telemotor link R is moved in end direction.

Fulcrumed on hunting shaft S, the telemotor shaft compresses the power shaft on bevel-wheel B or C, forming a connection and communicating rotation to the power shaft D, which, through spur wheels H and J, rotates the worm shaft. When the worm shaft is rotating, bevel-wheel K, which is keyed to the underside of the worm-wheel K, which is keyed to the underside of the worm-wheel, rotates bevel pinion L, mounted on a square thread on hunting shaft S, and returns the telemotor links R to the centre position freeing the clutch.

Therefore, when the wheel on the bridge is stopped, the control shaft E is immediately brought to neutral.

The direction of the gear depends wholly upon the movement of the steering-wheel on the bridge, so that the action is exactly the same either to port or starboard, and, according to the movement of the wheel on the bridge, power is communicated to the worm-wheel, which in turn transmits power to main pinion N, which is geared direct to the quadrant on the rudder-stock, and gives the required movement, speed and direction to the rudder.

If control is by shafting, the power shaft D, Figure 8.24, is turned in one direction or the other by bevel wheels B or C, in obedience to the movement of the control shaft E, which carries a pinion G and

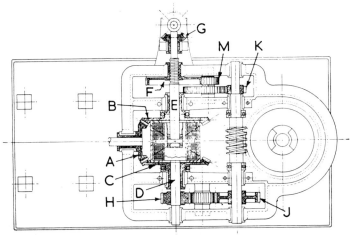

Figure 8.24 Reid electric steering gear (Thomas Reid & Sons (Paisley) Ltd)
F. Spur wheel; G. Mitre pinion

For other letters, see Figure 8.23

a square threaded screw working in the spur wheel F. When the mitre pinion G is turned by the control shafting, the spur wheel F being stationary, the shaft moves forward, pressing the clutch plates against either bevel wheels B or C, turning the power shaft which through its train of spur wheels turns the worm shaft and wheel.

When the worm shaft is turning, the train of spur wheels on the control shaft will turn in a direction tending to take the pressure off the clutch plates. When the control shafting ceases to turn, the spur wheel F having rotation, withdraws the control shaft to neutral.

Reversing the direction of rotation of the control shafting puts pressure on the opposite bevel wheel, the same action taking place in the opposite direction.

As in the case of a gear controlled by telemotor, the worm-wheel gives motion to the pinion N and to the rudder quadrant.

TELEMOTOR CONTROL

Figure 8.25 shows the Brown Telemotor. The system consists of a transmitter and steering-wheel on the bridge. Any movement of the transmitting gear displaces a non-freezing fluid, causing the receiving cylinder to move by a corresponding amount. In the transmitter the steering-wheel by means of a pinion on the hand-wheel shaft, spur wheel on pinion shaft, pinion on pinion shaft and rack, gives a reciprocating motion to the piston attached to the rack. With this piston in midships position there is an open connection between the top and bottom of the cylinder through the bypass; formed by drilling small holes into the annular space surrounding the centre of the cylinder. The replenishing tank is formed by a circular outer casing round the head of the transmitting cylinder. The annular space which forms the automatic bypass is enlarged and connected to the tank bottom, where self-contained suction and relief valves are fitted. The relief valve ensures that the fluid pressure does not exceed 8 bar on the whole system when the gear is at the normal or midship position, also on the non-pressure side when the gear is working, and is so arranged that any excess fluid is returned to the replenishing tank.

The suction or replenishing valve acts in the opposite direction to the relief valve, supplying fluid from the tank to the system to make up for external losses.

The *section through receiver* shows the two receiving cylinders in one casting with circuit pipes connected to the outer end of each ram. Any fluid displaced in the transmitter cylinder by the piston

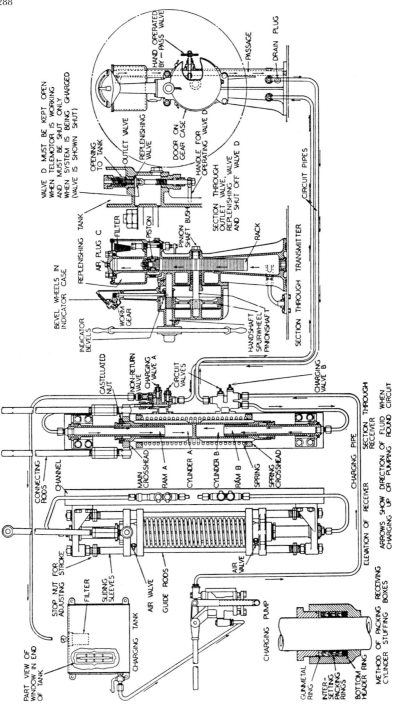

VALVE D MUST BE KEPT OPEN WHEN TELEMOTOR IS WORKING AND MUST BE SHUT ONLY WHEN SYSTEM IS BEING CHARGED (VALVE IS SHOWN SHUT)

HAND OPERATED BY-PASS VALVE

PASSAGE

DRAIN PLUG

OUTLET VALVE

REPLENISHING VALVE

DOOR ON GEAR CASE

HANDLE FOR OPERATING VALVE D

OPENING TO TANK

SECTION THROUGH OUTLET VALVE, REPLENISHING VALVE, AND SHUT OFF VALVE D

REPLENISHING TANK

AIR PLUG C

FILTER

PISTON

PINION SHAFT BUSH

RACK

BEVEL WHEELS IN INDICATOR CASE

WORM GEAR

INDICATOR BEVELS

HANDSHAFT SPURWHEEL PINIONSHAFT

SECTION THROUGH TRANSMITTER

CIRCUIT PIPES

CASTELLATED NUT

NON-RETURN VALVE

CHARGING VALVE A

CIRCUIT VALVES

CHARGING VALVE B

CONNECTING RODS

CHANNEL

MAIN CROSSHEAD

RAM A

CYLINDER A

CYLINDER B

RAM B

SPRING

SPRING CROSSHEAD

CHARGING PIPE

SECTION THROUGH RECEIVER

ELEVATION OF RECEIVER

ARROWS SHOW DIRECTION OF FLUID WHEN CHARGING UP OR PUMPING ROUND CIRCUIT

PART VIEW OF WINDOW IN END OF TANK

STOP NUT FOR ADJUSTING STROKE

FILTER

SLIDING SLEEVES

AIR VALVE

GUIDE RODS

CHARGING TANK

AIR VALVE

CHARGING PUMP

RECEIVING CYLINDER STUFFING BOXES

METHOD OF PACKING

GUNMETAL RING

INTER- SETTING PACKING RINGS

BOTTOM HEADER RING

Figure 8.25 Patent hydraulic steering Telemotor (*Brown Bros & Co Ltd*)

will therefore be forced through the pipes and circuit valve to the receiving cylinder. The receiver rams are fixed in the arrangement shown, and any displacement of the fluid causes the cylinder body to move along the rams against the compression of the spring, which brings the receiver, and with it the transmitter, back to midship position when the steering-wheel is let go. To one end of the cylinder body a cross-head is fitted to which are attached the connecting-rods for connecting in any suitable way to the control gear of the steering-gear.

THE FLOATING LEVER

Figure 8.26 shows diagrammatically three variants of the floating lever arranged as a control and cut-off gear. In each diagram, the control movement is applied at point A, point B is linked to the body whose movement is to be controlled and point C is linked to the control mechanism of the power source i.e. in a ship's steering gear, movement of point A is controlled by the steersman, point B, known as the hunting point, is linked to a point on the tiller and point C is linked to the control lever(s) on the pump(s). The floating lever pivots alternately about points B and A while executing the control and cut-off functions respectively so that the lever 'floats' in space and, as can be seen in the diagrams, the distance between one or more pairs of attachment points on the lever varies as movement takes place. In consequence, the motion of the lever is extremely

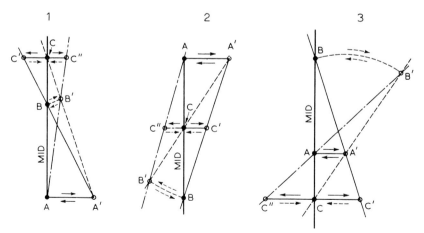

Figure 8.26 The floating lever
A. Control point   B. Hunting point   C. Pump control point
(Brown Bros & Co Ltd)

complex and is difficult to visualise as a continuous process. However, if the geometry of the system is examined step by step at each
of its rest positions, the principles of operation become clear.

To render the geometry of the system determinate and to ensure
that it functions correctly, it is prudent to impose the following limitations on the design:

(a) Either point A or point C, but not both, should occupy a
     fixed location in the length of the lever: the other must be left
     free to move longitudinally in the lever to accommodate the
     variation that occurs in the distance between them.
(b) Points A and C should be constrained to move on known loci
     to ensure the accuracy of the system.
(c) Point B should be free to move longitudinally in the lever.

In the three diagrams (Figure 8.26) the control point A has been
selected as the datum point and constrained to move only in a
straight line such as AA', the hunting point B is free to slide in a slot
cut in the floating lever and the pump control point C is arranged to
slide in another slot and constrained to move along a straight line
such as CC'.

Now, if we assume a ship to be proceeding on a straight course,
points A, B and C in each of the diagrams lie on the mid line since
the steering controls and the rudder are centralised and the pump is
in the no-stroke or neutral condition. Movement of point A to A',
which corresponds with the rudder angle required, causes the floating
lever to pivot about point B and point C moves to C' placing the
pump(s) on-stroke in the correct sense. As the rudder moves over
towards the angle 'ordered' by the movement of A to A', point B,
which is linked to the tiller, moves towards B' with the floating lever
now pivoting about A', being held there by the steersman, causing
point C' to return to C. Points B' and C' arrive at B and C simultaneously, placing the pumps in the no-stroke condition and bringing
the rudder to rest at precisely the angle ordered. If, now, the control
point is moved back from A' to the mid position A, the floating
lever pivots about B', point C moves to a new position C'' on the
opposite side of the mid line placing the pumps on-stroke to drive
the steering gear back towards the mid position. The lever again
pivots about A as the gear returns, point B' moves back to B, causing
C'' to move back to C thus placing the pump(s) in the no-stroke
condition and bringing the rudder to rest at the mid position. Movement of point A to a position A'' (not shown) on the opposite side
of the mid line and then back to A would have a similar effect
except, of course, that the rudder movements and all the hunting

movements would also occur on the opposite side of the mid line to that shown.

The mechanical arrangement of a floating lever system may not always be in one of the three forms illustrated but, if the principle is understood and it is kept in mind that its basic functions are to initiate movement of the steering gear and to stop movement when the rudder arrives at the angle ordered by the steersman, the reader should have no difficulty in recognising any floating lever arrangement he may encounter.

TESTING OF STEERING GEARS

Except in the case of ships regularly engaged on short voyages the steering gear should be thoroughly checked and tested within 12 hours before departure. These tests should include testing of power unit and control system failure alarms, the emergency power supply (when relevant) and automatic isolating arrangements.

Every three months an emergency steering drill should be held and should include direct control from within the steering compartment at which time the use of the communications procedure with the navigating bridge should be practised.

BIBLIOGRAPHY

1. *The International Convention for the Safety of Life at Sea* (1974) as amended in November 1981, Chapter 11-1: Regulation 29 and 30. Chapter V, Regulation 19.
2. Cowley, J., *Steering Gear: New Concepts and Requirements, I. Mar. E. Transactions,* Paper No 23 (January 1982)

9 Stabilisers and stabilising systems

A ship at sea has six degrees of freedom, i.e. roll, heave, pitch, yaw, sway and surge (Figure 9.1). Of these, only roll can effectively be reduced in practice by fitting bilge keels, anti-rolling tanks or fin stabilisers. A combination of fins and tanks has potential advantages in prime costs and effective stabilisation at both high and low speeds.

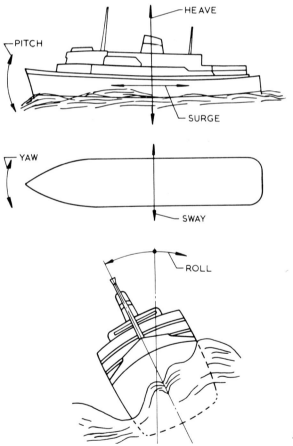

Figure 9.1 Ship motions

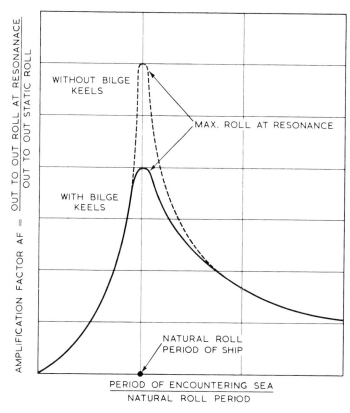

Figure 9.2 Amplification factor

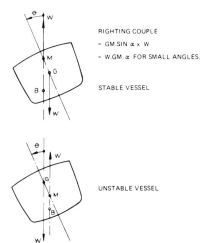

RIGHTING COUPLE

= GM.SIN α × W

= W.GM. α FOR SMALL ANGLES.

STABLE VESSEL

UNSTABLE VESSEL

Figure 9.3 Stable and unstable vessels
See text for key to letters

Since a ship is a damped mass elastic system, it has a natural rolling period and large rolling motions may be induced by resonance with relatively small wave forces. Large resonant rolls can be avoided by generating forces equal and opposite to the impressed sea force. Figure 9.2 shows that the roll amplitude at resonance is much greater than that at long wave periods. The ratio of these amplitudes is the dynamic amplification factor which is limited by the inherent damping of the ship, i.e. viscous damping and the action of bilge keels.

Natural roll period

In Figure 9.3, G is the centre of gravity, B is the centre of buoyancy, W is the weight or displacement of the ship, θ the roll angle. M, the point of intersection of a vertical line through B with the centre line of the ship, is known as the metacentre and the distance GM as the metacentric height which is of great importance in relation to the rolling characteristic and the stability of the ship. When a stable vessel is heeled over, the couple $W.GM. \sin \theta$ tends to restore it to the vertical whereas, in the unstable condition, the couple tends to turn the vessel further over. In the latter case, G is above M and GM is said to be negative, an unacceptable condition.

The natural rolling period, i.e. the time taken to roll from out to out and back again is

$$T = \frac{C \times \text{beam}}{\sqrt{GM}}$$

where C is a constant determined by experience on ships of similar hull form and usually falls within the range 0.72 to 0.80 for merchant vessels.

For example, where maximum beam is 16 m, GM 0.49 m and C is 0.80

$$T = \frac{0.80 \times 16}{\sqrt{0.49}} = 18.29 \text{ sec.}$$

If the GM changes to, say, 0.64 m

$$T = \frac{0.80 \times 16}{\sqrt{0.64}} = 16.0 \text{ sec.}$$

A 'stiff' vessel has a large GM and the natural period is short: a 'tender' vessel has a small GM and the period is long. It is important

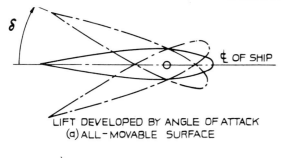

LIFT DEVELOPED BY ANGLE OF ATTACK
(a) ALL-MOVABLE SURFACE

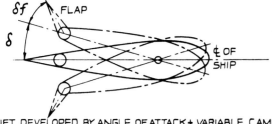

LIFT DEVELOPED BY ANGLE OF ATTACK + VARIABLE CAMBER
(b) ALL-MOVABLE SURFACE WITH TAIL FLAP

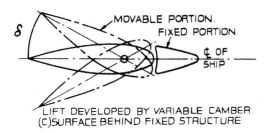

LIFT DEVELOPED BY VARIABLE CAMBER
(C) SURFACE BEHIND FIXED STRUCTURE

Figure 9.4 Fin arrangements

to note that, particularly in cargo vessels, the draught and dead weight may change appreciably during a voyage hence the natural rolling period may also change, varying inversely as the square root of the GM.

A ship in a beam sea will roll to an angle $\theta°$ if the rolling moment applied by wave action is equal to that required to heel the ship $\theta°$ in calm water. When θ is small, the heeling moment is $W.GM.\theta$, (θ now expressed in radians), which provides a basis for calculations of the stabilising force required.

Fin stabilisers

The stabilising power of fins is generated by the 'lift' on 'aerofoil' sections which may be all-moveable, with or without flaps or partly

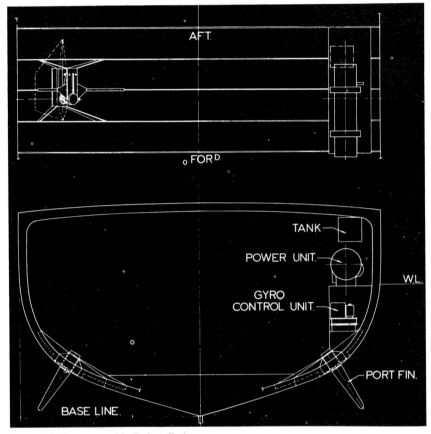

Figure 9.5 Non-retractable fin installations

fixed, partly moveable, (Figure 9.4). These fins are tilted, usually hydraulically, in phase with the roll at long wave periods, 90° out of phase at resonance and in phase with roll acceleration at short periods.

Non-retractable fins are commonly used where space within the hull is limited. They are usually fitted at the round of the bilge and do not project beyond the vertical line from the ship's side or the keel line, (Figure 9.5) to minimise the risk of contact with a quay wall or the sea bottom.

The fin shaft, to which the fin is rigidly attached, passes through a sea gland in a mounting plate welded or bolted to the hull and is supported by two substantial bearings. A double-ended lever keyed to the inner end of the finshaft is actuated by two hydraulic rams supplied from an electrically driven variable delivery pump (Figure 9.6).

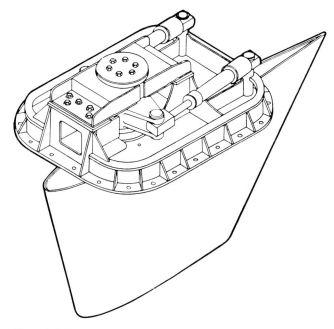

Figure 9.6 Non-retractable fin assembly

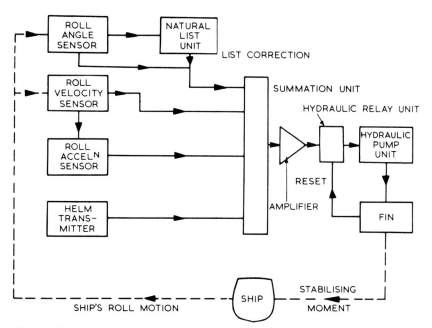

Figure 9.7 Principles of Multra control system

Control of fin movement is automatic and is usually derived from gyroscopic sensing gear which, in its simplest form, Velocity Control, is based on one small, electrically driven gyroscope mounted horizontally with its axis athwartships. The angular velocity of roll of the ship causes the gyroscope to precess against centralising springs to an amount proportional to the velocity and generates a small force which is hydraulically amplified by a hydraulic relay unit to provide power sufficient to operate the controls of the variable delivery pump via suitable linkage. Part of the linkage is coupled to the fin-shaft to transmit a cancelling signal to the pump control and to bring the fin to rest at the angle of tilt demanded by the sensing unit. This type of control is often fitted in small installations, usually for economic reasons, and is most effective against resonant rolling.

Ships seldom roll in a purely resonant mode: the sea state is often highly confused. More elaborate, and more expensive, control systems are required to deal with suddenly applied roll, rolling at periods off resonance and rolling in conditions arising from the combination of several wave frequencies. A sensing unit based on a vertical-keeping gyroscope and a velocity gyroscope coupled into differentiating and summation units enables fin movement to be controlled by a composite function derived from roll angle, roll velocity and roll acceleration. By adding a 'natural list' unit, stabilisation is achieved about the mean point of roll and so reduces both propulsion and stabilising power demand. This is known as a compensated control system, Figure 9.7, and is generally used in large installations.

Roll reduction in excess of 90%, typically 30° out to out reduced to less than 3° out to out, can be achieved at resonance and low residual rolls can be maintained over a wide range of frequencies. However, since the stabilising power varies as the square of the ship's speed, fins are least effective at low or zero speed where they function only as additional bilge keels.

Retractable fins fall into two classes, those that extend and stow athwartships and those that hinge into a fore-and-aft stowed position. In the athwartship-retracting type, Figure 9.8, the finshaft has a tapered outboard end to which the fin is keyed. The parallel inboard end passes through a sea gland on the inboard face of the fin box and is supported by two bearings. One, close to the inboard end of the fin, is carried in a heavy crosshead, arranged to slide in top and bottom guides within the fin box and the other in a crosshead slideably mounted on the extension guides, within the hull. The hollow bore of the parallel section of the finshaft houses a double-acting piston to act as housing and extending gear. Tilting of the fin is by two or four hydraulic cylinders which may be of the simple oscillating type or arranged on the Rapson-slide principle as described for steering gears. Power units, control and sensing equipment are as for non-retractable fins.

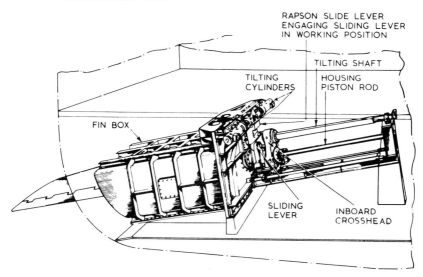

Figure 9.8 Arrangement of Denny-Brown athwartships retracting fin

In the Denny-Brown-AEG hinged or folding type, Figure 9.9, the finshaft is rigidly fixed into the crux which has two heavy trunnions disposed vertically and housed in bearings top and bottom to the fin box. The fin is free to oscillate on the finshaft and the tilting force is provided by a vane type motor the stator of which is secured to the crux and the rotor keyed to the fin through a flexible coupling. The van motor is housed in an oil-tight casing secured to the fin and is provided with a sea gland bearing on a sleeve fitted to the crux. The whole of the casing and the interior of the fin is full of oil under pressure to prevent the ingress of sea water.

Housing and extending the fin is achieved by a double acting oscillating cylinder connected to the upper trunnion. Power units, control and sensing equipment are generally similar to other types of stabiliser except that feed-back of fin angle is accomplished electrically by synchros.

Details of the variable-delivery pumps used to extend, house and operate stabilising fins, and also the vane motor will be found in Chapter 8 'Steering Gears'.

The fin and fin shaft are integral structures supported on taper roller bearings which are carried in the fin housing. Movement of the

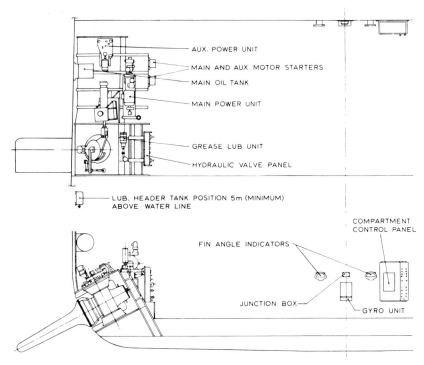

AUX. POWER UNIT

MAIN AND AUX. MOTOR STARTERS

MAIN OIL TANK

MAIN POWER UNIT

GREASE LUB. UNIT

HYDRAULIC VALVE PANEL

LUB. HEADER TANK POSITION 5m (MINIMUM) ABOVE WATER LINE

COMPARTMENT CONTROL PANEL

FIN ANGLE INDICATORS

JUNCTION BOX

GYRO UNIT

Figure 9.9 Arrangement of Denny-Brown-AEG forward folding fin

fin shaft to tilt the fin is effected by means of a cylinder tilting mechanism with a double acting piston. The fins are rigged out for operation from the stowed position by rotation of the fin housing about the rigging axis, in upper and lower bearings in the fin box. Figure 9.10 shows a sectional arrangement of the Sperry 'Gyrofin' stabiliser and Figures 9.11 and 9.12 shows respectively the Vosper retractable fin stabiliser and control system.

Tank stabilisers

Tank stabilisers are virtually independent of the forward speed of the vessel: they generate anti-rolling forces by phased flow of appropriate masses of fluid (water or reserve fuel, etc) in transverse tanks installed at suitable heights and distances from the ship's centre line. Fluid transfer may be by open flume or from and to wing tanks connected by cross ducts. The tank/fluid combination constitutes a damped mass elastic system having its own natural period and capable of developing large forces at resonance with the impressed wave motion.

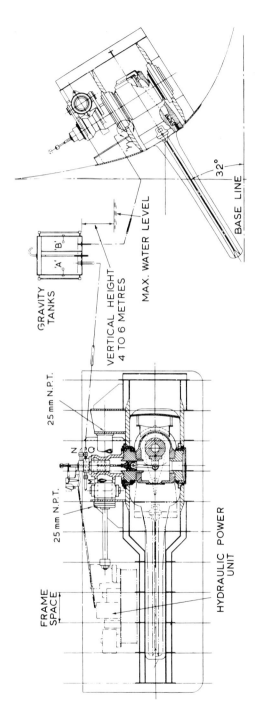

GRAVITY
TANKS

'A' 'B'

VERTICAL HEIGHT
4 TO 6 METRES

MAX. WATER LEVEL

25 mm N.P.T.

25 mm N.P.T.

FRAME
SPACE

HYDRAULIC POWER
UNIT

32°

BASE LINE

Figure 9.10 Sperry Gyrofin stabilisier (Sperry Marine Systems)

Since the fluid can only flow downhill and has inertia, it cannot start to move until the ship has rolled a few degrees, see Figure 9.13. The natural restoring forces limit the maximum roll angle and initiate a roll in the opposite sense. In the meantime the fluid continues to flow downhill, piles up on the still low side and provides a moment opposing the ship motion. As the ship returns and passes its upright position, fluid again flows downhill to repeat the process.

The fluid flow tends to lag quarter of a cycle behind the ship motion, a phase lag of approximately 90°, to generate a continuing stabilising moment. This is due, mainly, to the transfer of the centre of gravity of the fluid mass away from the centre line of the ship. The transverse acceleration of the fluid generates an inertia force and thereby a moment, about the roll centre, which reduces the gravity

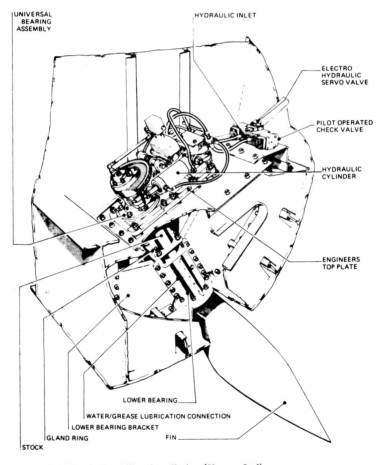

Figure 9.11 Typical stabiliser installation (Vosper Ltd)

303

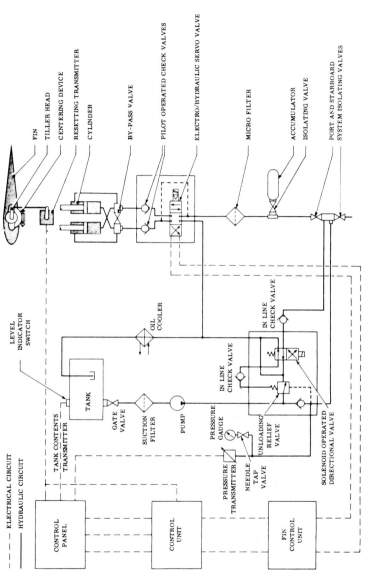

Figure 9.12 Schematic diagram for Vosper fin stabiliser

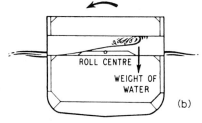

(a) *Stern view of ship with passive tank rolled to starboard. The water is moving in the direction shown.*

(b) *Ship rolling to port. The water in the tank on the starboard side provides a moment opposing the roll velocity.*

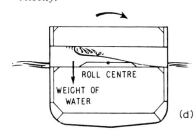

(c) *Ship at the end of its roll to port. The water is providing no moment to the ship.*

(d) *Ship rolling to starboard. The water in the tank on the port side provides a moment opposing the roll velocity.*

Figure 9.13 Brown-NPL passive tank stabiliser

moment when the tanks are below the roll centre and increases it when they are above. In practice, tanks may be placed 20% of the beam below the roll centre without serious loss of performance. Above the roll centre, other factors associated with the phase of fluid motion prevent augmentation of the gravity stabilising power being realised. The phase lag may be increased, within limits, by placing obstructions, e.g. orifice plates, grillages, etc. in the fluid flow path to increase the damping.

In the wing tank system the mode of operation is similar to the simple flume but the tank geometry combined with the dynamic amplification of the flow tends to make fluid pile up to a greater height at a greater distance from the ship's centre line to give more effective stabilisation. The wing tanks must be of sufficient depth to accommodate the maximum rise of the fluid without completely filling them. For purely passive action the tank tops are vented to atmosphere but in a controlled passive system, such as the Muirhead-Brown, Figure 9.14, they are connected by an air duct fitted with valves, controlled by a roll sensing device, which regulate the differential air pressure in the tanks to modify the natural fluid flow rate. This system will generate its full stabilising power from a

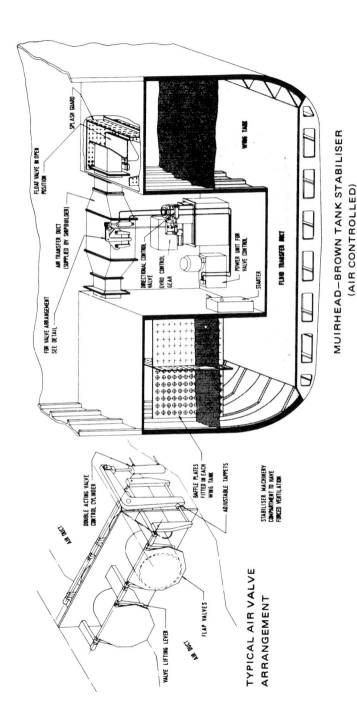

SPLASH GUARD

FLOAT VALVE IN OPEN
POSITION

AIR TRANSFER DUCT
(SUPPLIED BY SHIPBUILDER)

FOR VALVE ARRANGEMENT
SEE DETAIL

DIRECTIONAL CONTROL
VALVE

GYRO CONTROL
GEAR

POWER UNIT FOR
VALVE CONTROL

STARTER

WING TANK

FLUID TRANSFER DUCT

MUIRHEAD–BROWN TANK STABILISER
(AIR CONTROLLED)

AIR DUCT

DOUBLE ACTING VALVE
CONTROL CYLINDER

BAFFLE PLATES
FITTED IN EACH
WING TANK

ADJUSTABLE TAPPETS

STABILISER MACHINERY
COMPARTMENT TO HAVE
FORCED VENTILATION

VALVE LIFTING LEVER

FLAP VALVES

AIR DUCT

TYPICAL AIR VALVE
ARRANGEMENT

Figure 9.14 Muirhead-Brown controlled passive tank system

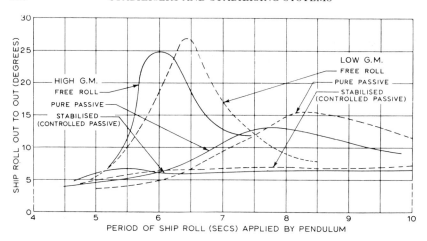

Figure 9.15 Typical performance curves for Muirhead-Brown tank stabiliser

residual roll of about 7° out-to-out at resonance, due to the fact that dynamic amplification of the fluid motion may be from twice to six times the long period effect. The natural period of the fluid is a function of tank geometry and the volume of fluid contained; it is arranged to be equal to or slightly less than the lowest natural roll frequency of the ship. Provided the system has little damping, maximum roll reduction is achieved at resonance and the roll amplitude/roll period characteristic is virtually a straight line at about the optimum residual roll characteristic, Figure 9.15.

Acknowledgement

All the illustrations in this chapter, with the exception of Figures 9.10, 9.11 and 9.12 are reproduced by courtesy of Brown Brothers and Co. Ltd.

10 Refrigerators

In a ship built solely for refrigerated cargo the value of the produce which could be lost in the event of serious failure of the refrigerating machinery may well exceed the value of the ship. Refrigeration is therefore, of prime importance.

Refrigeration is used principally in the carriage of perishable food-stuffs. Its purpose is to prevent or check spoilage, the more important causes of which are:

1. Excessive growth of micro-organisms, bacterial and fungal.
2. Changes due to oxidation, giving 'off' flavours and poor appearance.
3. Enzymatic or fermentive processes, causing rancidity.
4. Drying out (dessication).
5. The metabolism of certain produce, e.g. fruit and vegetables.

Table 10.1 Carrying temperatures, °C

Apples	1−2	Beef	
Bacon		frozen	−9
cured	1−5	chilled	−2
uncured	−9	Lamb	−9
Bananas	11−14	Pork	−9
Butter	−9	Lemons	12−13
Cheese	1−10	Oranges	3−7
Eggs		Pears	<0
frozen	−9	Plums	
chilled	0−1	4 days at	<0
Fish	−10	thereafter	7−10
Grapes	0−1		

Meat, fish, butter, cheese etc., which are usually carried frozen, may be regarded as dead cargoes. Fruit and vegetables are live cargoes until consumed, continuing their separate existences by absorbing oxygen, giving off CO_2 and so producing heat; the rate of respiration varying with the fruit and also, directly with the temperature. Apples may produce CO_2 at the rate of 0.06 m³/tonne/day at carrying

307

temperature, evolving heat at a rate of some 12 W tonne/h in the process. Each commodity has its own specific storage conditions for the best result obtainable. Table 10.1 gives some carrying temperatures commonly used satisfactorily.

Refrigeration has other important applications, namely:

1. Air-conditioning, perhaps employing the most machinery and consuming the most power.
2. The liquifaction of the boil-off which occurs in the transport of liquid natural gases.
3. To maintain the temperature of certain chemicals requiring carriage in bulk at temperatures below the ambient.
4. To cool liquid CO_2, carried for fire-fighting purposes, so that it may be stored in smaller, lighter containers than when stored at high pressure at ambient temperatures.

Choice of refrigerant

For practical purposes, shipboard refrigeration is brought about by the vaporisation of a liquid refrigerant. Just as a solid, changing its state absorbs heat from its surroundings, a liquid vaporising, absorbs heat similarly. For example, ethyl chloride at atmospheric pressure vaporises at 285.5K (54.4°F), taking up its latent heat of about 393 kJ/kg (169 Btu/lb); if placed in surroundings warmer than 285.5 K, it will cool these surroundings as it evaporates. If the pressure were reduced to say, 0.40 bar it would vaporise at 263.1 K and low temperature cooling would follow.

Mechanical refrigeration makes possible the control of the pressure and, therefore, the temperature of the boiling refrigerant and allows repeated use of the same refrigerant with little or no loss. Theoretically almost any liquid can be used as a refrigerant if its pressure/ temperature relationship is suitable for the conditions required. Although no perfect refrigerant is known, there are certain factors which determine a refrigerant's desirability for a particular duty and the one selected should possess as many as possible of the following characteristics:

1. Moderate condensing pressure, obviating the need for heavily constructed compressors, condensers and high pressure piping.
2. High critical temperature, as it is impossible to condense at a temperature above the critical, no matter how much the pressure is increased.

3. Low specific heat of the liquid. This is desirable as throttling at the expansion valve causes liquid refrigerant to be cooled at the expense of partial evaporation.
4. High latent heat of vaporisation, so that less refrigerant may be circulated to perform a given duty.
5. The refrigerant should be non-corrosive to all materials used in the construction of the refrigerating machinery and system.
6. It should be stable chemically.
7. It should be non-explosive and non-flammable.
8. World wide availability, low cost and ease of handling are most desirable.
9. Although there is some disagreement about the importance of this, the problem of oil return to the compressor crankcase is simplified when an oil-miscible refrigerant is used, by the admixture of the oil and the refrigerant in the system. With non-miscible refrigerants, once oil has passed to the condenser, its return to the crankcase can only be effected with difficulty.

Finally the refrigerant should preferably be non-toxic, have satisfactory heat transfer characteristics, and leakages should be easy to detect either by odour or by the use of suitable indicators.

It is not proposed to list or deal with all known refrigerants, but only those likely to be encountered on board. These refrigerants are referred to by their internationally recognised numbers, given in Table 10.2.

Table 10.2 List of commonly-used refrigerants

International Number	Chemical name	Boiling points, K
11	Trichloromonofluoromethane	298.0
12	Dichlorodifluoromethane	243.0
22	Monochlorodifluoromethane	233.0
501	An azeotropic mixture of 75% R22, 25% R12, by weight	232.0
717	Ammonia	240.0
744	Carbon Dioxide	195.0
50*	Methane	112.0
170*	Ethane	184.0
290*	Propane	230.0
600*	Butane	274.0
1150*	Ethylene	169.5
1270*	Propylene	225.0
1140*	Vinyl Chloride	259.0

*These refrigerants are used in the reliquifaction plants of ships where they form part or the whole of the cargo.

The most commonly used refrigerant in new construction is Refrigerant 22, followed by 12, 717, 501 and 744 in that order. However, the Azeotrope 501, only recently introduced, is rapidly gaining favour. This combines most of the desirable characteristics of R.12 and R.22. When as is likely, its price falls and its availability increases, it will be even more widely used.

Operation of the refrigeration system

The elements of a simple compression refrigeration system are shown in Figure 10.1. In the evaporator, the liquid refrigerant vaporising, absorbs heat from the circulating brine or directly from the space to be cooled. The low pressure vapour from the evaporator passes through a heat exchanger where it cools the high temperature liquid from the condenser and in doing so becomes super-heated before being drawn into the compressor. Here, its pressure and consequently its temperature, are raised for delivery to the condenser. With any refrigerant the condensing pressure of the system is determined by the temperature of the cooling water, or of the ambient air in the case of air.cooled condensers.

The refrigerant pressure must be sufficient to give it a saturation temperature above that of the available cooling medium, so that heat can be removed in the condenser. After removal of this heat, the liquid (which in water cooled installations is generally $1-2°C$ warmer than the circulating water) passes to the receiver thence through the heat-exchanger to the expansion valve, where its pressure is reduced

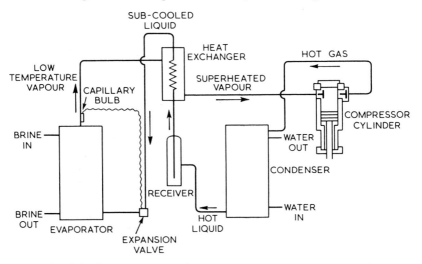

Figure 10.1 A simple vapour-compression system

to the evaporator pressure of the system by throttling. In passing through the expansion valve the liquid is cooled further by the inevitable partial evaporation which occurs.

COMPRESSORS

Compressors may be either reciprocating, centrifugal, screw, or rotary. The reciprocating type is still the most commonly used, but screw compressors for refrigeration of cargo and centrifugal compressors for air conditioning are becoming increasingly popular. Most modern reciprocating compressors have their cylinders arranged in either V or W formation with 4, 6, 8, 12, or even 16 cylinders. Rates of revolution have increased considerably in recent years: ten years ago, 500 rev/min put a compressor into the high speed category: speeds of 1500 to 2000 rev/min are common nowadays. The stroke/bore ratio has diminished to the point of becoming fractional because of improvements in valve design and manufacture.

Provision is made for unloading cylinders during starting and for subsequent load control, by holding the suction valves off their seats by suitable oil-pressure operated mechanisms. With this type of control the compressors can be run at constant speed; a great advantage when alternating current motors are used.

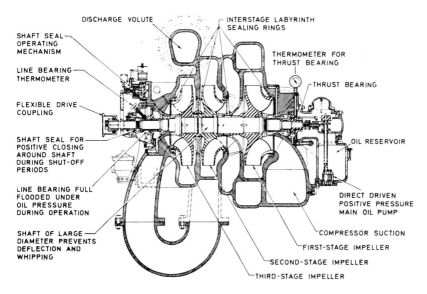

Figure 10.2 Three-stage centrifugal compressor

A bellows device actuated by suction pressure can serve to cut out one or more cylinders. Thus a falling suction pressure, indicating a reduced load on the system, can be used to reduce automatically the number of working cylinders to that required to deal with the existing load. Nearly all compressors of this type are fitted with plate type suction and delivery valves, whose large diameter and very small lifts offer the least resistance to the flow of refrigerant gas.

Centrifugal compressors

In these one or more impellers are mounted on a steel shaft and enclosed in a volute casing, usually of cast iron; the number of impellers depends on the compression ratio required and the number of stages necessary for this.

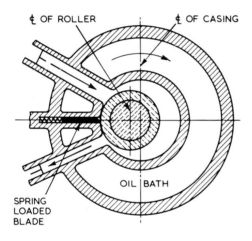

Figure 10.3 Rotary piston compressor

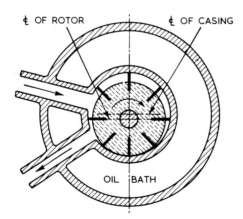

Figure 10.4 Rotary vane compressor

The use of centrifugal compressors (Figure 10.2) at sea is now limited generally to large air-conditioning installations, in which great capacity is required and using R11 or other refrigerants operating at atmospheric pressure, where the range of compression ratio is small. They are particularly well suited to this duty.

Rotary compressors

These are of two types, rotary piston (Figure 10.3) and rotary vane (Figure 10.4). The former consist of a cylindrical steel roller revolving on an eccentric shaft, mounted concentrically in a cylinder. A spring-loaded blade, free to move in a longitudinal slot in the cylinder wall bears against the revolving roller; as the shaft revolves and the roller moves round the cylinder wall in the direction of shaft rotation, it maintains continual contact.

Rotary vane compressors employ a number of vanes or blades carried in longitudinal slots machined in an eccentrically mounted rotor. This rotor is installed in the cylinder in such a way that it approaches closely the cylinder wall on one side. The vanes move radially in the rotor slots as the rotor revolves, being held against the cylinder wall by centrifugal force; in some cases springs are fitted behind the vanes.

Screw compressors

The capacity, range and use of screw compressors are increasing rapidly; their economy in space and weight and their capacity for long periods of uninterrupted running are inherent advantages. They may be oil-free, used within their limited pressure range when oil contamination of the gas cannot be tolerated or more generally, oil-injected for higher pressure range work. In both, two steel rotors are mounted in a gas-tight casing, usually of h.t. cast iron; the male rotors have a number of lobes (usually 4) and the female a number (usually 6) of flutes cut helically in cylindrical bodies, so dimensioned that they mesh like helical gears when mounted at appropriate centres.

Oil-free or dry machines are fitted with timing gears to ensure synchronisation of the rotors, out of contact; in oil-injected or wet machines, the female is driven by the male rotor, separated by the oil, which serves also for sealing and cooling. Capacity may be varied by throttling at the suction (at some expense of power) or by the incorporation of a slide valve which, moving axially within the casing, varies the effective length of the rotors. Figure 10.5 shows a

314

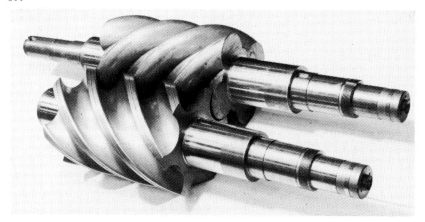

Figure 10.5 A pair of compressor rotors (Howden Group Ltd)

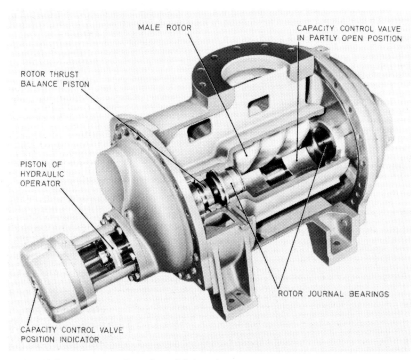

Figure 10.6 Cut-away section of an oil injected compressor (Howden/Godfrey Ltd)

pair of rotors and Figure 10.6 a cut-away section of an oil-injected compressor, with the female rotor removed to show the slide valve and the actuating piston in its cylinder.

As will be apparent, large, effective oil separators are necessary when rotary or oil-injected compressors are used.

Hermetic compressors

Hermetic and semi-hermetic compressors have been introduced fairly recently. In these, the mechanical elements of the compressor are conventional but the driving motor is enclosed within the crank chamber and is cooled by the passage over it of the refrigerant suction gas. The motor heat is, of course, added to the refrigeration load. This may be and is sometimes avoided by placing the motor in the oil-rich discharge of an oil-injected compressor, cool enough for the purpose.

These compressors are not intended to be overhauled on board but to be removed and replaced after a predetermined period of running. In this way, the ingress of moisture to the system may be prevented. This has always been a matter of great difficulty during the overhaul of marine refrigerators using halocarbon refrigerants.

EVAPORATORS

Evaporators are most commonly of the shell and tube type although submerged coil-in-casing evaporators may still be met in installations using Refrigerant 744 and in small domestic plants. In a shell and tube evaporator, the area of tube surface in contact with the liquid refrigerant determines its performance, and the present trend is to submerge all the tubes in the boiling liquid. This involves either a high liquid level in the shell, or the placing of the tubes in the lower part of the shell only, the upper part then forming a vapour chamber.

Modern flooded evaporators incorporate finned tubes, usually with integral fins, where the diameter of the fins is no greater than that of the holes in the tube plate. Shell and tube evaporators (Figure 10.7) may also be of the dry expansion type, in which the refrigerant passes through the tubes, and the brine is circulated through the shell. The advantages of this type are a smaller refrigerant charge and a more positive return of lubricating oil to the compressor.

In direct expansion systems, evaporation takes place in air coolers consisting of pipe grids, plain or finned, enclosed in closely fitting casings, though which air from the holds or chambers is circulated

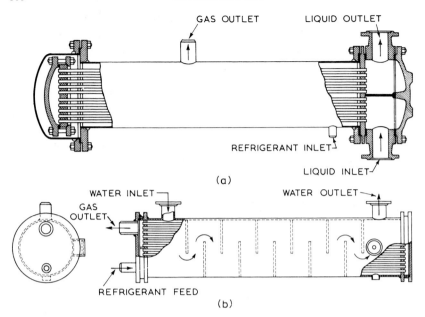

Figure 10.7
(a) Flooded-type evaporator   *(b)*   *Dry expansion-type evaporator*

by forced or induced draught fans. This type of evaporator can be operated either partly flooded, fully flooded, with incorporated accumulators, or even dry. In the latter, the refrigerant flow is controlled at the expansion valve in such a way that, as it passes through the grids, it is completely vaporised and slightly superheated.

CONDENSERS

Marine condensers are generally shell and tube, designed for high pressures. Although there are a few coil-in-casing and double-pipe condensers still in use, they are no longer made, except as replacements.

Condensers provide heat transfer surfaces in which heat from the hot refrigerant passes through the walls of the tubes to the condensing media; the vapour is cooled first to saturation point, then to the liquid state. The design of condensers is largely dictated by the quantity and cost of the circulating water, and, where water is plentiful as at sea, a large number of short circuits are used to keep pressure drop to a minimum. Water velocity must be restricted to prevent erosion of the tubes, and is usually kept below 2.5 m/sec.

Marine condensers are very susceptible to corrosion, and much research has been done to lengthen their useful life. With halo-carbon refrigerants, the use of aluminium brass or cupro-nickel tubes and tube plates has done much to reduce the rate of corrosion, but where Refrigerant 717 is employed, it has been found necessary to use bi-metallic tubes and clad tube plates, the ferrous metals being in contact with the refrigerant, and the non-ferrous with the sea water. The use of sacrificial anodes in the water ends of the condensers is common and sometimes a short length of ungalvanised pipe is fitted in a galvanised steel pipe system.

REFRIGERANT CONTROLS

There are six basic types of refrigerant controls which can be summarised as follows:

Manually operated expansion valves

These are suitable for fairly large installations, having an operator always on duty; they have the disadvantage of being unresponsive to changes in load and must be adjusted frequently. The valve itself is a screw down needle valve dimensioned to give fine adjustment.

Automatic expansion valves

These consist of a needle with seat and a pressure bellows or diaph-ragm, with a torsion spring capable of adjustment. Operated by evaporator pressure, their chief disadvantage is their relatively poor efficiency compared with other types. Constant pressure in the evaporator also requires a constant rate of vaporisation, which in turn calls for severe throttling of the liquid. There is also the danger of liquid being allowed to return to the compressor when the load falls below a certain level.

This type of valve is used principally in small equipment with fairly constant loads, such as domestic storage cabinets, freezers and the like.

Thermostatic expansion valves

These valves are similar in general design to automatic valves, but having the space above the bellows or diaphragm filled with the liquid refrigerant used in the main system and connected by capillary tube to a remote bulb. This remote bulb is fixed in close contact with the suction gas line at the outlet from the evaporator and is responsive to changes in refrigerant vapour temperature at this point. These valves are the most commonly employed and are suitable for the control of systems where changes in the loading are frequent.

Unlike the automatic valve, based on constant evaporator pressure, the thermostatic valve is based on a constant degree of superheat in the vapour at the evaporator outlet, so enabling the evaporator at any load to be kept fully charged with liquid refrigerant without any danger of liquid carryover to the suction line and thence to the compressor.

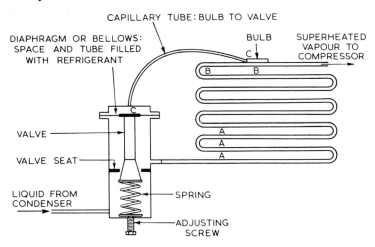

Figure 10.8 Diagrammatic arrangement of automatic expansion valve
At A: Liquid and vapour at saturated temperature and pressure
At B: Pressure as at A, but vapour superheated at evaporator outlet
At C: Liquid and vapour at the saturated pressure corresponding with temperature at B

Referring to Figure 10.8 if the temperature at B falls, the valve closes: if the temperature at B rises, the valve opens. For best results this type of valve should be placed as near to the evaporator as possible and the sensing bulb should be clamped firmly to a horizontal section of the suction line, preferable within the refrigerated space.

Low pressure float controls

The mechanisms are similar to most other float controls, and act to maintain a constant level of liquid refrigerant in the evaporator by relating the flow of refrigerant to the rate of evaporation. It is responsive only to liquid level which it will keep constant, irrespective of evaporator temperature or pressure.

This type of valve is usually provided with a bypass fitted with a manually operated valve, so that the system can be kept in operation in the event of a float valve failure or the float valve serviced without evacuating the system.

High pressure float valves

These valves are similar to low pressure valves in that they relate the flow of liquid into the evaporator to the rate of vaporisation. The low pressure valve controls the evaporator liquid level directly; the high pressure valve located on the high pressure side of the system, controls the evaporator liquid level indirectly, by maintaining a constant liquid level in the high pressure float chamber.

As vapour is always condensed in the condenser at the same rate as liquid is vaporised in the evaporator, the high pressure float valve will automatically allow liquid to flow to the evaporator at the same rate as it is being evaporated, irrespective of the load on the system.

Capillary tube control

This is the simplest of all refrigerant controls and consists of a length of small diameter tubing inserted in the liquid line between the condenser and the evaporator. For a given tube bore and length the resistance will be constant, so that the liquid flow through the tube will always be proportional to the pressure difference between the condensing and evaporating pressures of the system.

Although self-compensating to some extent, this type of control will only work at maximum efficiency under one set of operating conditions, and for this reason is principally employed on close coupled package systems using hermetic or semi-hermetic compressors.

PIPELINES AND AUXILIARIES

Refrigerant piping may be of iron, steel, copper and their alloys, but copper and brass should not be used in contact with Refrigerant 717.

The design of piping for refrigerating purposes differs a little from other shipboard systems in that the diameter of the piping used is determined principally by the permissible pressure drop and the cost of reducing this. However, any pressure drop in refrigerant suction lines demands increased power input per unit of refrigeration and decreases the capacity of the plant; therefore pressure drop in these lines should be kept to a minimum.

To ensure continuous oil return, horizontal lines are usually dimensioned to give a minimum gas velocity of 230 m/min and vertical risers to give 460 m/min. The pressure drop normally considered allowable is that equal to about 1°C change in saturated refrigerant temperature. This means a very small loss in low temperature systems as the pressure change at 244K for a one-degree saturation temperature change, is only one half of that consequent upon the same temperature change at 278 K.

Horizontal pipelines should be pitched downstream to induce free draining and where the compressor is 10 m above the evaporator level, U-traps should be provided in vertical risers.

Welding, or in the case of non-ferrous piping, soldering and brazing, are practically universal in pipe assembly, and except where piping is connected to removable components of the system, flanges are rarely used.

Oil separators

Oil separators are generally fitted in hot gas discharge lines and are either of the impingement or chiller types. The former are closed vessels fitted with a series of baffles or a knitted wire mesh through which the oil-laden vapour passes. The reduction in velocity of the vapour as it enters the larger area of the separator allows the oil particles, which have greater momentum, to impinge on the baffles, then draining by gravity to a float valve and to the compressor crankcase.

The chiller type is usually water cooled and is similar to a shell and tube condenser; water circulates through the tubes and hot vapour through the shell. The oil is separated from the vapour by precipitation on the cold tubes from which it drains into a sump, or is returned automatically to the compressor through a float valve. With this type of separator the water circulation must be controlled to prevent the refrigerant vapour being brought below its condensing temperature.

The chief disadvantage of separators placed in the hot gas discharge line is the possibility of liquid refrigerant passing from the separator to the compressor crankcase when the compressor is stopped. In order to minimise this, the separators should be placed in the warmest position available. It is good practice to drain the oil from the separator into a receiver containing a heating element, where the liquid refrigerant boils off to the compressor suction line and allows the oil to drain to the compressor crankcase through a float valve.

Liquid indicators

These can be either cylindrical or circular glasses installed in the liquid line, providing a means of ascertaining whether or no the system is fully charged with refrigerant. If undercharged, vapour bubbles will appear in the sight glass.

To be most effective indicators should be installed in the liquid line as close to the liquid receiver as possible. Some types incorporate a moisture indicator which, by changing colour indicates the relative moisture content of the liquid passing through it.

Driers

Where halocarbon refrigerants are used it is absolutely essential that driers are fitted in the refrigerant piping and most Classification Societies make this mandatory. The presence of even the smallest amount of water can have so disastrous an effect on plant performance that the greatest care should be used in the selection and installation of the driers. These are usually simple cylindrical vessels, the refrigerant entering at one end and leaving at the other. The vessel is filled with the drying agent, which may be activated alumina, silica gel, calcium oxide, or calcium chloride, the three first mentioned being the most acceptable and widely used.

If the drier is located in the liquid line it should be arranged so that the liquid enters at the bottom and leaves at the top. This is to ensure that there is uniform contact between the liquid refrigerant and the drying agent and that any entrained oil globules will be floated out without fouling the particles of the drying agent. If located in the suction line, the gas should enter at the top and leave at the bottom so that any oil can pass straight through and out.

In most large installations the driers have by passes so that they can be isolated without interfering with the running of the plant or the drying agent renewed or re-activated by the application of heat.

Pressure relief devices

These either take the form of simple spring loaded safety valves or bursting discs and generally, where the discharge is led to atmosphere, a combination of both is used, so arranged that the bursting disc has to be ruptured before the safety valve lifts. This is to ensure that there is no undetected leakage of refrigerant from a safety valve improperly seated.

Apart from safety valves fitted in such obvious places as between compressor discharge pipes and their shut-off valves, it is important that all refrigerant-holding vessels which can be completely filled with liquid refrigerant and then shut off and isolated, should be fitted with safety valves or bursting discs. Even so called 'low sides' will be subject to high pressure when the plant is shut down and the ship is in waters where the ambient temperature is high.

CHAMBER COOLING ARRANGEMENTS

Direct expansion grids

These provide the cheapest and the simplest means of cooling a refrigerated chamber. The pipe grids in which the refrigerant expands and vaporises are arranged so that they cover as much as possible of the roof and walls of the chamber; the greatest coverage being given to those surfaces which form external boundaries and the least to divisional bulkheads and decks. As the actual cooling of the cargo depends on movements of air by natural convection, this type of chamber cooling requires good, careful and ample dunnaging within the cargo stowage. This appreciably diminishes the amount of cargo that can be carried, and is not now favoured.

A further objection to this system is that multiple circuits of liquid refrigerant make it extremely difficult to control the flow of liquid, particularly when the movements of the vessel are lively.

Cold brine grids

These are similar to the foregoing except that brine, the cooling medium, is more easily regulated and controlled. The system is little used in modern installations except in refrigerated trawlers and domestic provision chambers.

Direct expansion batteries and air

This is perhaps the system most commonly used where the liquid refrigerant is vaporised in a multiple battery of grids, enclosed in a casing through which the air from the refrigerated chambers is circulated by forced or induced draught fans. Its great advantages are economy in space, weight and cost. There is also a growing tendency to use flooded direct expansion batteries, where the liquid refrigerant at suction pressure and temperature is pumped through the battery grids.

This system is almost universally adopted in all new cold stores; the rate of heat transfer between liquid refrigerant and air is much higher than that between refrigerant vapour and air. This more than compensates for the complexity and higher cost of flooded systems.

Brine battery and air

This system, in which brine, instead of primary refrigerant, is circulated through the batteries is still employed largely in fully refrigerated ships carrying such cargoes as chilled meat or bananas, where extremely close control of temperature is required. The flow of brine as a cooling medium is relatively easy to regulate with electric, electronic or pneumatic control valves. The brine commonly used is made with calcium chloride and fresh water to a specific gravity of about 1.25. Sodium dichromate or lime may be added to maintain the brine in an alkaline condition.

Recently systems have been designed in which brine is replaced by one of the Glycols, e.g. ethylene glycol. The glycols have the advantage of being non-corrosive, and may be used at much lower working temperatures than brine. Trichlorethylene is also used as a secondary refrigerant, but has the disadvantages of being toxic and a solvent of many of the synthetic rubbers and other materials normally used as jointing.

Air circulation systems

The design of an air circulating system is dictated principally by the allowable temperature spread in the cargo spaces and is not influenced by the type of air cooler in use: brine and direct expansion systems have similar air circuits.

In large ships, the frames and beams have generous dimensions and advantage has been taken of the otherwise waste spaces between them to accommodate the suction and delivery ducts. The greatest disadvantage of this is the difficulty of making these ducts completely air-tight since even the smallest leak has a direct effect on the efficiency of the plant. Cases have been reported where air leaks alone have accounted for loads as high as three times the theoretical calculated load.

For many years almost all large fruit carriers used side to side air circulation, and were fitted with deck to deckhead ducts that could be entered during the voyage to alter the delivery and suction apertures to suit the spread of temperature in and the ripeness of the fruit.

A very popular method of air circulation in smaller installations is one in which the air cooler and fan unit are mounted behind a floor-to-ceiling screen at one end of the chamber (Figure 10.9). Air is delivered at the lower end of the chamber under a false floor provided with openings and then passes through the cargo, returning to the cooler via the top of the cooler screen, which has a side to side grille. The delivery openings in the false floor are arranged with the largest in that part of the floor furthest away from the cooler, where the air pressure is at its lowest and the smallest nearest the cooler. With this system the cargo stowage is most important, as voids in the stow which would allow the delivery air to short circuit to the suction side of the cooler must not be allowed.

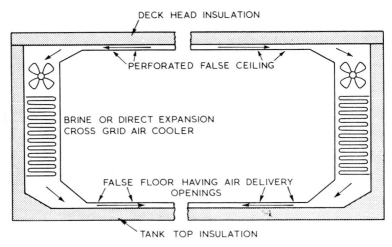

Figure 10.9 Typical system of vertical air circulation

Most modern installations employ a vertical air circulation; the air is delivered as described above and is returned to the cooler either by buried roof ducts or a false ceiling. In larger ships it is usual to have a cooler and fan on each side of the chamber discharging into tapered ducts running its full length. These ducts are connected to the false floor from which the air rises through the cargo to the false ceiling and so back to the cooler. This type of vertical air circulation allows the spread of temperature in the cargo to be at a minimum.

All the above systems are designed for the cooling of cargo holds or chambers, but before leaving this subject, some mention should be made of the refrigerated container ship. These ships are specifically designed to carry insulated containers, built to ISO dimensions. The containers are stowed in stacks between built-in guide rails and are connected to the suction and delivery air ducts of the ship's refrigeration plant by bellows pieces operated pneumatically. The air is cooled in either brine or direct expansion batteries and the containers are arranged so that one cooler can maintain several containers at a given temperature. A principal difference between container and conventional refrigerated ships is that the insulation of the latter is designed to retard the flow of heat into the cargo chambers; in the former the purpose of the insulation is to keep the decks and strength members of the ship's structure as near to ambient temperature as possible, so that the use of the costly steels suitable for low temperatures may be unnecessary.

Air cooler fans

Fans may be either centrifugal or propeller, but except in the cases of a few fruit carrying vessels, and of airconditioning plants, where high duct pressures are usual, the propeller type is most commonly used, the air circulation systems being based on a pressure requirement of about 50 mm W.G.

The capacity of the fans is determined by the number of air changes per hour required in the cargo chambers, and this in turn is influenced by the maximum calculated heat load. In a system using air coolers and fans, all the heat load must be carried away by the circulating air and the difference between delivery and suction air temperatures is directly proportional to the weight of air being circulated. Since the temperature difference is limited by the allowable temperature spread in the cargo chambers and the maximum temperature spread in the cargo chambers and the maximum load can be estimated, the selection of suitable fans is fairly easy.

In most installations the number of air changes per hour (based on empty chambers) varies between forty for 'dead' cargoes such as frozen meat and fish and eighty for fruit cargoes, such as bananas, which evolve heat freely.

As all the electrical energy input of the fan motors is dissipated in the form of heat and has to be removed by the refrigerating plant, it is essential that the fans designed for maximum load should be able to run at lower output when the heat load diminishes. There is no problem with d.c. motors; with a.c. either two-speed fan motors are used or each cooler has a number of fixed speed fans which can be switched off individually to suit the existing load: in the latter case provision must be made to blank off the stopped fans to prevent delivery air bypassing through them to the suction. Reversible fans can be extremely useful when trying to clear out 'hot spots' in the cargo and are commonly fitted.

Instruments

Perhaps the most important instruments in a refrigerated ship are those used to measure the temperatures in the cargo. In the past, these were mercury or spirit thermometers suspended in vertical perforated steel tubes passing through the decks into the cargo chambers, access to them being through screwed plugs on deck. Many ingenious devices were used to try to keep the temperature reading steady during the sometimes long and tedious operation of raising the thermometers for reading, the most useful being the immersion of the bulb of the thermometer in non-freezing oil contained in a heavy brass pot. Generally these have been superseded by either electric resistance or self balancing electronic thermometers or by data loggers, though direct reading and recording capillary tube instruments are sometimes still preferred.

Electrical resistance and electronic self balancing theremometers use the principle of the Wheatstone bridge, the former relying on a galvanometer to indicate a balance; in the latter the unbalanced current causes an electric motor to adjust the resistance.

Some authorities state that it is not essential to know the temperature of the cargo as it must lie between those of delivery and suction air; this is not strictly correct, as there have been many cases where chilled cargoes, stowed in chambers adjacent to frozen cargoes, have been adversely affected when stowed close to bulkheads separating the chambers. The temperatures are required also of refrigerant gas and liquid, cooling water, brine and the ambient. Most of these can be easily obtained from direct reading mercury or spirit thermometers but there is a growing tendency to incorporate them all on a data logger which, coupled to a typewriter gives an automatic record of all temperatures. Pressure gauges are required for refrigerant suction and discharge (combined pressure-temperature gauges are usual) compressor lubricating oil, brine and cooling water; the last named may be important if the condensed water speed has to be restricted. Ammeters provide valuable information on the functioning of compressors, pumps and fans.

Carbon dioxide measurement

The carbon dioxide concentration in the cargo chambers is of prime importance when, for instance, fruit or chilled beef is carried. Two common types of CO_2 indicator are in use, absorption and electric. In the former, a known volume of air-CO_2 mixture is passed through a caustic soda solution and the loss of CO_2 by absorption indicated on a scale, suitably calibrated. CO_2 is a better heat conductor than air and this characteristic is the basis of the electric CO_2 indicator. The sample of air-CO_2 mixture is passed over a wire carrying a constant heating current; the wire temperature falls progressively with increasing CO_2 content. This temperature change is detected on a Wheatstone Bridge circuit; a milliammeter, suitably calibrated, giving the CO_2 content (Figure 10.10).

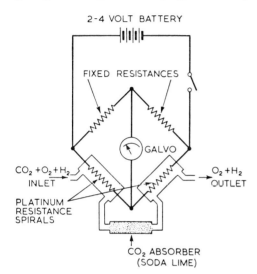

Figure 10.10 Wheatstone bridge circuit for CO_2 indicator

In addition to a halide torch leak indicator, it is usual to provide large installations using halocarbon refrigerants with a vacuum pump. This, used to create a very low pressure in the refrigerant system, brings about the evaporation and removal of any moisture present after opening out for maintenance or inspection.

Air conditioning

Except for the washing, filtering and heating elements, most of the components are very similar to those of cargo refrigerators but as the capacity required increases, centrifugal or positive displacement rotary compressors, using the so-called vacuum refrigerants, such as R11 and R21, tend to displace reciprocating types.

Duct-pressures between 200 and 250 mm W.G. may be required and centrifugal fans are used almost invariably, see Chapter 11.

Reliquifaction of natural gases

Here again the refrigerating machinery is no different to that used for cargo refrigeration, but because of the low temperatures required, a 'cascade' system is often installed employing two or more circuits using refrigerants with progressively lower boiling points. The evaporators of the higher stages act also as the condensers of the lower, the hot compressed gas from the lower stage compressors being condensed by the evaporating liquid from the higher. Oil contamination of some natural gases is wholly unacceptable and the oil-free screw compressor may be favoured.

In ships carrying these cargoes, the instrumentation is much more extensive and complex than in the conventional refrigerated ship, it having been found necessary in certain cases to provide for constant monitoring of the temperature of all the ship's steelwork in the vicinity of the cargo tanks and the gas leakage from the cargo into any other space. The use of refrigeration for this purpose is likely to increase with the exploitation of new sources of natural gas and the increasing transport of volatile products, where venting to atmosphere would affect adversely the physical characteristics of the liquids remaining in the tanks, by the loss of their more volatile constituents.

PLANT OPERATION

Because of the large number of refrigerants in common use, it is not proposed to specify here any specific operating temperatures and pressures for optimum performance. Under normal conditions, a fully charged plant will show a condenser gauge temperature between 6 and 10°C above the circulating water inlet, and an evaporator gauge temperature 4 to 6°C below that of the brine. In direct expansion installations the difference between chamber and evaporator temperatures may be slightly greater. It should be appreciated that condenser and evaporator gauges are actually pressure gauges; the temperatures shown are the saturation temperatures of the refrigerant in use.

Overcharged systems, or those where the system contains non-condensible gases, will show unduly high condensing pressures; undercharged systems will show lower than normal evaporator gauge readings. If high condenser and low evaporator pressures are indicated at the same time, the following test should be made for the presence of non-condensible gases. The unit suspected should be stopped and the expansion valve or regulator opened to allow the pressures in the high and low sides of the system to equalise, while the water circulation should continue. After a few minutes the temperature shown on the condenser gauge should be about that of the circulating water. A higher reading on the condenser gauge indicates that non-condensible gases are present and these should be purged to atmosphere by means of the purging valve, usually found on top of the condenser. Testing in this way, it has been found that if the condenser gauge temperature is 6° above that of the circulating water there can be 12° to 14°C difference under operating conditions.

The temperature of the liquid refrigerant leaving the condenser is a good indication of plant efficiency and in a well-designed clean condenser with an adequate supply of cooling water, the liquid temperature should be not more than 1–2°C above the sea.

In order to prevent fouling and corrosion in condensers, it is good practice to circulate the condenser from clean ballast tanks when the ship is in badly polluted waters. It goes without saying that if the plant is to be stopped for long periods in freezing weather, the condensers should be drained. A number of textbooks state that frosting of the compressor suction pipe in the vicinity of the compressor cylinders indicates satisfactory working, but where Refrigerant 12 is used, a high degree of superheat is desirable and best results are obtained when the frost line ends on the suction pipe some distance from the compressor. With this refrigerant, superheating increases the heat content of the gas more rapidly than it decreases the specific volume; therefore capacity actually increases progressively with superheat. In small compressors superheating also gives an improvement in volumetric efficiency, an added advantage.

Defrosting

This very necessary operation presents no difficulty when the cooling medium is brine. All that is required is a brine heater and the requisite pump and circuits to circulate hot brine in the coolers (Figure 10.11).

In direct expansion systems, defrosting has to be effected either by separate electric heaters (in some coolers, heaters are incorporated in the cooler grids) or by providing means of bypassing the condenser, so that hot gas from the compressor goes directly to the coolers, where it condenses, giving up its latent and sensible heat to rid the cooler piping of ice. To prevent the condensed refrigerant returning as liquid to the compressor, it is necessary to re-evaporate it, either in another air cooler or a specially designed re-evaporator.

Although a large build-up of frost and snow on the cooler is undesirable and will reduce its efficiency, a slight film of frost can be beneficial, increasing as it does, the heat transfer coefficient of the cooler pipe surfaces.

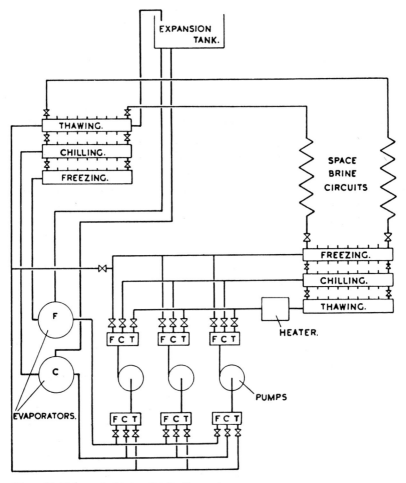

Figure 10.11 Layout of brine distribution system

HEAT LEAKAGE AND INSULATION

This chapter on refrigeration would not be complete without some mention of how the load is made up, and the steps taken to keep it at the lowest practicable level. The total load on a cargo refrigerating plant is the sum of:

1. Surface heat leakage from sea and ambient air.
2. Deck and bulkhead edge leakage from the same sources.
3. Heat leakage into brine and refrigerant piping from their surroundings.

4. Heat equivalent of fan and brine pump power.
5. Cooling of cargo not precooled at loading. Respiratory heat of live cargoes.
6. Heat introduced by air refreshment generally necessary with live cargoes.

The load arising from 1, 2 and 3 can be much reduced by the efficient use of insulation. A number of materials are used for this, slab and granulated cork, glass and mineral wools, expanded plastics, aluminium foil, and polyurethane. The latter, although generally the most costly, is the best insulator, having the lowest coefficient of conductivity, with the further advantages of being completely impervious to air leaks and almost impervious to the passage of water vapour, when the material is foamed in situ.

As all the materials mentioned have to be enclosed by linings, the prevention of air leakage is most important and the design and construction of the linings may make a greater contribution to the efficiency of an installation than the selection of the insulant.

Insulation

The insulation of a ship's structure is necessarily a complicated matter, as all structural material connected to the external plating must be enclosed in order to reduce the flow of heat into the refrigerated spaces from external sources. For many years, the design, capacity and the quality of insulation in an installation were judged by its performance in what was known as the temperature rise test. In this, the spaces were cooled to a specified temperature, the machinery stopped, the space temperature allowed to rise and the rate of temperature rise measured. This was not a useful exercise; in installations having excess capacity, the space temperatures were reduced so rapidly that there was insufficient time for the insulation and the structural members to cool correspondingly so that when the machinery was stopped the space temperatures rose rapidly. On the other hand in an installation having limited capacity or poor machinery, the cooling period was long, the cooling more complete and the rate of temperature rise correspondingly low. In 1947 the unsatisfactory nature of this test was recognised and the heat balance test, which is now a requirement of the major Classification Societies, was introduced.

In this test the temperatures of the refrigerated spaces are first reduced to those specified and then, after a lapse of time sufficient to remove all the residual heat from the insulation and structure, the spaces are maintained at constant temperature for at least six hours by varying the compressor output. During this period all temperatures, pressures, speeds and electrical consumption of compressors, fans and pumps are carefully logged and the compressors output noted from appropriate tables.

From this information it is possible to compare the efficiency of the insulation with the theoretical estimate made during the design stage and also to decide whether or no the installation can maintain these temperatures in maximum tropical sea and ambient conditions. Obtaining the theoretical estimate entails taking each external surface of each individual chamber separately and considering all factors affecting the heat leakage. These factors include the pitch, depth, and width of face of all beams, frames, stiffeners etc., buried in the insulation, the type of grounds securing the linings; the presence of which have their effect in reducing the effective depth of the insulation.

Hatches, access doors, bilge limbers, air and sounding pipes also have their effect on heat leakage and must come into consideration. It should be noted that in these calculations the laboratory value of the insulation is generally increased by about 25% to allow for deficiencies in fitting.

It has been found that the overall co-efficient of hat leakage in well insulated installations can vary between 0.454 $W/m^2/°C$ for 'tweendecks in small lightly framed ships and 0.920 $W/m^2/°C$ for fully refrigerated moderate-sized ships having deep frames with reverse angles. Where there are also buried air ducts, the effective depth of the insulation may reduce to little more than zero.

11 Air conditioning, ventilating and heating

Air conditioning is vital to the comfort of those on board ship. Comfortable conditions depend mainly upon the temperature and the humidity but are also sensitive to air movement and air freshness and purity. The application of air conditioning to ships has become almost universal, because of the increasing proportion of vessels, particularly tankers, journeying through tropical waters. Even in temperate zones, the heat generated within the ship often requires the provision of air cooling.

AIR CONDITIONING

A very significant factor affecting an air conditioning system is the rapidly changing climatic conditions. The equipment has to perform within these variations and has to meet the differing requirements of the occupied spaces of the ship.

The essentials of air conditioning systems have been known for many years but were not applied to ships mainly because the standards of comfort were not sufficiently high to justify the installation and running costs. In most cases, relief from excessive heat was obtained by providing high velocity directional air outlets in the spaces. For some years early air conditioning systems were rather bulky because designs were based on low air velocities throughout, particularly in the distributing ducts, in which the velocities were of the order of 10 m/sec or less. In later years there was extensive standardisation, with very substantial increases in air velocities, reaching a maximum of about 22.5 m/sec in the ducts, producing a large reduction in the space occupied by the equipment. Increased operating costs as a result of higher velocities have to be set against reduced installation costs and the value of space saving but the owner is usually disposed favourably towards the high velocity system with its lower initial cost.

Basic standards

The designer and user of air conditioning plant must study the physiological factors involved. The terms used to define the atmospheric conditions are fairly well known, but are reviewed here since it is essential to know exactly what they mean before proceeding further:

Dry bulb temperature (d.b.). The temperature as measured by an ordinary thermometer which is not affected by radiation.

Wet bulb temperature (d.b.). When moisture evaporates from a surface, i.e. the skin, the large amount of heat required is drawn from the surroundings, causing them to be cooled. If a thermometer bulb is covered by a wetted fabric and exposed to the air, the rate of evaporation will depend upon the dryness of the air. As the heat required must come from the bulb, this results in a lower temperature reading than if the bulb was dry. The difference between wet bulb and dry bulb readings at the same location in the air is therefore a measure of the humidity of the atmosphere.

It is important to note that the wet bulb temperature is a measure of the heat contained in the air (or its enthalpy), so it is an essential factor in all cooling load calculations.

Relative humidity (r.h.). The capacity of the atmosphere to hold water vapour is dependent upon its temperature; at higher temperatures this is much greater than at the lower temperatures. When the maximum is reached at a given temperature, the air is said to be saturated (100% relative humidity).

Dewpoint temperature (d.p.). If an unsaturated mixture of air and water vapour be cooled at constant pressure, the temperature at which condensation of water vapour begins is known as the dewpoint temperature. As the air is further cooled, more moisture is deposited.

Temperature conditions

Apart from air conditions, temperature, humidity, movement and radiation, affecting the body heat balance, other factors have an influence. An obvious one is the amount of clothing being worn, while acclimatisation can have an effect. The degree of exertion is not usually a significant factor in cabins and public spaces, but can have a vital influence, e.g. a crew member carrying out some heavy duty.

The ideal conditions for comfort vary considerably between one person and another, so it is only possible to stipulate a fairly wide zone. In this connection, it would obviously be of great value if a single index could be used to define the physiological reaction to

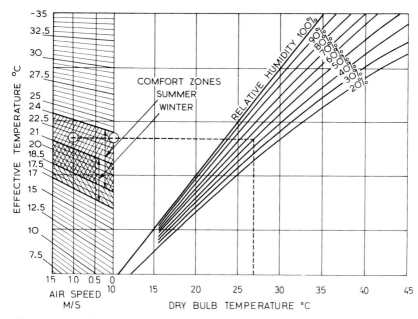

Figure 11.1 Comfort chart compiled from tests carried out by the American Society of Heating and Air Conditioning Engineers.

the various combinations of factors involved. Among other suggestions the most satisfactory has been the effective temperature index. It is the temperature of still, saturated air which would produce the same feeling of warmth.

The American Society of Heating and Air Conditioning Engineers carried out a comprehensive series of tests on a large number of people, from which they were able to draw up an effective temperature chart. The comfort chart, Figure 11.1, is based on the results. Taking as an example 27°C d.b. and 50% r.h., for zero air speed, the effective temperature is found to be 23.4°C, which is within the summer comfort zone. With an air speed of 1.0 m/sec, the effective temperature is reduced to 22.2°C.

The higher limit of comfort is more critical than the lower, since it is closely associated with the essential process of getting rid of body heat. Also it is of considerable significance to the air conditioning engineer, who has to fix the capacity of the refrigerating plant so that it can provide conditions within this limit when the outside conditions are at their most onerous. Protracted investigations have been made by Hall-Thermotank Ltd. to determine the effective temperature which could be said to represent the upper threshold of the comfort zone. The reactions of a large number of persons, mostly mariners, some of whom had just completed a voyage, were analysed. For a person dressed in tropical clothing and at rest, the threshold value was found to be about $25.6°C$ effective temperature.

The tests showed that most persons tended to sweat when the temperature rose a degree or so above this value, and to cease sweating at the same value as the temperature fell again. The conclusion was therefore reached that there could be a relationship between the threshold of comfort and the onset of sweating. Any difference in the thermal sensations of men and women can normally be explained by the different clothing worn, and the comfortable level of warmth for acclimatised persons of all races is very similar, in spite of certain differences in the reaction to heat stress.

At room temperatures above $21°C$ an air velocity of $0.15-0.20$ m/sec is desirable, to avoid any feeling of stuffiness, and to provide proof that the space is being ventilated. On the other hand, velocities higher than 0.35 m/sec are usually classed as draughts, to be avoided particularly when the person is at rest. When a space is heated or cooled, it is impossible to ensure an absolutely uniform distribution of the effect throughout the space. Because warm air rises, the air can be appreciably warmer at the higher levels than at the deck, giving rise to discomfort, unless the air terminals are designed to counteract this effect.

Air purification

Radiation, planned or otherwise, can contribute to the effective temperature of a space, but in ships' accommodation it is not significant, even in cabins adjacent to engine room bulkheads, given high standards of insulation.

Outside air must be introduced to all living spaces, although the amount of fresh air necessary to sustain life is very small indeed. Space conditions can vary greatly in a short time; they are governed by sources of contamination, e.g., body odours and smoking, which may require a fresh air supply of 12 litre/sec per person or more. The actual quantity of fresh air supplied is usually governed by the provision that it must be not less than the total capacity of accommodation exhaust fans (excluding galley). On this basis the minimum D.O.T. requirement of 7 litre/sec per person is usually exceeded.

It is desirable that outside air should be cleaned before being introduced to the spaces, but this is less essential than on land, except when in port. Of greater influence are the impurities such as lint carried in the recirculating air, which must be filtered out to prevent eventual choking of the heating and cooling elements in the conditioning plant.

It can be shown that a large proportion of the total cooling load

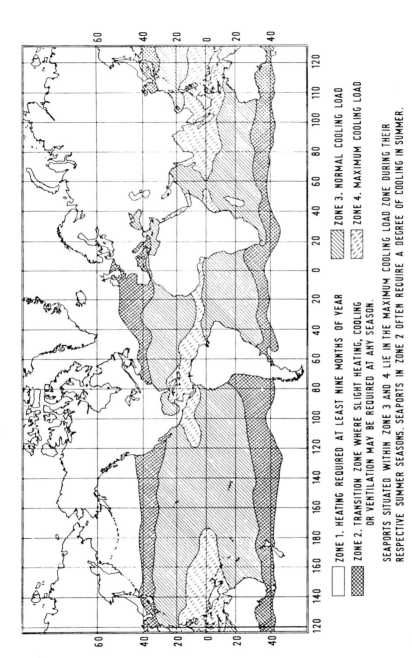

Figure 11.2 Phases of air conditioning required during voyages.

ZONE 1. HEATING REQUIRED AT LEAST NINE MONTHS OF YEAR

ZONE 2. TRANSITION ZONE WHERE SLIGHT HEATING, COOLING OR VENTILATION MAY BE REQUIRED AT ANY SEASON.

ZONE 3. NORMAL COOLING LOAD

ZONE 4. MAXIMUM COOLING LOAD

SEAPORTS SITUATED WITHIN ZONE 3 AND 4 LIE IN THE MAXIMUM COOLING LOAD ZONE DURING THEIR RESPECTIVE SUMMER SEASONS. SEAPORTS IN ZONE 2 OFTEN REQUIRE A DEGREE OF COOLING IN SUMMER.

is required to reduce the temperature and moisture content of the incoming fresh air to the space conditions, and as the wet bulb temperature is a measure of the total heat of this air, the outside wet bulb temperature can be regarded as of the greatest significance in the specification of outside conditions for design. Surveys have determined the maximum wet bulb temperatures over the main trade routes of the world. If peak temperatures of short duration are excluded, it appears that the highest wet bulb recording is 30.5°C, which occurs in the Persian Gulf. A compromise figure of 29.0°C wet bulb has been accepted internationally for many years for oil tankers; for other classes of ship, not normally trading in the Persian Gulf, a lower figure might be taken.

The figure commonly taken for outside dry bulb temperature is 32.2°C; much higher temperatures, of the order of 50°C and more, may be experienced in tropical zones with offshore winds, but these winds are invariably very dry, and do not impose as high a load on the cooling plant. Figure 11.2 has been prepared from data and from operating experience with ships in service, to indicate the phases of air conditioning normally called for during voyages.

It is usual to express the inside conditions in terms of a depression in dry bulb temperature relative to the outside temperature, coupled with a value for the relative humidity. A differential of about 3.3°C is commonly chosen for the most humid outside conditions. Although people are not sensitive to the degree of humidity over quite a wide range, it is usual to design for between 40% and 60% r.h. For 32.2°C dry bulb outside temperature, the inside conditions could therefore be 29.0°C dry bulb, 50% r.h.

The inside design conditions have a very important bearing on the cooling plant power. Another very significant factor is the degree of insulation of the surfaces bounding the airconditioned spaces. At one extreme, the only insulation may be provided by e.g., the decorative lining of the bulkheads and by the treatment of the engine room bulkhead, while some insulating value would lie in the deck coverings.

At the other extreme, some shipowners specify generous insulation enclosing all sections of the accommodation. A reasonable mean standard would be obtained by assuming the equivalent of 25 mm of high class insulating material, having an insulating value of 1.5 W/m^2/deg. C, over all surfaces normally treated.

EVALUATION OF HEATING AND COOLING LOADS AND AIR QUANTITIES

The heating load comprises the transmission losses through the structure, calculated with the aid of the requisite coefficients for the various materials involved, together with the heat required to raise the outside air temperature to the space temperature. The latter is evaluated from the formula:

$$H = 1.21 \; Q(t_i - t_o)$$

where:

H = heat required, kW,
Q = airflow, m^3/s,
t_i = inside temperature, °C,
t_o = outside temperature, °C,

and the density of the air is 1.2 kg/m^3 at 20°C.

The outside temperature chosen is not the extreme minimum for the trading routes of the vessel, usually of short duration; the actual value chosen can range from -20°C to 0°C, and the inside condition from 18°C to 24°C depending upon the type of accommodation.

The cooling load has a greater influence on the design of the equipment since it not only determines the size of the refrigerating plant, but fixes the quantity of air to be circulated. The following sensible heat gains must be balanced to maintain the required inside temperature, when cooling is in operation:

1. *Heat transmission through the structure.* This is dependent on the physical properties of the materials surrounding the conditioned spaces and the dry bulb depression maintained inside. Allowance has also to be made for the effect of sun heat on exposed surfaces. This is very difficult to define with any accuracy, and is usually computed with the aid of tables and charts based on experience.
2. *Body heat.* Account must be taken of the heat gain in the space due to the occupants.
3. *Lighting heat.* This can be a significant factor on board ship, where lighting is in use almost continuously.
4. *Fan heat.* The energy applied to the air is converted to heat in the passage of the air through the system.

The air delivered conveys the cooling effect to the spaces. This air must be delivered at a temperature below that desired in the accommodation fixed by the required dew point of the air. The air passing through the cooling coils becomes saturated on cooling and gives up moisture as its temperature falls; the temperature at which it leaves the cooler is a measure of its moisture content, which remains practically unchanged when the air enters the accommodation. Once inside the accommodation, the temperature of the air rises due to the heat gains already discussed. Any specified combination of dry bulb temperature and relative humidity has a specific dew point and this is the temperature at which the air is delivered to the space.

The quantity of air must be so arranged that the temperature rises to the specified inside conditions. The quantity is given by the formula:

$$Q = \frac{H}{1.21 \, (t_i - t_e)}$$

where:

    Q = total volume of air circulated, m^3/s,
    H = total heat gain in the spaces, kW,
    t_i = inside temperature, $°C$,
    t_e = temperature of entering air, $°C$.

It invariably happens that this quantity is considerably greater than the fresh air requirements discussed previously, so that the balance is recirculated in order to economise in cooling load. In practice, usually about two-thirds of the air delivered to the space is recirculated.

TYPES OF AIR CONDITIONING SYSTEM

Air conditioning systems may be divided into two main classes; the central unit type in which the air is distributed to a group of spaces through ducting, and the self contained type, installed in the space it is to serve.

The central unit type is the most widely applied, in one or other of a number of alternative systems characterised by the means provided to meet the varying requirements of each of the spaces being conditioned. The systems in general use are as follows:

    1. Zone control system.
    2. Double duct system.
    3. Reheat system.

Zone control system

This is the most popular because of its basic simplicity. The accommodation is divided into zones, having different heating requirements.

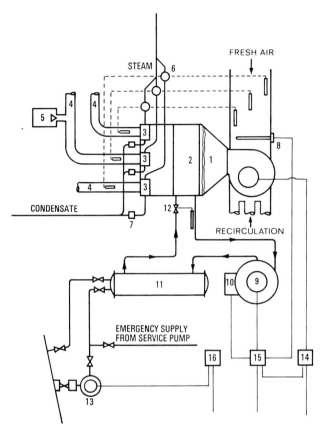

Figure 11.3 Zone control system

1. Filter
2. Cooler
3. One, two or three-zone heaters as required
4. Pre-insulated pipes delivering air to zones
5. Sound attenuating air terminal, with volume control
6. Automatic steam valves. One per zone heater
7. Steam trap. One per zone heater

8. Multi-step cooling thermostat
9. Compressor
10. Automatic capacity control valves
11. Condenser
12. Thermostatic expansion valve
13. Sea water pump
14. Fan starter
15. Compressor starter
16. Sea water pump starter

Separate air heaters for each zone are provided at the central unit, as shown in Figure 11.3.

The main problem is to obtain a typical sample of air for thermo-static control of the heaters, for it may not be possible to choose a location which is uninfluenced by local factors. This has led to the general adoption of a compromise solution, which is to vary the temperature of the air leaving the heater in accordance with the outside temperature prevailing. This can be effectively performed by a self-actuating regulator, controlled by two thermostat sensors, one in the air leaving the heater, the other outside. Air quantity control in each room served gives individual refinement. In summer, air temperature is controlled by a multi-step thermostat in the recirculating air stream, which governs the automatic capacity control of the refrigerating plant.

The regulation of temperature by individual air quantity control in this system can give rise to difficulties unless special arrangements are made. For instance, a concerted move to reduce the air volume in a number of cabins would cause increased air pressure in the ducts, with a consequent increase in air flow and possibly in noise level at other outlets. This can be avoided but economic factors usually place a limit on this. Some degree of control is possible through maintaining a constant pressure at the central unit, but since most of the variation in pressure drop takes place in the ducts, the effect is very limited. A pressure-sensing device some way along each branch duct, controlling a valve at the entry to the branch, strikes a reasonable mean, and is fairly widely applied.

Double duct system

In this system, two separate ducts are run from the central unit to each of the air terminals, as shown in Figure 11.4. In winter, two warm air streams, of differing temperatures, are carried to the air terminals, for individual mixing. The temperatures of both air streams are automatically controlled. In summer, the air temperature leaving the cooler is controlled by a multi-step thermostat in the recirculating air stream, which governs the automatic capacity control of the refrigerating plant, as with zone control. Steam is supplied to one of the heaters, so that two air streams are available at the air terminals for individual mixing.

Reheat system

In winter, the air is preheated at the central unit, its temperature being automatically controlled. The air terminals are equipped with electric or hot water heating elements, as shown in Figure 11.5. These raise the temperature of the air to meet the demand of the individually set room thermostats.

In the case of electric reheat, fire protection is provided by over-heat thermostats which shut down the heaters in the event of air

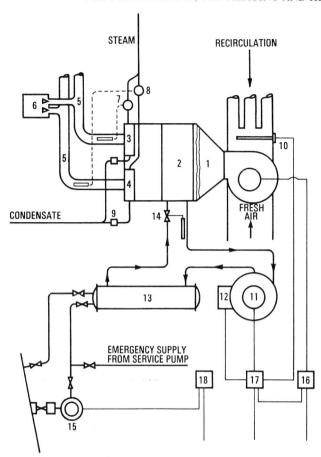

STEAM

RECIRCULATION

CONDENSATE

FRESH
AIR

EMERGENCY SUPPLY
FROM SERVICE PUMP

Figure 11.4 Double duct system

1. *Filter*
2. *Cooler*
3. *Low-duty heater*
4. *High-duty heater*
5. *Pre-insulated air pipes*
6. *Sound attenuating air terminal with volume and temperature control*
7. *Automatic steam valve for tempered air stream*
8. *Automatic steam valve for warm air system*
9. *Steam traps*
10. *Multi-step cooling thermostat*
11. *Compressor*
12. *Auto-capacity control valves*
13. *Condenser*
14. *Thermostatic expansion valve*
15. *Sea water pump*
16. *Fan starter*
17. *Compressor starter*
18. *Sea water pump starter*

starvation, while a fan failure automatically cuts off the power supply. In summer, the air temperature is controlled by a multi-step thermostat in the recirculating air stream, which governs the automatic capacity control of the refrigerating plant, as in the other system.

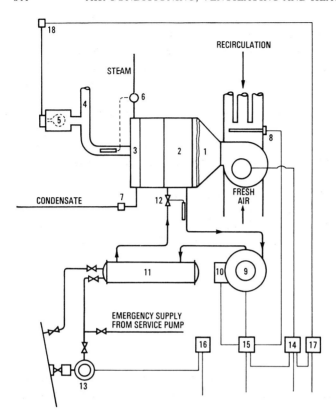

Figure 11.5 Electric reheat system

1. *Filter*
2. *Cooler*
3. *Pre-heater*
4. *Pre-insulated air pipe*
5. *Sound attenuating air terminal containing electric re-heater and overheat thermostat*
6. *Automatic steam valve*
7. *Steam trap*
8. *Multi-step cooling thermostat*

9. *Compressor*
10. *Auto capacity control valves*
11. *Condenser*
12. *Thermostatic expansion valve*
13. *Sea water pump*
14. *Fan starter*
15. *Compressor starter*
16. *Sea water pump starter*
17. *Heater contractor*
18. *Room type thermostat*

Self-contained air conditioners

In the early days of air conditioning, there was a demand for self-contained units to serve, e.g. hospitals, public rooms, etc., where the advantages of air conditioning were very obvious. At first these units were rather cumbersome, but with the advent of hermetically sealed components and other developments associated with land applications, the modern cooling unit can usually be accommodated within the space it is to serve. By taking full advantage of the available height, the deck space required is relatively small.

Self-contained units may be used instead of a central unit system in new or existing ships where space is not available for the latter. The term 'self-contained' is only relative, since fresh air and cooling water are required and provision made for removing condensate.

The S-type Thermo-Unit is widely used on board ship. It combines a compact arrangement of the elements with the accessibility which is essential for marine use. To facilitate installation, the unit is divided into an upper and a lower section which can be taken apart readily.

The self-contained unit is ideally suited to the engine control rooms of automated ships. With the additional heat load coming from the equipment housed within the room, cooling may be required when the accommodation is calling for heat from the central unit system.

Conversion units

For a ship fitted originally with mechanical ventilation only, a good case may be made out for the provision of full air conditioning, if the ship has a reasonable span of life ahead of it. This can be done by mounting a conversion unit on the deck, embracing the essential features of a central cooling plant. The unit is so designed that it can be coupled to the existing fan, heater and air distribution system.

The central unit

The elements of a central unit are fan, filter, cooler, heaters, and plenum chamber. Normally these are all housed within a single casing, with the possible exception of the fan. It is possible to carry this further by including the refrigerating plant in a single assembly, thus providing a complete package. Apart from the obvious saving in space and economy in pipework, the possibility of refrigerant leakage is minimised by having the circuit sealed in the factory. Figure 11.6 shows a central unit of this type.

The filter, which is essential to keep the heat transfer elements clean, is usually formed of a terylene fibre mat, easily removable for periodic cleaning. The cooler is of the fin tube type, as are the heaters, usually steam. The air passes from the heaters into a plenum chamber, and from there into the pipes or ducts leading to the various spaces. The plenum chamber, acoustically lined, acts as a very

Figure 11.6 Central unit comprising fan, filter, cooler, heaters and plenum chamber (Hall Thermotank International Ltd)

effective silencer for the fan noise otherwise transmitted along the ducts.

AIR DISTRIBUTION

Friction and eddy losses in the ducts make up the greater part of the pressure required at the fan, hence the design of the duct system affects the fan power very considerably. The fan power is a function of the air quantity and the pressure, and is expressed as follows:

$$\text{Fan power (kW)} = \frac{\text{volume (m}^3/\text{s)} \times \text{pressure (mbar)}}{10 \times \text{efficiency}}$$

The efficiency is static or total, depending on whether the pressure is static or total. The total pressure is the algebraic sum of the static and velocity pressures.

The system is sized for the longest duct branch, so that artificial resistances must be inserted in other branches to balance the air distribution. In designing the system, account is taken of static pressure regain to reduce the rate of fall along the ducts. This regain results from a reduction in velocity when the volume of air in the duct is reduced after an outlet is passed, and can amount to about 75% of the fall in velocity pressure.

High velocity distribution

Over the years, the most significant development has been the introduction of high velocity air distribution, made possible by the reduced air quantity required. In other words, the inevitable increase in fan power associated with higher velocities (and hence higher pressures) has been kept within reasonable limits by a reduction in air quantity.

High velocity distribution has a number of clear advantages, among these being:

1. Ducting costs much reduced.
2. Standardising on a few diameters of round ducting up to about 175 mm instead of a great variety of widths and depths of rectangular ducting.
3. Standardised bends and fittings, having improved aerodynamic efficiencies.
4. Use of automatic machines for fabrication of ducts, with a spiral joint giving great stiffness.
5. Greatly reduced erection costs resulting from light weight and small bulk of ductwork.
6. Considerable space saving in the ship.
7. Possible reduced fire risk with smaller duct sections.

Against all these advantages must be set the increase in fan power already referred to. Thus an older system with duct velocities of the order of 8 m/s might require a fan pressure of no more than 50 mm w.g. whereas with high velocities around 22.5 m/sec the fan pressure could exceed 230 mm. The ratio of fan power increase is not so great as this, however, since high efficiency centrifugal fans of the backward-bent blade type are suited to high pressure operation, but for low pressures the relatively inefficient forward-bent blade type of fan must be used, since the high efficiency type fans would be far too bulky if designed for low pressures.

With the increase in friction loss due to high velocity, the reheating of the air can result in an appreciable increase in the cooling load, when compared with a low velocity system, and this could be a limiting factor in the choice of the duct velocity.

The design of the air terminals is very important with high velocity distribution, in order to minimise noise and prevent draughts.

Duct insulation

Duct insulation is standard practice, being particularly necessary in installations where the policy has been to reduce the volume of air handled to a minimum, resulting in greater temperature differentials. The ideal is to integrate the insulation with the duct manufacture, or at least to apply it before the ductwork is despatched to the ship.

There are a number of high class fire resistance insulating materials on the market, such as mineral wool and fibreglass. These, of course, must have a suitable covering to resist the entry of moisture and protect the material from damage. Jointing of the duct sections is usually by sleeves, with external adhesive binding.

Air terminals

The best designed air conditioning system is only as good as the means of delivering the air to the spaces. The main function of the air terminal is to distribute the air uniformly throughout the spaces without draughts. It is not possible to provide ideal conditions for both heating and cooling from the same outlet. Too low a discharge velocity in the heating season can result in stratification, the air at ceiling level remaining warmer than the air at the floor. Even when cooling, a low velocity stream could fall through to lower levels in localised streams, without upsetting the stratification.

Careful selection of the discharge velocity and direction of flow in the design stages can provide an acceptable compromise between good distribution and draught-free conditions. Generally, it is found that the ceiling is the most convenient location for the air terminal, although in large public spaces extended slot-type outlets on the bulkhead, with near-horizontal discharge are satisfactory, and blend well with the decorative features. The usual recirculation outlet at the bottom of the door normally ensures a good distribution of the air in the space.

With high air velocities, some control of the noise level in the system becomes essential, and it is true to say that equipment design, particularly as applied to the terminals, has been influenced more by this than by any other factor.

Figure 11.7 shows a typical layout of ducts and terminals for the accommodation in a tanker or bulk-carrier.

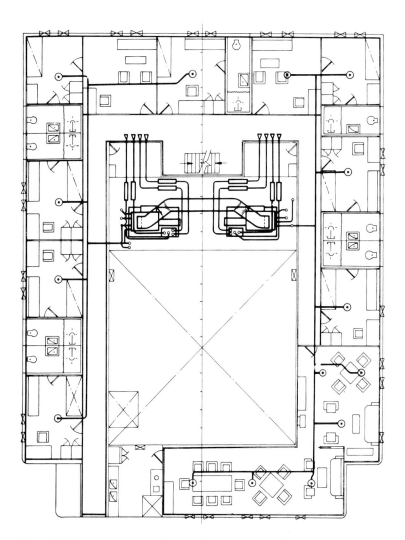

Figure 11.7 Layout of air conditioning ducts and terminals in a tanker or bulk carrier

Ventilation of boiler and engine rooms

Due to the large amount of heat picked up by the air in these spaces, it would be impracticable to maintain ambient conditions within the comfort zone by air conditioning or any other means. The normal practice is to provide copious mechanical ventilation; in boiler rooms, the quantity is equated to the combustion requirements, while in a motor ship engine room the supply may be 25—50% in excess of the requirements of the engines.

The axial flow fan is particularly suited to handle these large air volumes at the moderate pressures required, while of course the 'straight-through' flow feature places it at an advantage over the centrifugal fan.

The increasing adoption of automation, with the provision of a separate control room makes less significant the fact that comfort conditions cannot be maintained in the engine room all the time.

TYPICAL SPECIFICATION FOR AIR CONDITIONING INSTALLATION IN A TANKER OR BULK CARRIER

Installation serving the accommodation aft

The officers', engineers' and crew accommodation is served by two air conditioning units, with a direct expansion refrigerating plant of the Freon 22 type, capable of maintaining an inside condition of 26.7°C d.b. and 20.0°C w.b. (55% relative humidity) when the outside condition is 32.2°C d.b. and 28.9°C w.b. (78% relative humidity). Finned tube type steam heating coils to maintain 21°C in the spaces when the outside temperature is minus 20.5°C. The schedule in Table 11.3 indicates the rates of air changes to be provided.

A proportion of the air would be recirculated, except from the hospital, wheelhouse, etc. The recirculated air is to be withdrawn through wire mesh grids mounted on mild steel ducting at ceiling level, located in the passageways adjoining the treated spaces.

With heating in operation, automatic temperature regulating valves in the steam supply lines give independent temperature control to the following spaces

Officers:
  Navigating bridge deck.
  Bridge deck.
  Boat deck.
  Part poop deck.

Crew:
  Part poop deck.
  Upper deck.

The refrigerating plant comprises a compressor driven by a marine type motor with automatic starter, shell and tube type condenser, evaporator/air cooling coils within the air conditioning unit casings, gas connecting piping and fittings, safeguards, automatic cylinder unloading gear for cooling capacity control, and first charge of refrigerant. Mechanical and electrical spare gear is included. Air conditioning units are installed on boat deck aft, with refrigerating machinery remotely situated in engine room at middle flat level port side, approximately 7 m above ship's keel.

A sea water circulating pump to serve the condenser is provided. The shipbuilder would supply and fit sea water piping, valves, etc. between ship's side and pump, from pump to condenser and from condenser overboard.

Air delivery to the spaces is by means of distributors, with sound attentuating chambers and air volume regulators, mounted on mild steel ducting at high level.

The cargo control room is served by a branch duct delivering 0.18 m^3/sec through the distributors. Cooling at selected spots in the galley is provided by the conditioning units.

Non-return valves are fitted in ducts serving hospital and laundries. This prevents odours reaching accommodation should the fan unit be stopped for any reason.

Mechanical supply ventilation. Gyro and electronics room and motor generator room on bridge deck, switchboard room and telephone exchange on boat deck, motor room, storerooms and bedding store on upper deck, galley on poop deck are ventilated at atmospheric temperature by two axial flow supply fans.

Air is delivered through diffusing type punkah louvres and domed diffusers fitted on mild steel ducts at ceiling level.

Mechanical exhaust ventilation. Galley, pantry, laundries, drying rooms, oilskin lockers, overall lockers, gyro and electronics room, motor gear room, switchboard room, telephone exchange and motor room and all private and communal toilets, washplaces, bathrooms, and w.cs. are ventilated by five axial flow exhaust fans.

Hospitals, w.c., medical locker and bathroom must be indpendently ventilated by an axial flow exhaust fan.

Vitiated air is withdrawn through domed extractors and regulating type grid openings mounted on mild steel ducting at ceiling level. Canopies over main galley range, etc., should be supplied and fitted by Shipbuilder.

Air filtering equipment. Filter screens of washable nylon fibrous material are supplied with air conditioning and supply fan units. Spare screens would be provided.

Technical data

No. of air conditioning units	2
No. of axial flow supply fans	2
No. of axial flow exhaust fans	6
Total fan motor power of above	27.5 kW
Approx. steam consumption for heating and humidification purposes (with 50% recirculation)	0.13 kg/s @ 4.15 bar gauge
No. of direct expansion refrigerating plants	1
Capacity of plant	208 kW
Compressor power	63.3 kW
No. of sea water circulating pumps	1
Capacity of pumps	16.4 litre/s at 1.72 bar gauge
Power of above pump	7.25 kW
Approx. weight of installation	12 000 kg

General. All natural ventilation, together with shut-off valves as required, operating in conjunction with the mechanical system and serving spaces not mechanically ventilated, should be supplied and fitted by shipbuilder.

The shipbuilder provides the necessary air inlet and outlet jalousies to the unit compartment.

To prevent an excessive leak of conditioned air, all doors leading from the conditioned spaces to the outside atmosphere, machinery casing, etc. should be of the selfclosing type and reasonably airtight.

Cooling load calculations have been based on the understanding that there would be no awnings and that all accommodation would have airtight panelling on deckheads and house sides, and ship's surfaces would be insulated by shipbuilder as follows:

Approximate amount of insulation to be supplied and fitted around ducting by shipbuilder:

In accommodation	300 m² of 25 mm thick glass fibre slabs with vapour seal (or equivalent).
Exposed deckheads	68.5 mm thick wood deck or equivalent thickness of insulation.

Exposed sides	25 mm thick insulation suitably finished fitted on back of panelling or on steel-work around beams and stiffeners.
Galley surfaces adjacent to conditioned spaces	Suitable insulation applied.
Surfaces of machinery spaces adjacent to conditioned spaces	50 mm thick insulation suitably finished.

All unit rooms in proximity to accommodation should be suitably sound-insulated.

The above proposal is exclusive of any ventilation whatsoever to the forward pump, paint and lamp rooms and stores also the emergency generator room and battery room, book stores and CO_2 bottle room.

From Table 11.3 it will be observed that in certain of the air conditioned spaces, the air changes are less than those required by D.O.T., First Schedule Regulation No. 1036, 1953, for ventilated spaces. Since the conditions achieved by air conditioning would be superior to any under ventilation without artificial air cooling, it is understood from communication with the D.O.T., that relaxation from the above requirements would be made by them under the terms of Regulation 38(3) which provides for such relaxation in this event.

The fitting of an installation as described above would, therefore, be subject to final approval of the Department of Trade.

Installation serving the engine control room

The engine control room would be served by three selfcontained room conditioners, to operate on 440 V, 3-phase, 60 Hz supply. Power consumption 1.2 kW per unit.

At least 7 litres/sec of fresh air per occupant of room should be taken from machinery space system. The shipbuilder should pipe condenser cooling water from ship's services. Erosion of condenser piping may be prevented by fitting a 'Constaflo' control valve in each unit to limit flow to 0.15 litre/sec per unit.

This proposal is exclusive of heating. The shipbuilder should supply and fit suitable bases for units, provide switches, wire up to motors and switches, and supply and fit cooling water drain piping and fittings.

Table 11.1 Fan particulars

Unit No.	Volume m³/s	Static pressure mbar	Dia. mm	Power kW	Speed rev/sec	Encl. with Class 'B' insulation	Spaces served
Air conditioning unit No. 1	3.40	26.2	–	14.9	30	T.E.F.C.	Officers' accommodation
Air conditioning unit No. 2	1.87	22.4	–	7.5	30	T.E.F.C.	Crew's accommodation
Axial flow supply fan No. 1	0.57	4.0	240	0.6	60	T.E.	Motor generator room, electronic and switchboard rooms, etc.
Axial flow supply fan No. 2	1.08	4.2	380	0.8	60	T.E.	Galley and stores
Axial flow exhaust fan Nos. 1, 2, 3 and 6	0.57	4.0	240	0.6	60	T.E.	Toilet spaces
Axial flow exhaust fan No. 4	1.65	3.0	610	0.8	30	T.E.	Galley and handling space
Axial flow exhaust fan No. 5	0.16	1.9	152	0.14	60	T.E.	Hospital, W.C., bathroom and medical locker
Axial flow exhaust fan Nos. 1–2 P.	1.41	3.1	445	1.1	20	T.E.F.C.	Cargo pump room

All the above electric motors would be suitable for operation on a ship's voltage of 440 V, 3 ph, 60 Hz.

Table 11.2 List of motor and starter spares

Unit	Motors	Starters
Air conditioning unit No. 1	1 14.9 kW motor 1 set bearings	1 set contacts, coils and springs
Air conditioning unit No. 2	1 7.5 kW motor 1 set bearings	1 set contacts, coils and springs
Supply fan No. 1	1 fan unit, complete	1 Auto-West switch
Supply fan No. 2	1 0.8 kW motor 1 set bearings	1 set contacts, coils and springs
Exhaust fan No. 1	1 0.6 kW motor	1 set contacts, coils and springs
Exhaust fan No. 2 No. 3 No. 4 No. 5 No. 6	1 fan unit, complete	
Exhaust fan No. 1 P Exhaust fan No. 2 P	1 set bearings	1 set contacts, coils and springs

Note All starters 0.75 kW and above would have running lights and in addition those 3.75 kW and above would have ammeters fitted. The above spares would be suitably packed for stowage on board.

Table 11.3 Schedule of air changes

		Air changes/hour	
Spaces served	*Deck*	*Supply*	*Exhaust*
Wheelhouse	Nav. Bridge	6	—
W.C.		—	15
Captains, chief officer's and deck officers' accommodation		9	—
Radio office		12	—
Motor gen. room, gyro and electronics room	Bridge	15/30*	15/30
Officers' toilets and drying room		—	15
Baggage room		—	10
Officers' smoke room and bar, games room		12	—
Chief engineer's and engineers' accommodation, pilot and owners representatives		9	—
Laundry	Boat	6	15
Switchboard room, telephone exchange		30*	30
Engineers' toilets and change room, pantry, drying room, equipment locker		—	15
Lockers		—	10
Cadets, catering officer, P.O.'s and junior engineers		9	—
Cargo control room, dining saloon, crew's mess room, crew's smoke room	Poop	12	—
Hospital		12	0.14m³/s†
Medical locker		—	10
Galley		20*	40
Cadets', engineers', catering officer's and P.O.'s toilets		—	15
Games room		12	—
Crew accommodation		9	—
Laundry		6	15
Motor room		30*	30
Bedding, beer, dry provision and bonded and equipment stores	Upper	10*	—
Crew toilets, change room, laundry, drying rooms		—	15
Handling space		—	12.5

*Air at atmospheric temperature † Through W.C. and bathroom.

Installation serving the cargo pump room

The cargo pump room is ventilated by two axial flow exhaust fans each to extract 1.4 m^3/sec against 3.1 mbar static, requiring total of 2.25 kW. The fans should be of split case bevel gear driven type, enabling the fan motor to be mounted in the adjacent engine room. Motor and controller spare gear must be provided. Air is extracted from two low level points through wire mesh grids fitted on galvanised steel ducting.

The shipbuilder must supply and fit air outlet jalousies with hinged W.T. covers, and natural ventilation to operate in conjunction with the mechanical system.

12 Deck machinery

The range of deck machinery is extensive and varied, it can be divided broadly into:

Anchor handling (windlasses and capstans).
Mooring (winches and capstans).
Cargo handling (winches and cranes).

The basic requirement of all the above is to control loads associated with chain cable or wire rope and whilst each type of equipment has its own operational requirements, certain aspects of design and operation are common.

Most deck machinery is idle during much of its life and due to this intermittent duty requirement, gears and drives are normally designed to a limited rating of one half to one hour. Despite long periods of idleness, often in severe weather conditions, the machinery must operate immediately, when required.

It is essential that deck machinery should require minimum maintenance. Totally enclosed equipment with oil bath lubrication for gears and bearings is now standard but maintenance cannot be completely eliminated and routine checking and greasing should be carried out on a planned basis. Prime movers may be used to perform more than one basic duty. For example, mooring winches are often combined with windlass units so that one prime mover drives both. The port mooring winch motor can thus be used to drive the starboard windlass and vice versa. This applies also to the power supply where generators or hydraulic pumps are also cross-connected.

There are many instances where remote or centralised control is of great advantage, for example, the facility for letting go anchor from the bridge under emergency conditions, the use of shipside controllers with mooring winches or the central control position required for the multiwinch slewing derrick system.

DRIVE AND CONTROL ARRANGEMENTS

The three most common forms of main drive used on deck auxiliaries are steam, electric and hydraulic.

When fitted, steam auxiliaries are frequently of the totally enclosed type using forced or splash lubrication or a combination of both. A typical arrangement has an oil pump driven from the crank-shaft supplying pressure oil to the crankshaft and connecting rod bearings whilst crossheads, eccentrics and gearing are splash lubri-cated from the sump or drip trays. The steam cylinders and valves are not normally lubricated as, due to the low working pressure and condensation, steam temperatures encountered on deck rarely exceed 180°C. A superheated steam supply creates problems with cylinder lubrication and in any event has little effect on non-expansive working. Inlet steam temperatures are limited for use with cast iron cylinders by Classification Society rules. Full load crankshaft speeds are normally between 90—160 rev/min increasing to twice this figure for light line duties.

An alternative form of steam drive is the reversing steam turbine which is illustrated in Figure 12.1. This machine requires less main-tenance and it is able to accept higher steam pressures and temper-atures, (up to 24 bar, 290°C) directly through a reducing valve from the main boilers, thus increasing turbine efficiency and minimising its size and weight. On large equipment, even though the turbine has to be geared down from a normal full load speed of 2000—2500 rev/min there is a saving in total weight when compared with engine driven equipment.

As illustrated in Figure 12.1, the turbine shaft and rotor are supported in anti-friction bearings; the coupling end bearing is grease lubricated and the governor mechanism and location bearing are splash oil lubricated. Although the bearings warm up rapidly due to conduction of heat along the shaft from the exhaust casing, normal lubricating oils and greases have proved satisfactory.

Both reciprocators and turbines, as used on supertankers, drive the largest deck machinery in service. This equipment is normally locally controlled at the machinery position, however, it is of great advantage to have shipside control for both windlass and mooring winch operation and this can be simply achieved by the use of push-pull hydraulic mechanisms which are effective up to approximately 45 m.

Electric drives

Electric drives are most commonly used for deck machinery. The

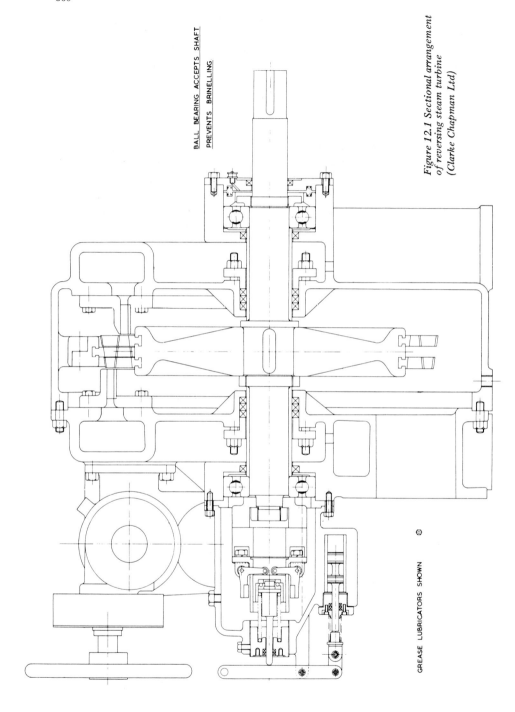

BALL BEARING ACCEPTS SHAFT

PREVENTS BRINELLING

GREASE LUBRICATORS SHOWN

Figure 12.1 Sectional arrangement of reversing steam turbine (Clarke Chapman Ltd)

motors are generally totally enclosed, watertight and in most cases embody a spring applied, magnetically released, fail safe disc brake.

The direct current motor, although it is relatively costly and requires regular brushgear maintenance, is frequently used where the characteristic flexibility of control may be used to good advantage. The control of d.c. motors by contactor-switched armature resistances, common in the days when ships' supplies were mainly d.c., has now largely been superseded by a variety of Ward Leonard type control systems which give a better, more positive control particularly for controlled lowering of loads. The Ward Leonard generator may be driven by either a d.c. or an a.c. motor.

The most important feature of d.c. drive is its efficiency, particularly in comparison with a.c. drives, when operating at speeds in the lower portion of its working range. The d.c. motor is the only electric drive at present in production which can be designed to operate in a stalled condition continuously against its full rate torque and this feature is used extensively for automatic mooring winches

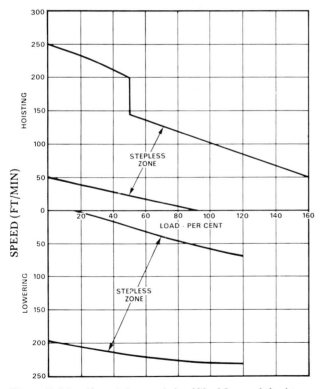

Figure 12.2 Load/speed characteristic of Ward Leonard thyristor controlled winch. See Figure 12.3 for conversions to m/sec   *(Clarke Chapman Ltd)*

of the 'live motor' type. The majority of d.c. winch motors develop full output at speeds of the order of 500 rev/min and wherever necessary are arranged to run up to two to four times this speed for light line duties. Windlass motors on the other hand do not normally operate with a run up in excess of 2 : 1 and usually have a full load working speed of the order of 1 000 rev/min.

D.C. motors may also be controlled by static 'thyristor' converters which convert a.c. supply into a variable d.c. voltage of the required magnitude and polarity for any required armature speed. These converters must be of a type capable of controlled rectification and inversion with bi-directional current flow if full control is to be obtained (Figure 12.2).

A.C. motors, either 'wound rotor' or 'cage', are also in common use. With these the speed may be changed by means of pole changing connections or, in the case of the wound rotor induction motor, by

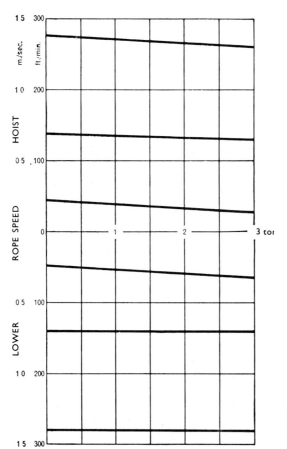

Figure 12.3 Performance curves of a 3 tonne winch. AC pole-changing 'cage' motor.

(Clarke, Chapman Ltd)

changing the value of the resistance connected to the rotor. Both methods involve the switching of high currents at medium voltage in several lines simultaneously and the use of multi pole contactors is common. These drives offer a very limited choice of only two or three discrete speeds such as 0.65, 0.325 and 0.1025 m/sec corresponding to 4, 8 and 24 pole operation. The wound rotor type is slightly more flexible in 'hoist' control but, as with the resistance controlled d.c. motor, difficulty is experienced in providing effective control of an overhauling load e.g. lowering a suspended load. These disadvantages are often outweighed by lower cost, particularly of the cage induction motor, in comparison with the more flexible d.c. Typical performance curves are shown in Figure 12.3.

Another form of induction motor control system is based on the relationship between output torque and applied voltage, the torque being proportional to the voltage squared. The controller takes the form of a three-phase series regulator with an arm in each supply line to the motor. A stable drive system can only be achieved by this means if a closed loop servo control system is used in conjunction with a very fast acting regulator which automatically adjusts the output torque to suit the load demand at the set speed. Control of an overhauling load is possibly by means of injection braking techniques. A combined system employing both these control principles can provide full control requirements for all deck machinery.

The a.c. drives described operate at the supply frequency and consequently rapid heating of the motor will occur if the drive is stalled when energised.

The majority of a.c. motors on deck machinery run at a maximum speed corresponding to the 4 pole synchronous speed of 1800 rev/min on a 60 Hz supply. These speeds are similar to the maximum speeds used with d.c. drives and the bearings and shaft details tend to be much the same. The motor bearings are normally grease lubricated; however, in some cases where the motor is flange mounted on an oil bath gearcase, the driving end bearing is open to the gearcase oil and grease lubrication is not required.

Hydraulic drives

Hydraulic drives can be broadly sub-divided into constant pressure, constant volume and variable displacement systems.

Constant pressure systems for use with prime movers on deck machinery are rare, control and oil cooling problems being difficult to overcome with large equipment. Constant volume systems are more common, one example being the high torque vane type system which operates between 25—45 bar. As the pump (see under 'Positive displacement pumps' in Chapter 3) delivers a constant volume of oil, speed control of the hydraulic motor is obtained by throttling the required amount of oil to the motor through a control valve, the remainder being by-passed to the pump suction. The pump discharge pressure is determined by the load, speed and direction of rotation being controlled at the hydraulic motor by a reversing lever positioning a balanced spool valve.

A third form of hydraulic drive consists of a variable displacement axial piston pump (see Chapter 8) which is driven at constant speed, normally by an a.c. electric motor, and which supplies oil to a fixed displacement hydraulic motor coupled to the mechanical portion of the machinery (Figure 12.4). The piston stroke and hence the pump

Figure 12.4 Norwinch hydraulic winch motor (open)

delivery is controlled by servo motor, the operation of which is dependent on the movement of a pilot valve in either direction from the neutral position. As the motor speed is directly related to the amount of oil available, should light line duties be required the full load speed of the motor is obtained when the pump is on some fraction of full stroke. This type of system incorporates a constant horsepower device which limits the maximum amount of power absorbed from the pump drive motor under loaded conditions and provides an automatic speed/load discrimination feature similar in performance to the steam drive or Ward Leonard electrical system. Full load system pressures normally fall within the range 140–240 bar, though some work at lower pressures, using a vane-type motor.

Many of the hydraulic systems, fitted to deck machinery are of the 'unit' type, one pump driving one motor, but there are great advantages to be gained by the use of a 'ring main' system. With this system one centrally located hydraulic pump is able to cater for the needs of a number of auxiliaries working simultaneously at varying loads. As the equipment powered from this central pumping station need not be restricted to deck machinery, the system offers considerable savings on capital cost.

It is important with all hydraulic systems to ensure that interlocking arrangements provided for pump or motor control levers are in the neutral position before the pump driving motor can be started in order to avoid inadvertent running of unmanned machinery. Overload protection on hydraulic systems is provided by the use of pressure relief valves set between 30–50% in excess of rated full load pressures.

ANCHOR HANDLING

The efficient working of the anchor windlass is essential to the safety of the ship. An anchor windlass can expect to fulfill the following:

1. The windlass cablelifter brakes must be able to control the running anchor and cable when the cablelifter is disconnected from the gearing during 'letting go'. Average cable speeds vary between 5–7 m/sec during this operation.
2. The windlass must be able to heave a certain weight of cable at a specified speed. This 'full load' duty of the windlass varies (it may be as high as 70 tonne; figures between 20 and 40 tonne are not unusual) but is commonly between 4 and 6 times the weight of one anchor, the speed of haul being at least 9 m/min and up to 15 m/min.
3 The braking effort obtained at the cable lifter must be at least equal to 40 per cent of the breaking strength of the cable.

Most anchor handling equipment incorporates warpends for mooring purposes and light line speeds of up to 0.75 to 1.00 m/sec are required. The most conventional types of equipment in use are as follows.

Mooring windlasses

This equipment is self contained and normally one prime mover drives two cablelifters and two warpends, the latter may not be declutchable and, if so, rotate when the cablelifters are engaged. There is some variation in detail design of cable lifters and in their drives. Figure 12.5 shows a typical arrangement. Due to the low speed of rotation required of the cablelifter whilst heaving anchor, (3–5 rev/min) a high gear reduction is needed when the windlass is driven by a high speed electric or hydraulic motor. This is generally obtained by using a high ratio worm gear followed by a single step of spur gears between the warpend shaft and cablelifters, typically as shown in Figure 12.6. Alternatively, multi steps of spur gears are used.

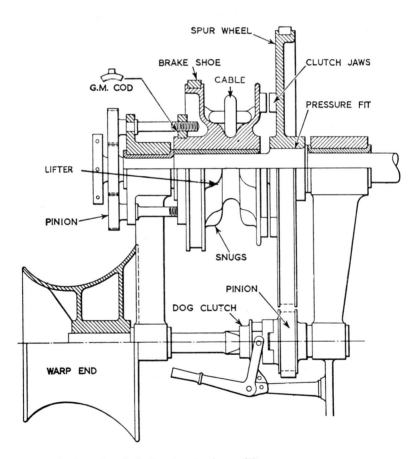

Figure 12.5 Part plan of windlass dog-clutch-type lifter

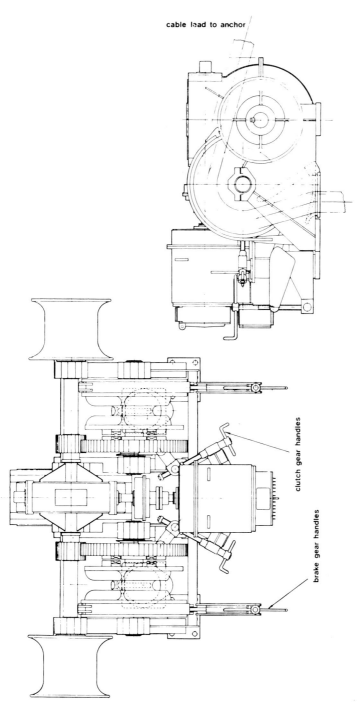

cable lead to anchor

clutch gear handles

brake gear handles

Figure 12.6 Typical electrically driven mooring windlass

Anchor capstans

With this type of equipment the driving machinery is situated below deck and the cablelifters are mounted horizontally being driven by vertical shafts as shown in Figure 12.7. In this example a capstan barrel is shown mounted above the cablelifter (not shown) although with larger equipment (above 76 mm dia cable) it is usual to have only the cablelifter, the capstan barrel being mounted on a separate shaft.

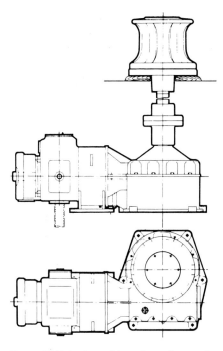

Figure 12.7 Anchor cable and warping capstan

Winch windlasses

This arrangement utilises a forward mooring winch to drive a windlass unit thus reducing the number of prime movers required. The port and starboard units are normally interconnected, both mechanically and for power, in order to provide standby drive and to utilise the power of both winches on the windlass should this be required.

Control of windlasses

As windlasses are required for intermittent duty only, gearing is designed with an adequate margin on strength rather than wear.

Slipping clutches are commonly fitted between the prime mover and gearing to avoid the inertia of the prime mover being transmitted to the machinery in the event of shock loading on the cable when, for example, the anchor is being housed (see Figure 12.8).

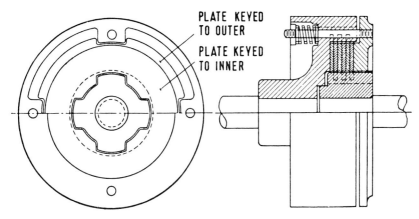

Figure 12.8 Slipping clutch

Windlasses are normally controlled from a local position, the operator manually applying the cablelifter brake as required to control the speed of the running cable, and whilst heaving anchor the operator is positioned at the windlass or at the shipside so that he can see the anchor for housing purposes. It is quite feasible, however, to control all functions of the windlass from a remote position. The spring applied cablelifter brakes are hydraulically released and to aid the operator, the running cable speed and the length paid out are indicated at the remote position during letting go. The cablelifter can also be engaged from the remote position so that the anchor can be veered out to the waterline before letting go or heaved in as required.

The windlass is in the most vulnerable position so far as exposure to the elements is concerned and maintenance demands should be an absolute minimum. Normally primary gearing is enclosed and splash lubricated, maintenance being limited to pressure grease points for gunmetal sleeve bearings. However, due to the large size of the final of bevel or spur reduction gears, and the clutching arrangements required, these gears are often of the open type and are lubricated with open gear compounds.

MOORING EQUIPMENT

Full load duties of warping capstans and mooring winches vary between 3—30 tonnes at 0.3 to 0.6 m/sec and twice full load speed is normally provided for recovering slack lines.

The size of wire rope used on mooring winch barrels is governed by the weight of wire manageable by the crew; this is currently accepted as 140 mm circumference maximum. The basic problems associated with the use of wire ropes is that they are difficult to handle, do not float and when used in multi-layers, due to inadequate spooling, the top, tensioned layer cuts down into the underlying layers causing damage. To counteract this latter problem a divided barrel can be used such that the wire may be stored on one portion and a single layer of wire transfered to the second portion when tensioned. Low density, high breaking strength synthetic ropes (polypropylene, nylon, terylene etc.) offer certain advantages over wire, its main disadvantage being a tendency to fuse if scrubbed against itself or the barrel.

Winches

Mooring winches provide the facility for tensioning the wire up to the stalling capacity of the winch, usually 1.5 times full load, thereafter the load is held by the prime mover brake or barrel brake when the power is shut off. The winch cannot pay out wire unless the brake is overhauled or recover wire unless manually operated, thus wires may become slack.

Automatic mooring winches provide the manual control previously described but in addition incorporate control features such that, in the 'automatic' setting, the winch may be overhauled and wire is paid off the barrel at a pre-determined maximum tension; also wire is recovered at a lower tension should it tend to become slack. Thus there is a certain range of tension, associated with each step of automatic control, when the wire is stationary. It is not practical to reduce this range to the minimum possible as this results in hunting of the controls.

It should be noted that the principal reason for incorporating automatic controls with the features described is to limit the render value of the winch and avoid broken wires; also to prevent mooring wires becoming slack. Load sensing devices are used with automatic mooring winches, e.g. spring-loaded gearwheels and torsion bars are widely used with steam and electric winches; fluid pressure sensing, either steam or hydraulic oil pressure, is also used where appropriate.

Mooring winches are usually controlled at the local position, i.e. the winch; for vessels of unusually large beam or where docking operations are a frequent occurrence e.g. in ships regularly traversing the St. Lawrence Seaway, remote and shipside controllers are of great advantage. As mooring techniques vary widely, the position and type of control must be engineered to suit the application. It is considered, especially on vessels where mooring lines may be long and ship position critical, that the greatest asset to the operator is knowledge of the wire tensions existing during the mooring operation coupled with an indication of the amount of wire paid off the barrel. It is quite feasible to record these at a central position and mooring lines would then only have to be adjusted periodically as indicated by the recording instruments.

The majority of automatic mooring winches are spur geared to improve the backward efficiency of the gear train for rendering, the gearing and bearings being totally enclosed and lubricated from the oil sump. On larger mooring winches were a barrel brake is fitted, it is now common practice to design the brake to withstand the breaking strength of the mooring wire. Worm geared automatic mooring winches are uncommon as the multistart feature required to improve gear efficiency reduces the main advantage of the worm gear i.e.: the high gear ratio.

CARGO HANDLING

The duty of a deck winch is to lift and lower a load by means of a fixed rope on a barrel, or by means of whipping the load on the warp ends; to top or luff the derricks, and to warp the ship. In fulfilling these duties it is essential that the winch should be capable of carrying out the following requirements:

(a) lift the load at suitable speeds;
(b) hold the load from running back;
(c) lower the load under control;
(d) take up the slack on the slings without undue stress;
(e) drop the load smartly on the skids by answering the operator's application without delay;
(f) allow the winch to be stalled when overloaded, and to start up again automatically when the stress is reduced;
(g) have good acceleration and retardation:

In addition when the winch is electrically driven the requirements are:

(a) prevent the load being lowered at a speed which will damage the motor armature;

(b) stop the load running back should the power supply fail;

(c) prevent the winch starting up again when the power is restored, until the controller has been turned to the correct position.

Hydraulic winch systems are now quite common but electric drives for cargo winches and cranes are most widely used. For the conventional union purchase cargo handling arrangement or for slewing derrick systems handling loads up to 20 tonne, standard cargo winches are normally used for hoist, topping and slewing motions, the full load duties varying from 3–10 tonne at 0.65 to 0.3 m/sec. For the handling of heavy loads, although this may be accomplished with conventional derrick systems using multipart tackle, specially designed heavy lift equipment is available. The winches used with these heavy lift systems may have to be specially designed to fit in with the mast arrangements and the winch duty pull may be as high as 30 tonne.

Cargo winches

It is usual to select the number and capacity of and to group the winches in such a way that within practical limits, all hatches may be worked simultaneously and having regard to their size (and the hold capacity beneath them) work at each is carried out in the same period.

Reduction of the cycle time during cargo handling is best accomplished by the use of equipment offering high speeds say from 0.45 m/sec at full load to 1.75 m/sec light, the power required varying from 40 kW at 7 tonnes to 20 kW at 3 tonnes; this feature is available with electro-hydraulic and d.c. electric drives as they offer an automatic load discrimination feature. However, the rationalisation of electrical power supply on board ship has resulted in the increased use of a.c. power and the majority of winch machinery now produced for cargo handling utilises the pole-changing induction motor. This offers two or more discrete speeds of operation in fixed gear and a mechanical change speed gear is normally provided for half load conditions. Normally all modern cargo handling machinery, of the electric or electrohydraulic type, is designed to 'fail safe'. A typical example of this is the automatic application of the disc brake on an electric driving motor should the supply fail or when the controller is returned to the 'OFF' position.

Derricks

Most older ships and some recent ones use winches in conjunction with derricks for working cargo. The derricks may be arranged for fixed outreach working or slewing derricks may be fitted. A fixed outreach system uses two derricks, one 'topped' to a position over the ship's side and the other to a position over the hold. Figure 12.9 shows the commonest arrangement adopted, known as Union Purchase rig. The disadvantages of the fixed outreach systems are that firstly if the outreach requires adjustment cargo work must be interrupted, and secondly the load that can be lifted is less than the safe working load of the derricks since an indirect lift is used. More-over considerable time and man power is required to prepare a ship for cargo working.

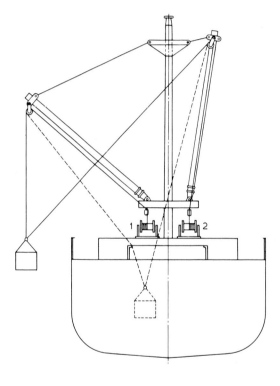

Figure 12.9 Union purchase rig (Clarke, Chapman & Co Ltd)

The main advantages of the system are that only two winches are required for each pair of derricks and it has a faster cycle time than the slewing derrick system.

The slewing derrick system, one type of which is shown in Figure 12.10, has the advantages that there is no interruption in cargo work for adjustments and that cargo can be more accurately placed in the hold; however in such a system three winches are required for each derrick to hoist luff and slew.

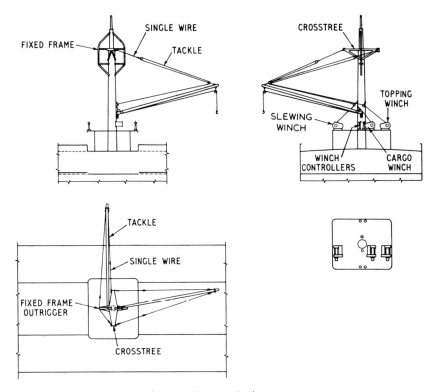

Figure 12.10 Slewing derrick (Clarke, Chapman Ltd)

Deck cranes

A large number of ships are fitted with deck cranes. These require less time to prepare for working cargo than derricks and have the advantage of being able to accurately place (or spot) cargo in the hold. On container ships using ports without special container handling facilities, cranes with special container-handling gear are essential.

Deck-mounted cranes for both conventional cargo handling and grabbing duties are available with lifting capacities of up to 50 tonnes. Ships specialising in carrying very heavy loads, however, are invariably equipped with special derrick systems such as the Stulken (Figure 12.11). These derrick systems are capable of lifting loads of up to 500 tonnes.

Although crane motors may rely upon pole changing for speed variation, Ward Leonard and electro-hydraulic controls are those most widely used. One of the reasons for this is that pole-changing

Figure 12.11 Stulken derrick (Blohm and Voss)

motors can only give a range of discrete speeds, but additional
factors favouring the two alternative methods include less fierce
power surges since the Ward-Leonard motor or the electric drive
motor in the hydraulic system run continuously and secondly the
contactors required are far simpler and need less maintenance since
they are not continuously being exposed to the high starting currents
of pole-changing systems.

Deck cranes require to hoist, luff and slew and separate electric or
hydraulic motors will be required for each motion. Most makes of
crane incorporate a rope system to effect luffing and this is

commonly rove to give a level luff — in other words the cable geometry is such that the load is not lifted or lowered by the action of luffing the jib and the luffing motor need therefore only be rated to lift the jib and not the load as well.

Generally, deck cranes of this type use the 'Toplis' three-part reeving system for the hoist rope and the luffing ropes are rove between the jib head and the superstructure apex which gives them an approximately constant load, irrespective of the jib radius. This load depends only on the weight of the jib, the resultant of loads in the hoisting rope due to the load on the hook passes through the jib to the jib foot pin (Figure 12.12(a)). If the crane is inclined 5° in

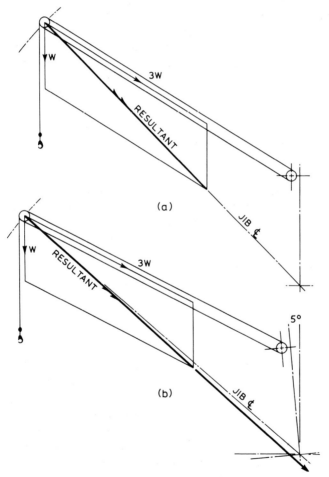

Figure 12.12 Rope lift cranes — resultant loads when hoisting. (Stothert & Pitt Ltd)

the forward direction due to heel of the ship the level-luffing geometry is disturbed and the hook load produces a considerable moment on the jib which increases the pull on the luffing rope (Figure 12.12(b)). In the case of a 5 tonne crane the pull under these conditions is approximately doubled and the luffing ropes need to be over-proportioned to meet the required factor of safety. If the inclination is in the inward direction and the jib is near minimum radius there is a danger that its weight moment will not be sufficient to prevent it from luffing up under the action of the hoisting rope resultant. Swinging of the hook will produce similar effects to inclination of the crane.

In the Stothert & Pitt 'Stevedore' electro-hydraulic crane the jib is luffed by one or two hydraulic rams. Pilot operated leak valves in the rams ensure that the jib is supported in the event of hydraulic pressure being lost and an automatic limiting device is incorporated which ensures that maximum radius cannot be exceeded. When the jib is to be stowed the operator can override the limiting device. In the horizontal stowed position the cylinder rods are fully retracted into the rams where they are protected from the weather.

Some cranes are mounted in pairs on a common platform which can be rotated through 360°. The cranes can be operated independently or locked together and operated as a twin-jib crane of double capacity, usually to give capacities of up to 50 tonnes.

Most cranes can, if required, be fitted with a two-gear selection to give a choice of a faster maximum hoisting speed on less than half load. For a 5 tonne crane full load maximum hoisting speeds in the range 50–75 m/min are available with slewing speeds in the range 1–2 rev/min. For a 25 tonne capacity crane, maximum full load hoisting speeds in the range 20–25 m/min are common with slewing speeds again in the range 1–2 rev/min. On half loads hoisting speeds increase two to three times.

Drive mechanism and safety features

In both electric and electro-hydraulic cranes it is usual to find that the crane revolves on roller bearings. A toothed rack is formed on the periphery of the supporting seat and a motor-driven pinion meshes with the rack to provide drive. Spring-loaded disc or band brakes are fitted on all the drive motors. These are arranged to fail safe in the event of a power or hydraulic failure. The brakes are also arranged to operate in conjunction with motor cut-outs when the crane has reached its hoisting and luffing limits, or if slack turns occur on the hoist barrel.

In the case of the electro-hydraulic cranes it is normal for one electric motor to drive all three hydraulic pumps and in Ward-Leonard electric crane systems the Ward-Leonard generator usually supplies all three drive motors.

CARGO ACCESS

Although not strictly a 'machinery' item, the mechanical complexity of present-day cargo hatch covers — whose periodic maintenance may fall in the domain of the ship's Engineering Department — warrants some mention in this chapter. Many types of mechanically operated hatch covers can now be found at sea. The principle ones are listed in Table 12.1.

Table 12.1 Types of mechanically operated hatch covers

Operating mode	Type
Rolling and tipping	single pull
Rolling	end rolling
	side rolling
	lift and roll ('piggy-back')
Roll stowing	rolltite
Folding	hydraulic folding
	wire-operated folding
	direct pull
Sliding/nesting	tween-deck sliding

The most common type of hatch cover is the 'single pull'. The complete cover consists of a number of transverse panels which span the hatchway and are linked together by chains. In the closed position, the panel sides sit firmly on a horizontal steel bar attached to the top of the hatch coaming. Just inside the side plates is a rubber gasket housed in a channel on the under-side of the hatch cover and which rests on a steel compression bar to form a weathertight seal (Figure 12.13). When closed the covers are held onto the seals by a series of peripheral cleats. Rollers are arranged on the sides of the covers to facilitate opening/closing. The opening arrangements are shown in Figure 12.14.

To open a single pull cover the securing cleats are first freed and each panel is raised off its compression bars by hydraulic jacks. The cover wheels, which are arranged on eccentrics, are rotated through 180° and locked into position. The jacks are then removed and the cover can be pulled backwards or forwards as required.

Instead of the opening/closing arrangements shown in Figure

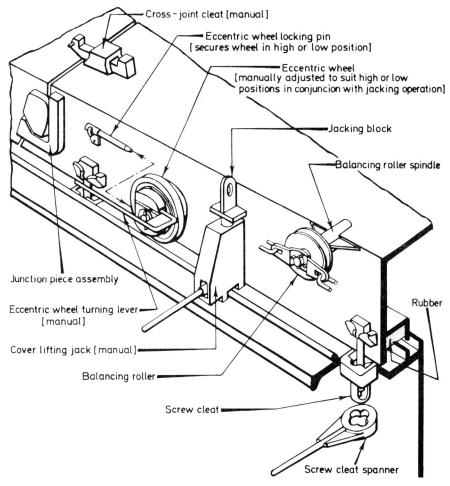

Cross - joint cleat [manual]

Eccentric wheel locking pin
[secures wheel in high or low position]

Eccentric wheel
[manually adjusted to suit high or low
positions in conjuncion with jacking operation]

Jacking block

Balancing roller spindle

Junction piece assembly

Eccentric wheel turning lever
[manual]

Cover lifting jack [manual]

Balancing roller

Screw cleat

Rubber

Screw cleat spanner

Figure 12.13 Detail of single-pull cover showing scaling arrangement and jacking system
(The Henri Kummerman Foundation)

12.14 the hatch may be fitted with a fixed chain drive on the periphery of the hatch, complete with its own electric or hydraulic motor.

Folding covers

These may be wire operated or hydraulically operated. A multi-panel end-folding hydraulic cover is shown in Figure 12.15, while Figure 12.16 shows an interesting hydraulic hinge arrangement. Known as the Navire Hydratorque hinge it incorporates a pair of

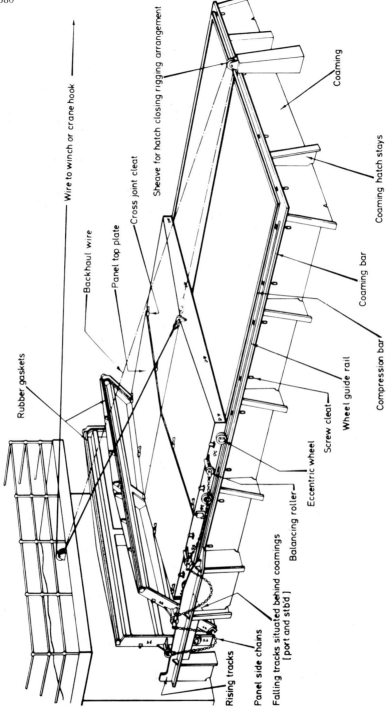

Figure 12.14 Single pull cover showing fittings and opening arrangements (The Henri Kummerman Foundation)

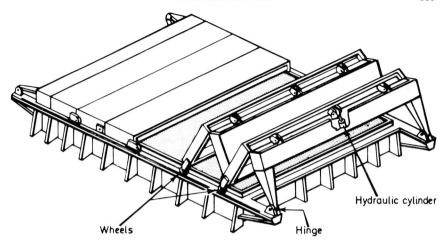

Figure 12.15 A multi-panel end folding hydraulic cover for weather deck use

helixes attached to two pistons. When hydraulic pressure is applied between the two pistons it forces them apart thus rotating the helixes. Pressure applied to the outside of the pistons creates a torque in the opposite direction.

Maintenance

Hatch cover equipment has to exist in a very hostile environment. The importance of regular maintenance cannot be over-emphasised. Drive boxes and electrical enclosures should be checked regularly for water-tightness. Drive chains, trolleys and adjusting devices such as peripheral and cross-joint cleats should be cleaned and greased regularly. Seals, compression bars and coamings should be inspected and cleaned at each port. Drain channels should be cleared regularly.

On the subject of seals and cleats it is important not to over-tighten cleats. The seal should be compressed but not beyond the elastic limit of the gasket material. Standard rubber gaskets can be expected to last from four to five years of normal service. In freezing conditions special grease or commercial glycerine should be spread over the surface of all gaskets to prevent them from sticking to their compression bars. Quick-acting cleats are fitted with thick neoprene washers arranged to exert the correct degree of compression. After a time these lose their elasticity and the cleat must be adjusted or replaced.

382

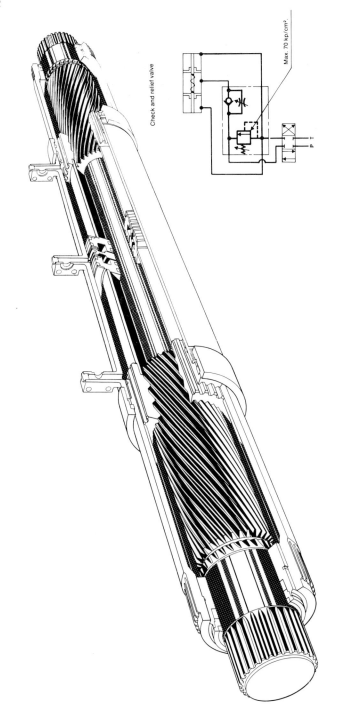

Check and relief valve

Max. 70 kp/cm²

P T

Figure 12.16 The Navire Hydratorque rotary actuator and hinge (Navire Cargo Gear International AB, Sweden)

Hydraulic systems

The most important thing about any hydraulic system is to make sure that the hydraulic oil remains clean (regular inspection of filters). Any protective boots fitted over rams etc. should be periodically examined as also should flexible hoses. Hydraulic hoses should have their date of manufacture printed on them and can be expected to have a life of about five years.

13 Safety equipment

Safety equipment, in the context of this chapter, constitutes the following:

1. Fire detection and fire-fighting equipment.
2. Inert gas system and gas detection meters.
3. Emergency generator sets.
4. Emergency pumping and watertight doors.
5. Lifeboat engines and davits.
6. Whistles.

FIRE DETECTION AND FIRE-FIGHTING EQUIPMENT

Out of the many hazards that can occur on ships, fire is by far the most frequent, and results in more total losses than any other casualty. Most fires start through negligence. A number of fatal accommodation fires have been started by people falling asleep smoking in bed; engine room fires have been started by neglected oil leaks dripping on to hot pipes or exhaust manifolds whose insulation has not been replaced after maintenance. Again, thoughtless smoking habits in machinery spaces have been the cause of many severe fires.

Nearly all fires can be easily controlled by one man with a portable fire extinguisher providing that the fire is detected soon enough and handled in the correct manner. A fire starts when the following three factors come together:

1. Combustible material — fuel.
2. Ignition source — heat.
3. Oxygen.

By removing one or more of these factors the fire will be subdued, always providing action is taken early enough (this may be within seconds).

Table 13.1 Types of fire extinguisher

Type	Class of fire			Operation	Range and duration	Yearly maintenance	Remarks
	A	B	C				
Soda and acid	$\checkmark$	X	X	Turn upside down. Direct stream at base of fire	Not less than 6 m. 60–80 sec (9 litres)	Discharge Clean Re-charge	Should not be kept at a temp. below 5° C
Gas cartridge and water	$\checkmark$	X	X	Release gas. Hold upright. Direct stream at base of fire	Length of jet 10.6 m. spray 6 m. Coverage 37 m² (9 litres)	As above	
Foam	$\checkmark$	$\checkmark$	X	Turn upside down. Direct at bulkhead just above fire	Jet of foam 6 m. 60 sec (9 litres)	As above	Should not be kept at a temp. below 5° C
CO₂	X $\checkmark$	$\checkmark$	$\checkmark$	Release gas. Direct discharge at base of fire and maintain slow sweeping motion	Over 2 250 litres of gas from 4.5 kg cont'r.	Weigh or check with isotope. Recharge if loss more than 10%	Can be of value on Class A fires. Not allowed in Accom.
Dry Powder	X $\checkmark$	$\checkmark$	$\checkmark$	Release gas and control discharge with squeeze grip nozzle. Direct at base of fire with a sweeping motion	In still air 7.6 m long, 1.8 m wide, 1.2 m high.	Weigh gas cart. Renew if 10% loss. Check that powder is free running	Can be of value on Class A Fires. Needs careful aim

For reference purposes fires are designated three classes:

A. This class covers fires in solid materials such as wood shavings and soft furnishings. Fires in this class are best extinguished by quenching or cooling with a water spray i.e. heat removal.
B. This class covers fires in fluids, such as petrol, lubricating oil and grease. It is dangerous to attempt to extinguish such fires with water although a very fine water spray can be beneficial as a heat screen. These fires are best extinguished by smothering, i.e. deprival of oxygen.
C. This class covers fires in electrical equipment and any extinguishing agents used must be non-conductive.

The first line of defence against fire is the portable fire extinguisher. The most usual portable extinguishers found at sea are:

1. The soda-acid type.
2. The foam extinguisher.
3. The dry powder type.
4. The carbon dioxide (CO_2) extinguisher.

The soda-acid and foam extinguishers are usually of 9 litres capacity and are listed in Table 13.1.

The soda-acid extinguisher

A 9 litre soda-acid extinguisher is shown in Figure 13.1. It consists of a 1.63 mm thick steel shell approximately 180 mm dia. and 530 mm high. The shell ends are dished and welded or riveted to the wrapper plate. The shell must be capable of withstanding pressures of up to 14 bar in the event of a blockage occurring in the discharge nozzle. When new the containers are thus hydraulically tested to a pressure of 25 bar, this pressure being maintained for five minutes. The container is also subjected to a pressure test of 21 bar at four-yearly intervals.

Suspended from the neck of the container is a glass phial containing sulphuric acid. The main body of the extinguisher contains an alkali solution, sodium bicarbonate. To operate the extinguisher a brass knob located on the top of the container is pressed sharply inwards, thus breaking the acid phial. When the acid and alkali mix, carbon dioxide is formed which forces the liquid out of the discharge nozzle. The chemical reaction is as follows:

$$H_2SO_4 + 2NaHCO_3 = Na_2SO_4 + 2H_2O + 2CO_2$$

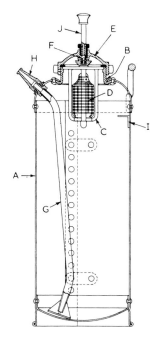

Figure 13.1 Portable soda acid extinguisher
A   Steel container   F   Atmospheric valve
B   Neck ring   G   Internal pipe
C   Acid bottle cage   H   Nozzle
D   Acid bottle   I   Alkali level indicator
E   Gunmetal top cap   J   Plunger

All portable extinguishers must be discharged and refilled yearly. When refilling a soda-acid extinguisher the screwed brass cap which offers internal access should be carefully examined, firstly to see that the cap seal is in good order and secondly to check that the small holes in the threads are clear. The purpose of the holes is to release any internal pressure before the cap is fully removed. It is also important to check that the ball valve F is free. This prevents liquid rising up the internal pipe and dribbling from the nozzle in hot weather. When the appliance is operated the internal pressure closes this valve. If it is jammed open the appliance will not function correctly. Before refilling the container, the internal pipe G should be checked to see that it is firmly attached to discharge nozzle H and that it is not perished. With all extinguishers chemicals are provided in kits, complete with instructions on re-charging. While still common the soda acid extinguisher has largely been replaced by the CO_2/ water type. (See Figure 13.3(a)).

Foam extinguisher

A portable foam extinguisher is shown in Figure 13.2 The outer container of this extinguisher is similar in construction to that of the soda-acid appliance and is required to withstand the same initial test pressure of 25 bar and is retested at 21 bars every four years.

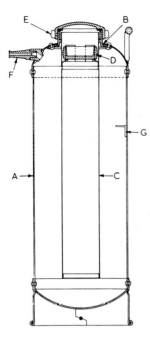

Figure 13.2 Portable foam extinguisher
A   Steel container   E   Gunmetal top cap
B   Neck ring   F   Removable nozzle
C   Inner acid container   G   Alkali level indicator
D   Lead sealing weight

This extinguisher has a long inner container of polythene suspended from the neck ring and filled with aluminium sulphate solution. The outer container is filled to a marked level with sodium bicarbonate solution. A lead disc sits on top of the inner container and acts as a stopper. By inverting the extinguisher (and giving it a bit of a shake) the disc is dislodged and the two solutions mix. The following reaction takes place:

$$Al_2 (SO_4)_3 + 6NaHCO_3 = 2Al(OH)_3 + 3Na_2 SO_4 + 6CO_2$$

The chemicals react in much the same way as those in the soda-acid extinguisher but the action is slower, giving time for bubbles to form. Foam-making substances added to the sodium carbonate determine the nature of the foam formed. The ratio of the foam produced to liquid is in the order 8 : 1 to 12 : 1.

Since the extinguisher has to be turned upside down to bring it into use, no internal pipe is fitted. When being recharged the cap seal should be examined and the pressure relief holes in the cap thread checked.

An alternative type of portable foam extinguisher is shown in Figure 13.3(b). The appliance is filled with water and contains a phial of liquid carbon dioxide, surrounded by a collar holding a

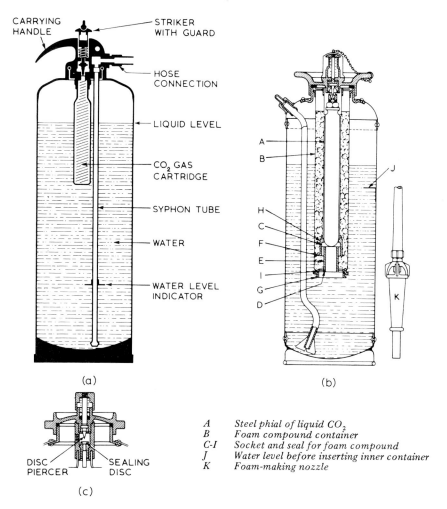

A   *Steel phial of liquid CO₂*
B   *Foam compound container*
C-I   *Socket and seal for foam compound*
J   *Water level before inserting inner container*
K   *Foam-making nozzle*

Figure 13.3
(a) Carbon dioxide and water extinguisher
(b) Air foam extinguisher
(c) Details of sealing disc and disc piercer

foam-inducing compound. To operate this extinguisher a plunger on the cap is depressed, thus bursting the phial. The CO_2 gas then ejects the water and foam compound through a special nozzle which agitates the mixture and creates a mechanical foam. In this type of extinguisher the ratio of foam to liquid is only about 8 : 1. The extinguisher has an internal pipe and is operated in the upright position.

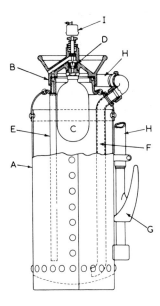

Figure 13.4 Dry powder extinguisher
A Steel container
B Top cap
C CO_2 capsule
D Sealing ring and disc
*   pieces*
E CO_2 injector tube
F Discharge tube
G Controlled discharge
H Flexible hose
I Safety cap

Dry powder extinguisher

This type of extinguisher contains a charge of sodium bicarbonate powder. As can be seen from Figure 13.4 the container holds a steel capsule containing liquid carbon dioxide. On breaking the capsule, by depressing the knob, the gas forces all the powder out of the discharge nozzle. The powder blankets the fire in a cloud of dust, cutting off the supply of oxygen. If the fire is hot enough the sodium bicarbonate will decompose to form sodium carbonate, carbon dioxide and water all of which will aid the smothering operation and give some cooling effect. The extinguisher discharges in only 15 sec and it is thus important to aim the appliance accurately. The operating pressure of this container is higher than that of the soda-acid type. It is consequently hydraulically tested at a pressure of 35 bar.

The carbon dioxide extinguisher

While carbon dioxide is used in some extinguishers as an inert propellant the gas is also used extensively as a blanketing agent. Figure 13.5 shows a carbon dioxide portable extinguisher. The carbon dioxide is in liquid form and is at a pressure of 6 bar at 20°C, necessitating a far stronger container. This type of extinguisher can only be recharged ashore. To check for leakage a record should be kept of the weight of the extinguisher. Alternatively the liquid level can be determined by using a special instrument which uses a radioactive source and a Geiger-Muller counter to detect the gas/liquid interface although this method is usually only used on large fixed CO_2 installations.

This could be lethal if discharged accidentally in a confined space and for this reason this type of extinguisher is not allowed in the accommodation.

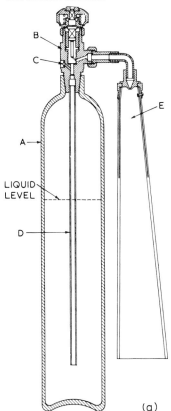

(a)

Figure 13.5(a) Liquid carbon dioxide extinguisher
A Steel cylinder
B Bronze cap with screw-down valve
C Safety bursting disc
D Liquid discharge tube
E Gas discharge horn

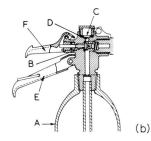

(b)

Figure 13.5(b) Liquid carbon dioxide extinguisher —
alternative method of operation
A Steel cylinder   D Liquid outlet
B Liquid release valve   E Lifting handle
C Safety bursting disc   F Operating trigger

FIXED INSTALLATIONS

While the portable extinguishers form the front line of attack, if fire becomes established it becomes necessary to deploy more extensive means of extinguishing it. This is the purpose of the fixed fire-fighting installations.

The most universal medium for fighting fires is water although due attention should be paid to the inadvisability of using a solid jet of water to attempt to extinguish an oil fire — a fine spray should be used — an on no account should an attempt be made to extinguish an electrical fire with water.

Because water is freely available aboard ships, in all but the smallest vessels a water hydrant system will be installed for fire-fighting duties. This will be supplied by one or more pumps located in the engine room. The number of hydrants fitted and the size of pumps required will be stipulated by the classification society or national legislative body (Department of Trade for U.K.-registered ships).

To ensure that an adequate water supply will be available even when the machinery space is on fire or flooded, an emergency fire pump is normally required to be fitted elsewhere in the vessel. This pump may be driven by a hand started diesel engine or an electric motor supplied from the emergency power circuit. The usual location of the emergency fire pump is the steering flat and because of the high suction lift involved a priming pump is fitted. This is usually friction driven from the fire pump shaft and once the fire pump is running the priming pump drive wheel, normally held away from the fire pump shaft by a spring, must be held against the shaft until the fire pump is primed. Emergency fire pumps are invariably centrifugal pumps in which case a throttling valve will be found on the discharge branch. This is shut while the pump is being primed and opened gradually as the suction is taken up.

If the fire pump is driven by a water-cooled diesel engine supplied by cooling water from the fire pump a header tank will be provided to ensure that the engine is cooled while the fire pump is being primed. It is more normal however to find the engine cooled from a closed circuit fresh water system, the water being cooled in a radiator. Alternatively an air-cooled diesel engine may be fitted.

Where a closed-circuit fresh-water cooled engine is installed it is as well to remember to put anti-freeze in the radiator should the ship be on a cold run.

In addition to the water hydrant systems the following fixed fire-fighting installations will be found. The particular system or systems will depend on the type and size of vessel, the application (e.g. machinery space or cargo tank protection) the requirements of SOLAS 74 and or National legislation and of course the choice of the owner.

1. Fixed foam systems (also called froth systems), either
(a) low expansion foams with an expansion ratio (the ratio of the volume of foam to the volume of the water and foam-making concentrate supplied) usually not more than 12 : 1. These are engineered to cover protected decks in machinery spaces to a depth of 150 mm of foam within five minutes using a system of distribution pipes and foam making sprinkler heads. The sprinklers are usually backed up by portable foam making branch pipes or in the case of the open decks of tankers by foam monitors.
(b) high expansion foams with expansion ratios of not more than 1000 : 1. These are engineered as total flooding systems with sufficient capacity to fill the largest space to be protected at a rate of $1m^3$/min per m^2 of the maximum horizontal surface of the space. The foam is distributed by a system of ducts and the quantity of concentrate carried must be capable of producing enough foam to fill five times the volume of the largest space to be protected.

2 Fixed gas fire-extinguishing systems which may be
(a) CO_2 stored under pressure;
(b) CO_2 store in low-pressure refrigerated tanks;
(c) Halon stored under pressure of nitrogen;
(d) Inert gas generated by combustion and suitably dried and cleaned;
(e) Steam but this is rarely used today. It is only permitted by SOLAS 74 in restricted areas of new ships (e.g. cargo spaces) and only as a back-up system to one of the above.

3. Water sprinkler systems which are used in the accommodation of passenger ships and are usually operated automatically. The system may be extended to the machinery space of the vessel and in the case of roll-on/roll-off vessels to the car decks.

Mechanical foam systems

Foam installations for ship's machinery spaces are designed specifically to suit the vessels concerned, and must therefore vary in arrangement and capacity. The following illustrations show typical foam liquid tank assemblies of both pump-operated and pressurised self-contained foam systems now installed in ships.

Pump-operated mechanical foam system

The pump-operated Chubb Fire mechanical foam system is installed to extinguish oil fires on the tank top in the machinery space, or above the machinery and over such vulnerable auxiliaries as the oil fuel plant and any other surface in the machinery compartment over which oil is liable to spread. Figure 13.6 shows a twin tank venturi proportionator unit. This installation is typical of the kind of equipment used in the machinery spaces of older ships, or for the cargo oil tank decks of older tankers. The two tanks (one being a reserve) are seen together with the foam liquid venturi fed by a dual water supply from the ship's pumps. Sometimes single tank units are fitted. The foam liquid tanks are located at a suitable and approved position outside the machinery space, the capacity of the tank depending upon the surface area to be covered with foam.

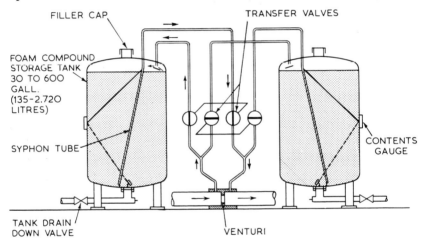

FILLER CAP   TRANSFER VALVES

FOAM COMPOUND
STORAGE TANK
30 TO 600
GALL.
(135–2.720
LITRES)

SYPHON TUBE

CONTENTS
GAUGE

TANK DRAIN
DOWN VALVE

VENTURI

Figure 13.6 Automatic foam compound proportioning by venturi

Under the tank is fitted the venturi to which the dual water supply from the ship's pumps is led, the capacity of which also depends upon the area of the tank top surface, the operating pressure at the venturi must be at least 7 bar. The dual water supply ensures operation of the system should one pump be out of order and is a usual requirement of the Survey Authorities.

From the venturi two small bore pipes are led to the foam liquid tanks, the outlet pipe terminating at the top of the tank, the inlet pipe connected to the internal syphon tube. Water passing through the venturi takes with it the correct proportion of foam liquid induced from the tank. The resultant solution of water and foam-making liquid is led to the foam maker sprinkler network in the machinery space or to the foam monitors on the cargo tank decks of the oil tankers, where the foam is produced.

The capacity of the foam liquid tank is determined by the area of the surface to be covered with foam and is sufficient to deposit foam to a depth of 150 mm over this area in a machinery space or a specified quantity over the cargo oil tank deck of a tanker to comply with Survey Authority requirements.

A foam liquid level gauge at the side of each tank, indicates when the tank is empty of foam liquid and filled with replacement water, so necessitating the closing of the water control valve on the inlet side of the venturi.

For vessels having two machinery spaces, one foam-making liquid tank and venturi may be fitted, with distribution valves to discharge the foam to either space, or twin tanks with venturis can be fitted (one for each space). In these systems the units are usually cross-connected so that foam can be discharged into either space from either tank.

In recent years many new foam concentrates have been introduced, including Protein, Fluoroprotein and Aqueous film-forming foam solutions (AFFF) many of which have a specific gravity close to that of water and a modification to the tank proportioning system has become necessary. In these modified units the foam liquid is stored in a flexible rubber bag within the steel tank. Water, by-passed via the venturi, feeds into the space between the tank and the bag thus exerting an indirect pressure on the foam liquid and forcing it into the low pressure side of the venturi (see Figure 13.7). The system

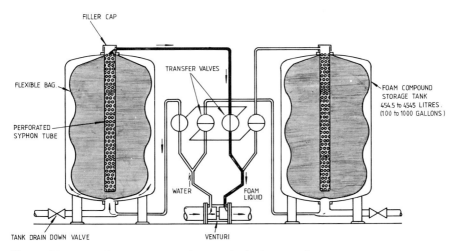

Figure 13.7 Automatic foam compound proportioning by venturi

may have a single or twin tank unit and may be used to feed a variety of low or high-expansion foam concentrates although in the latter case the foam generator is somewhat different.

When replenishing any foam concentrate it is essential to use a product compatible with the system.

Automatic foam liquid induction

An alternative and more flexible foam liquid induction unit is shown in Figure 13.8. This system is used in many tankers to operate deck foam-monitor equipment. The automatic inductor on the suction side of the pump maintains the appropriate foam liquid concentration induced into the water stream at the water flow rate demanded by the apparatus in use at any one time.

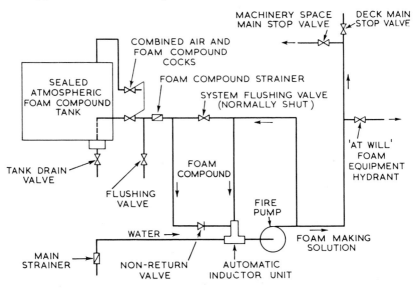

Figure 13.8 Automatic foam compound induction system

Water pressure and foam liquid suction ports are provided at the side of the automatic foam inductor unit. Water under pressure from the discharge side of the pump enters the pressure port inducing foam liquid from the tank through the foam liquid suction port. A swing paddle fitted in the body of the inductor, in way of the main water flow, moves backward and forward, according to the rate of flow. The paddle rotates a water metering vane in the water pressure port, by-passing water into the foam liquid port, thus diluting the foam liquid entering the water stream at the correct concentration to meet all flow conditions. The atmospheric type foam liquid tank enables the system to be replenished during operation.

Figure 13.9 shows the Jetmaster foam monitor as installed on the decks of tankers. This monitor requires a minimum water pressure of 8 bar in the deck main, and foam output is approximately 11 800 litre/min. A foam jet radius up to 33—37 m can be achieved at the working pressure of 8 bar.

As an alternative to introducing the foam-making compound from a bulk storage tank, foam-making nozzles capable of use with the normal fire main but drawing the compound into the nozzle from portable tanks are also used. Branchpipes for portable application of foam are made in sizes capable of generating foam up to 9000 litre/min. Normally an operating pressure of 7 bar is required from the

Figure 13.9 Marine jet-master foam monitor
(Chubb Fire Security Ltd)

fire hydrant, but lower pressures can be tolerated with slight drop in foam output. The Chubb Fire FB5X foam branchpipe, illustrated in Figure 13.10(a) will deliver up to 2200 litre/min of foam at a foam jet range of 16/18 m.

Pressure water from the fire main passes through a jet orifice assembly in the water head, inducing foam liquid and air into the water stream in the correct proportion to form foam. Foam liquid is induced from a 20-litre drum placed next to the operator. For installation in ships with no pump suitable for operating the above system a self-contained pressurised pre-mix type may be fitted. This unit contains the water and foam liquid in one tank mixed together in the correct proportion to form the foam making solution. One or more CO_2 cylinders are connected to the tank to expel the solution. A diagrammatic drawing of this type of installation is reproduced in Figure 13.11. When the CO_2 is released into the headspace of the tank, the solution is expelled to the foam sprinkler

Figure 13.10(a), left, Chubb fire FB5X foam making branch pipe (Chubb Fire Security Ltd)

Figure 13.10(b), above, Chubb Fire foam sprinkler (Chubb Fire Security Ltd)

network in the machinery space, where the foam is produced. The capacity of the tank and the number of CO_2 cylinders will depend upon the tank top area of the protected machinery space and must be sufficient to deposit a foam carpet of at least 150 mm depth to Survey Authority and Classification Society Requirements.

CO_2 FIRE EXTINGUISHING INSTALLATIONS

Fire extinguishing installations using CO_2 stored under pressure at ambient temperature are extensively used to protect ships cargo compartments, boiler rooms and machinery spaces. When released the CO_2 is distributed throughout the compartment, rendering the atmosphere inert.

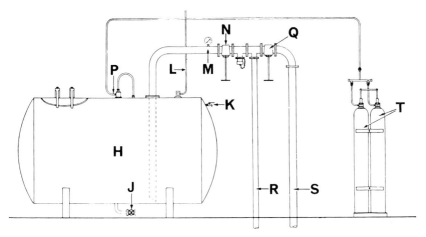

Figure 13.11 Layout of foam maker sprinkler system
H Solution storage tank
J Drain valve
K Level cock
L Safety valve
M Pressure gauge
N Stop valve
P CO_2 supply from cylinders
Q Distribution valve
R Solution supply to branchpipe in machinery space
S Solution supply to foam-makers
T CO_2 cylinders

The quantity of CO_2 required is calculated from the gross volume of the largest cargo space or machinery compartment, whichever is the greater of the two. Additional CO_2 may be required for machinery spaces containing large air receivers. This is because air released from the cylinders during a fire might seriously affect the efficiency of the fire-fighting system.

The CO_2 is stored in liquefied form in 45 kg solid drawn steel cylinders. In systems designed for cargo space and machinery space protection the cylinders will be arranged for a bulk discharge to the machinery space and ganged discharge of one or more cylinders (usually by manual release) to individual cargo spaces, depending on the volume of each space. Instructions on how many cylinders to be released for each space will be displayed at the control station. The cylinders are stored vertically, their discharge valves at the top, and internal tubes are arranged so that discharge is always from the liquid phase.

A variety of gas release mechanisms will be found. In some systems the cylinders are closed by metal discs which are cut or pierced by the release mechanism. All systems discharging to spaces normally accessible to personnel (e.g. machinery spaces and pump rooms) must be fitted with automatic audible alarms. These must operate for a suitable period before the gas is released.

Figure 13.12 is a schematic layout of a Walter Kidde CO_2 system

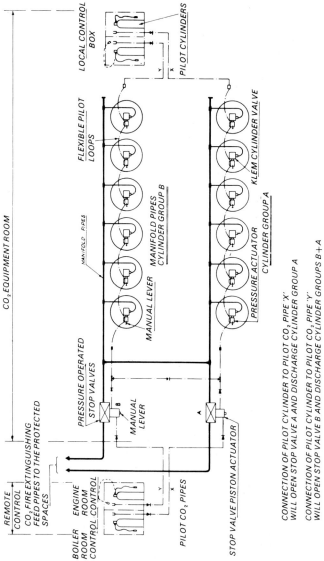

Figure 13.12A CO_2 fire extinguishing and smoke detector installation (The Walter Kidde Co. Ltd)

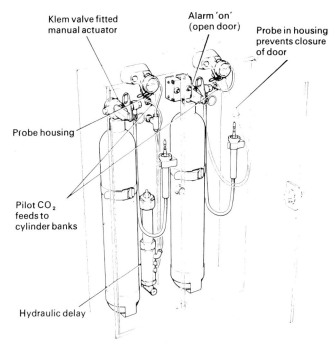

Klem valve fitted
manual actuator

Alarm 'on'
(open door)

Probe in housing
prevents closure
of door

Probe housing

Pilot CO_2
feeds to
cylinder banks

Hydraulic delay

Figure 13.13 Marine control box (The Walter Kidde Co. Ltd)

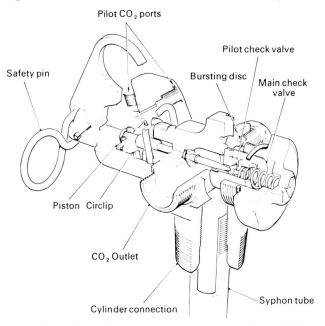

Pilot CO_2 ports

Safety pin

Pilot check valve

Bursting disc

Main check
valve

Piston Circlip

CO_2 Outlet

Cylinder connection

Syphon tube

Figure 13.14 Klem valve with pressure actuator (The Walter Kidde Co. Ltd)

in which pilot CO_2 cylinders are used to open the distribution system main stop valve and subsequently the valves on the individual cylinders. The system shown has two banks of cylinders. The pilot CO_2 cylinders are contained in a control box (Figure 13.13) and are normally disconnected from the pilot system. To operate the system a flexible pipe fitted with a quick action coupling is plugged into a corresponding socket. When the valve on the pilot cylinder is opened the pilot CO_2 will open the system main stop valve. The stop valve actuator is a piston device and when the piston is fully depressed a second port is exposed which allows the pilot gas to flow to the CO_2 cylinder bank and to operate the cylinder valves. As soon as the control cabinet door is opened an alarm is initiated and the position of the hoses in the quick-coupling housings prevents the door from being closed.

The pilot CO_2 cylinders and the main CO_2 cylinders are fitted with Klem valves. An isometric sketch of a Klem valve fitted with a CO_2 actuator is shown in Figure 13.14. The safety pin shown is for transporting the cylinders. When installed the safety pins are removed from the valves, allowing them to be operated manually or by pilot pressure. As soon as mechanical or pilot pressure is removed the valve will close again. In this system each of the cylinders is fitted with a CO_2-operated actuator.

Smoke detection

The CO_2 fire extinguishing installation may incorporate a smoke sampling system so that fire can be detected in the early stage at the smoke detecting cabinet, which is situated in the wheelhouse. Each cargo compartment is individually connected to the smoke detecting cabinet with small bore piping and, by a suction fan mounted on the wheelhouse roof, samples of the air in each cargo space are continually drawn up to the smoke detecting cabinet and discharged into the wheelhouse or the atmosphere through a diverting valve. Thus smoke from a fire in any cargo space can be immediately detected.

If desired a smoke detecting cabinet of combined visual and audible pattern can be incorporated. The audible smoke detecting unit, integral with the visual cabinet will automatically operate an alarm bell at the cabinet and in the staff accommodation as soon as smoke emerges from any of the smoke tubes in the cabinet.

A three-way valve is fitted on each pipeline below the smoke detector cabinet which is normally kept open between the cargo space and the smoke detecting cabinet. When smoke is detected, the requisite valve is turned, thus isolating the smoke detecting cabinet and opening the pipe to the cylinder battery ready to convey CO_2 to the cargo space on fire. In this way the same system of piping is used for smoke detection as well as fire extinguishing.

Low pressure CO_2 storage plants

In some vessels the CO_2 is stored in low pressure refrigerated tanks. The Distillers storage vessels (Figure 13.15) are normally arranged as cylindrical tanks fabricated to the requirements of various Pressure Vessel Codes such as Lloyds Class 1, Bureau Veritas and Merkblatte. Low temperature steels, fully tested and stress relieved, are used and the vessels are mounted on supports capable of withstanding collision shocks. The insulation to limit heat ingress into the CO_2, which is stored at a temperature of $-17°C$, is polyurethane with fire resistant additives foamed in situ.

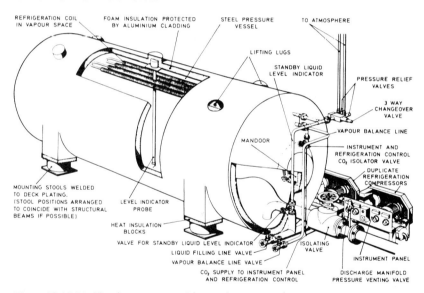

Figure 13.15 Distillers low-pressure refrigerated storage tank

Two refrigeration compressors connected to separate coils mounted inside the pressure vessel maintain the CO_2 temperature at $-17°C$. One unit is sufficient to deal with heat ingress into the CO_2 and is rated to operate for no more than 18 hours per 24 hours when the ambient temperature is $+38°C$. The other unit is standby. Either water or air cooling can be arranged. The duty from main to standby unit can be reversed to equalise running hours. The instrument panel contains tank contents gauge, pressure gauge and alarms to indicate high or low CO_2 liquid level, high or low CO_2 pressure and the refrigeration controls.

To ensure a dangerous high pressure condition does not exist if a serious refrigeration fault develops, pressure relief valves are fitted, discharging directly to atmosphere. The valves are mounted on a changeover valve and are set to discharge CO_2 gas if the pressure in the vessel rises above the design pressure of 23.8 bar. Each valve can in turn be isolated for removal and periodic testing.

The vessels are fitted with a capacitance type continuous indicator, together with a standby liquid level indicator which ensures that the CO_2 liquid level can always be checked approximately by opening the standby liquid level indicator valve which will flood the pipe to the same level as the pressure vessel. A frost line will appear due to the low temperature of the liquid CO_2. Closing the valve will cause the CO_2 to vaporise back into the pressure vessel.

The filling and balance lines are normally run to the main deck port and starboard sides for hose connections to be made to a road tanker. The balance line is used to equalise pressure with the road tanker during the filling operation.

The liquid CO_2 discharge is through a 150 mm bore pipe fitted with an isolating valve but the quantity of CO_2 discharged into the various spaces is controlled by timed opening of a discharge valve. A relief valve is fitted which will relieve excess pressure in the discharge pipe should the isolating valve be closed with liquid CO_2 trapped in the discharge manifold. Automatic or remote operation can be achieved by utilising CO_2 gas pressure from the top of the tank as the operating medium.

Because of the considerably reduced amount of steel, the storage tank compared with cylinders gives an approximate 50% weight saving and because low pressure CO_2 has a greater density than CO_2 at ambient temperature, the volume it occupies is considerably less in terms of deck space. Also, low pressure CO_2 usually costs considerably less than CO_2 supplied in cylinders.

A periodical survey of the refrigeration compressors is required by the various classification societies and this is limited normally to the exchange of pressure relief valves. At intervals of up to 10 years, an internal inspection is required via the man-door provided.

HALON SYSTEMS

Many machinery spaces are fitted with fixed fire-fighting installations which use Halon 1301 as an extinguishing medium. Halon 1301 is a group of halogenated hydrocarbon compounds with the chemical formula CF_3 Br — also known as bromotrifluoremethane. It is a colourless, odourless gas with a density five times that of air and extinguishes fire by breaking the combustion chain reaction.

Other halogenated hydro-carbons such as methylbromide and carbon tetrachloride have been used in the past as fire extinguishing agents but have been banned by various authorities because of their extremely toxic nature. Halon 1301 however is classed by Underwriters Laboratories as 'least toxic' (Group 6) and properly applied discharges of the gas allow people to see and breathe permitting them to leave the fire area with some safety.

It must be pointed out however that when Halon 1301 vapours are exposed to flame or hot surfaces above 480°C halogen acids and free halogens having a higher level of toxicity are produced. A self-contained breathing apparatus or a fresh air mask is therefore essential equipment when entering a space which has been flooded with Halon 1301.

In some small machinery spaces Halon 1301 systems may be found in which the Halon is stored in a sphere within the machinery space. In larger installations the storage battery is similar to that used in CO_2 systems, i.e. the gas is stored in 67-litre cylinders and at a pressure of 40 bars. This equates to 75 kg of Halon 1301.

Walter Kidde Halon systems utilise the CO_2 pilot release system shown in Figure 13.13 to open the Halon cylinder valves. The Klem valve shown in Figure 13.14 is not however fitted to the Halon cylinders; instead the valve shown in Figure 13.16 is used which provides a much greater through bore and hence a lower pressure drop than the Klem valve. The valve is operated by CO_2 acting on a piston actuator which dislodges the porous plug located above the main valve. This plug allows the space above the valve, normally maintained at bottle pressure, to evacuate.

Figure 13.17 shows the release gear used by Fire-Fighting Enterprises (UK) Ltd. In this system a cam on the end of the horizontal pin pushes the vertical valve rod downwards and when the pin is actuated. The pin may be moved by a pull wire or by one of a range of electrically or pneumatically operated actuators. As soon as the valve rod opens the cylinder valve gas from the cylinder is admitted onto the top side of the piston which holds the valve open. When the gas pressure falls away a spring below the piston re-seats the valve. Only one cylinder needs to be fitted with the horizontal pin, shown in the illustration. The valves on the other cylinder can be linked to it with small bore copper pipe so that gas from the master cylinder opens all of the other valves simultaneously. Most Halon cylinders are fitted with pressure gauges so that leakage can be more readily detected. The cylinders are also fitted with bursting discs.

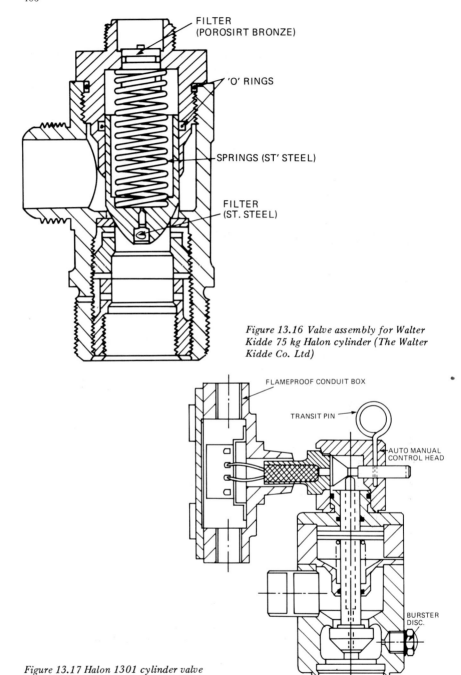

FILTER
(POROSIRT BRONZE)

'O' RINGS

SPRINGS (ST' STEEL)

FILTER
(ST. STEEL)

Figure 13.16 Valve assembly for Walter Kidde 75 kg Halon cylinder (The Walter Kidde Co. Ltd)

FLAMEPROOF CONDUIT BOX

TRANSIT PIN

AUTO MANUAL CONTROL HEAD

BURSTER DISC.

Figure 13.17 Halon 1301 cylinder valve with electric actuator (Fire Fighting Enterprises (UK) Ltd)

AUTOMATIC SPRINKLER SYSTEMS

Accommodation may be fitted with an automatic water sprinkling system which detects and extinguishes fires. A network of sprinkler heads is arranged throughout the spaces to be covered. Each sprinkler

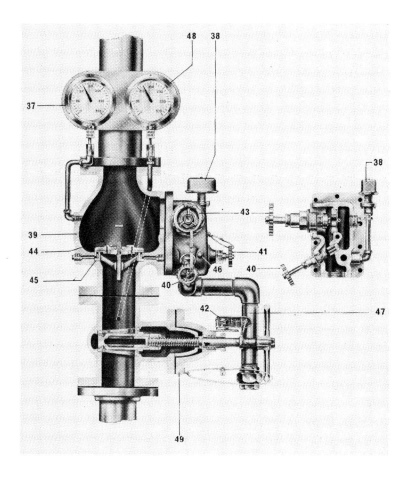

Figure 13.18 Grinnel alarm valve   *(Mather & Platt Ltd.)*

37. *Pressure gauge (above valve)*
38. *Diaphragm alarm switch (Installation)*
39. *Alarm valve*
40. *Test valve*
41. *Alarm isolating valve*
42. *Stop valve alarm switch*
43. *Drain valve*

44. *Non return valve*
45. *Alarm valve seat with annular groove*
46. *Drip plug including orifice*
47. *Drain pipe*
48. *Pressure gauge (below valve)*
49. *Stop valve*

head is normally kept closed by a quartzoid bulb which is almost filled with a liquid having a high expansion ratio. When the liquid is exposed to abnormal heat it expands rapidly to completely fill the bulb. Further expansion is sufficient to shatter the bulb. Water, maintained under pressure by compressed air, is then expelled from the sprinkler head or heads in the form of a fine spray. Each head adequately covers a floor are of 16 m² and the heads are arranged such that every part of each space requiring protection will be covered by a spray of water.

The system is shown in Figure 13.19. The pressure tank is kept half-filled with fresh water. The pressure in the tank is maintained at 8 bar by compressed air. When the pressure drops below 5.5 bar, a salt water pump cuts in automatically so that if the sprinklers come into operation a supply of water is maintained as long as is required.

Each installation is divided into sections containing up to 200 sprinkler heads and each section has an alarm valve as shown in Figure 13.18. When a sprinkler comes into operation water flows through the section alarm valve. The water lifts the non return valve 44 exposing an annular groove which connects to diaphragm alarm switch 38. This switch is coupled to an alarm bell and indicator panel on the bridge which gives audible and visual warning that fire has broken out in the section indicated. A stop valve 49, normally locked open, may be closed to carry out maintenance on the section. A test valve 40, normally closed, can be opened causing the non return valve 44 to open, thus testing the alarm system.

Key to Figure 13.19

1. Air compressor starting panel
2. Pressure tank
3. Air compressor
4. Compressor safety valve
5. Compressor check valve
6. Compressor stop valve
7. Pressure gauges
8. Water level indicator
9. Drain valve
10. Safety valve
11. Fresh water supply stop valve
12. Fresh water supply check valve
13. Connection from fresh water supply
14. Shore connections
15. Stop valves
16. Alarm valve
17. Diaphragm alarm switch
18. Stop valve
19. Check valve
20. Pressure tank alarm bell

21. Hose coupling
22. Trunk main
23. Pump test valve
24. Pump delivery stop valves
25. Pump delivery check valve
26. Automatic pressure relay
27. Pump
28. Installation control valves
29. Automatic relay isolating valve of wedge pattern
30. Release valve
31. Indicator pilot lamp
32. Indicator
33. Alarm bell
34. Automatic pump starting panel
35. Diagram of ship
36. Sprinklers
52. Trunk main drain valve
53. Air release valve
54. By-pass check valve

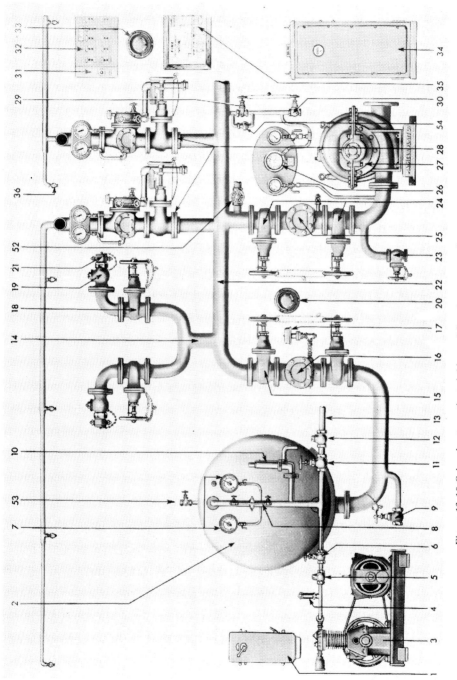

Figure 13.19 Grinnel automatic sprinkler and fire alarm system (Mather and Platt Ltd)

Weekly maintenance of the system consists of greasing the various valves and checking their freedom of movement, logging the pressure gauge reading, before and after each alarm valve (thus checking the tightness of the non-return valves) and checking the alarm systems. This latter is done by opening the test valves and checking that the audible and visual alarms work. The main stop valve should also be closed slightly which should also activate audible and visual alarms on the bridge. On completion the pressure tank should be recharged with fresh water and air. The salt water pump should also be checked by closing the isolating valve 1 and opening release valve 2 and test

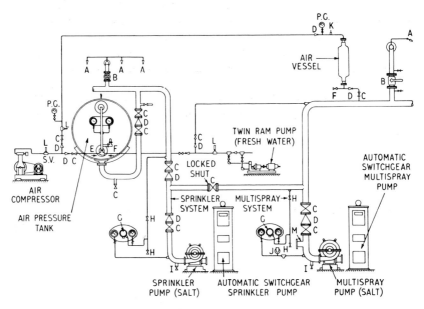

(left) Sprinkler system for accommodation spaces

(right) Machinery space system

Figure 13.20 Water spray systems for accommodation and machinery spaces

A	Sprinkler heads in accommodation	A	Sprayer head in engine room (open)
B	Section control, alarm and test valve	B	Section control valve
C	Stop valves	C	Stop valves
D	Non-return valves	D	Non-return valves
E	Air supply to pressure tank	F	Fresh water supply to pressure bottle
F	Fresh water supply to pressure tank	G	Pressure switch to start pump
G	Pressure switch to start pump	H	Pressure release valves to test pump automatic start
H	Pressure release valves to test pump automatic start	I	Water outlet for S.W. pump testing
I	Water outlet for S.W. pump testing	J	Pressure switch for alarm
L	Safety valves	K	Fusible plug
P.G.	Pressure gauge	L	Safety valve
		M	Connection to ship's fire main (optional)

valve 3. On the fall-off of line pressure the pump should start automatically. The pump delivery pressure should be logged. In the event of a fire, when a normal situation is recovered the system should be flushed out and recharged with fresh water and air.

For machinery space protection quartzoid bulbs are not fitted. Sections of sprinklers are brought into operation by manually opening isolating valves. As with the automatic systems the lines are kept pressurised by compressed air and a sea water pump cuts in automatically to provide a continuous supply of water in the event of a fire. The two systems may well be combined with a cross-over valve, normally locked shut, separating the two (Figure 13.20).

Where an automatic system is not fitted in accommodation spaces it is necessary to install an automatic fire alarm system. The usual system fitted consists of an electric circuit in which a number of bi-metallic temperature sensors are located. Any one of these temperature sensors, in the event of a sudden temperature rise, breaks the electrical circuit causing an alarm to sound. Visual indication showing the section in which the fire has occurred is also provided.

DETECTORS

A variety of devices exist for detecting fires namely:
1. Smoke detectors based on the ionization principle or on the photo-electric cell principle
2. Heat sensors;
3. Rate of temperature rise sensors;
4. Flame detectors.

Each type has its merits and most systems will deploy two or more of the above types of devices depending on the space being protected.

Within machinery spaces, ionization smoke detectors are generally considered the first line of defence. These devices monitor the electrical charge which occurs when combustion particles reach the charged space in the detector. So that the units can be readily tested an indicator light is provided on the sensor body. This lights up in the presence of gas from a test canister or even from the fumes off a cigarette.

The appropriate number of detector heads are sited in strategic positions at deckhead level in the protected compartment or arranged in zones if the compartment is extensive.

The control equipment associated with the installation may be accommodated in the wheel house or in a fire control centre and comprises a fire indicating cabinet to which the detector heads are connected, a power unit, to convert the incoming ship's supply to the voltage appropriate to the equipment and a standby battery unit. The cabinet will indicate in which space a fire has been detected and will also monitor the system and indicate whether a fault has developed. It will also instigate an audible alarm.

INERT GAS SYSTEMS

Mention has already been made of the necessity for oxygen to be present for combustion to take place and of how carbon dioxide can be used to displace air from a compartment until combustion can no longer be supported. A gas such as CO_2, which is incapable of supporting combustion, is said to be inert. Another example of an inert gas is nitrogen and, under certain conditions, boiler flue gas containing less than 11% of oxygen can also be considered inert. The latter is being increasingly used aboard tankers to prevent explosions occurring in cargo tanks.

Safety

Combustible gases and vapours, such as petroleum vapour, when mixed in the correct proportion with air in an enclosed vessel will burn so rapidly that an explosion occurs. The burning can be initiated quite easily, ignition often being caused by a relatively small spark. For each gas or vapour there is, however, an upper and lower concentration of vapour in air between which an explosion can occur. These limits are referred to as the lower flammable and upper flammable limits (LFL and UFL) or, alternatively, the lower and upper explosive limits (LEL and UEL). Petroleum vapour for instance, has an LEL of 1.4% and a UEL of 6.4% while hydrogen has an LEL of 4% and a UEL of 75%. Crude oil contains a variety of hydrocarbon mixtures and the vapour contained in crude oil tanks will exhibit a wide variety of upper and lower explosive limits.

By ensuring that the vapour/air mixture in the tanks is maintained outside the explosive limits an increased measure of safety can be achieved. Boiler flue gas can be used to maintain the environment below the LEL. In designing the inerting system the following must however be borne in mind:

1. Boiler flue gas contains highly corrosive elements which must be removed.

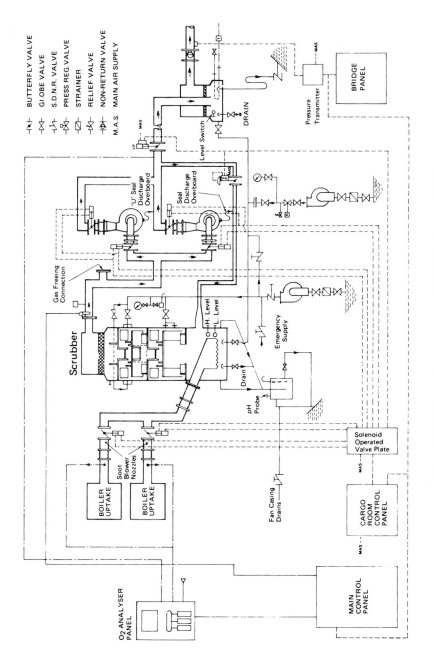

BUTTERFLY VALVE
GLOBE VALVE
S.D.N.R. VALVE
PRESS REG.VALVE
STRAINER
RELIEF VALVE
NON-RETURN VALVE
M.A.S.   MAIN AIR SUPPLY

Figure 13.21 Schematic arrangement of a Peabody inert gas plant

2. The gas must be cooled.
3. Measures must be taken to prevent the back flow of hydro-carbon gases to unsafe areas.
4. Interlocks must be provided to ensure that the inert gas system is not fed from a boiler undergoing soot-blowing.
5. On products tankers, which carry a variety of cargoes, it is necessary to ensure that contamination of one cargo by another cannot occur.

Operation of the inert gas system

An example of an inert gas system is shown in Figure 13.21 and while plant might differ in detail the principles remain the same for all boiler flue gas inerting systems.

The boiler flue gas is drawn through a scrubbing tower which uses sea water to scrub out some of the impurities; the sea water also cools down the gas from the boiler uptake temperature (upwards of 135°C) to only 1 or 2°C above the sea water temperature.

In the F.A. Hughes system the scrubbing tower (Figure 13.22) is a rectangular mild steel structure containing a number of polypro-pylene trays. Each tray is pierced by a number of shrouded slots. The shrouds are called tunnel caps and this type of scrubber is referred to as a tunnel cap scrubber.

The sea water enters the scrubber at the top and flows across each tray, a series of weirs being arranged to ensure that each tray is flooded to a depth of about 20 mm. Downcomers are arranged to conduct the water from one tray layer to the next.

The gas enters the pack at the bottom through a water seal and rises through the pack, bubbling up onto the trays through the tunnel caps which distribute the gas across the free water surface on the trays. At the top of the tray-pack a polypropylene mattress demister is arranged which removes the bulk of any entrained water. Above the demister a number of water sprays are located. These are used to wash down the scrubber after use.

The amount of sea water used varies to some extent with the design of the scrubber but units with sea water flows of from 100 tonnes/hr to 350 tonnes/hr are in common use.

Typical volumetric analyses of the gas before and after the scrubber are as follows:

		Before	*After*
Carbon dioxide	CO_2		12%
Oxygen	O_2		4.5%
Sulphur dioxide	SO_2	0.2%	0.02%
Nitrogen	N_2	84%	77%

remainder — water vapour and solids

415

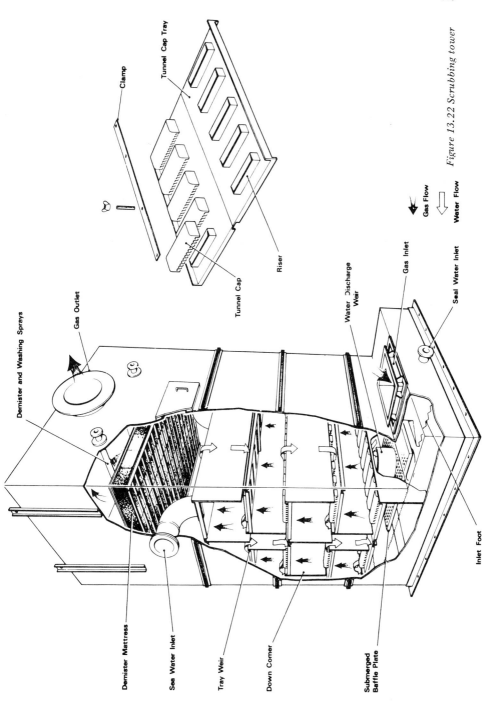

Tunnel Cap Tray

Clamp

Gas Outlet

Demister and Washing Sprays

Tunnel Cap

Riser

Water Discharge Weir

Gas Inlet

Seal Water Inlet

Gas Flow

Water Flow

Inlet Foot

Demister Mattress

Sea Water Inlet

Tray Weir

Down Comer

Submerged Baffle Plate

Figure 13.22 Scrubbing tower

Because of the sulphur dioxide absorbed by the sea water its pH value is changed from around about 7 to about 2.5 as it passes through the scrubber — hence the selection of polypropylene for the trays' tunnel caps and demister mattress. The tower itself is lined with ebonite rubber.

Other designs of scrubber used at sea include the impingement and agglomerating type such as the Peabody tower shown in Figure 13.23. In this type the incoming flue gas is first wetted by sea water sprayers and then passes upwards through a venturi slot stage which agglomerates the solid particles. The gas then rises through slots in a series of trays. Above the slots are a number of baffle plates. The trays are covered in water introduced at the top of the tower in much the same way as in the tunnel cap tower previously described. A mesh type demister is arranged at the top of the tower.

Fans

The fans used in the inert gas system must be capable of providing a throughput equivalent to about 1.33 times the maximum cargo

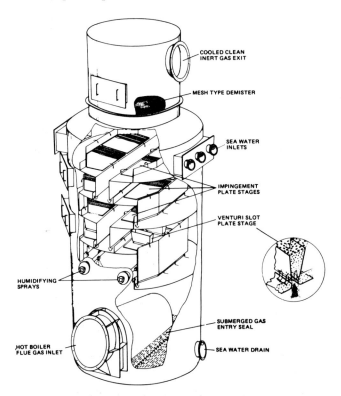

Figure 13.23 Peabody tower circular scrubber

pumping rate since the tanks must be fed with inert gas during cargo discharge. At full output the fans must be capable of delivering at an over-pressure of 670–1000 mm w.g. which with pipeline losses equates to a static pressure at the fan of up to 1600 mm W.G.

Both electric motor-driven and steam turbine driven fans are used and it is usual to provide one running and one stand-by unit. These are normally both 100% duty units although some installations with a 100% duty and a 50% duty fan have been used. Because of the corrosive nature of the gas the fan materials used must be carefully selected. Fan impellers of stainless steel or nickel-aluminium bronze are frequently used and the mild steel casings are internally coated with, for example, coal tar epoxy. Some problems have been encountered with bearing failures on inert gas fans and these have frequently

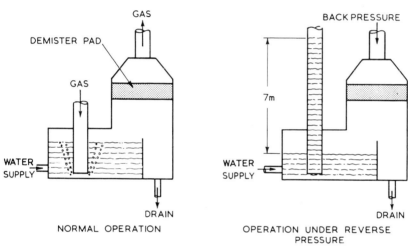

Figure 13.24(a) Wet type deck seal

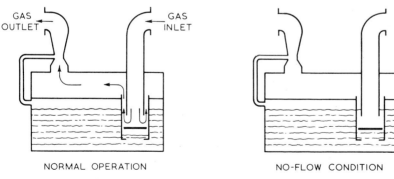

Figure 13.24(b) Dry type deck seal in which the water-trough is by-passed at high gas flow rates

been caused by blade imbalance brought about by solids deposits. It is common therefore to find cleaning water nozzles installed on fans and these should be used from time to time to clean the fan blades. The impeller may be supported in plain or anti-friction bearings. Where the latter are used it is normal to mount them on resilient pads.

The fans discharge to the deck main via a seal which prevents the back flow of gases. The seals used can be classified as wet or dry seals. Both types involve feeding the inert gas through a flooded trough but in the dry type seal a venturi gas outlet is used which effectively pulls the water away from the end of the gas inlet at high flows allowing the inert gas to by-pass the water trough (Figures 13.24a and b). The reason for developing this type of seal was because early wet-type seals frequently caused water carry-over into the system. As with other components in the inert gas system the internal surfaces of the deck seal must be corrosion protected, usually by a rubber lining.

Topping up systems

Diesel engine exhaust gas has too high an oxygen content for use as an inert gas. In motor tankers, the exhaust gas from an auxiliary boiler must therefore be used. This is usually plentiful when discharging cargo since the auxiliary boiler will be supplying steam to cargo pumps and perhaps a turbo-alternator. At sea however the auxiliary boiler might well be shut down and an alternative supply of inert gas must be found to make good losses from the system, or to supply gas prior to tank washing.

The W.C. Holmes oil-fired inert gas generator can be used for this purpose and indeed a number of inert gas generators of this make have also been fitted in dry cargo ships for fire-fighting duties. A Holmes vertical chamber oil-fired inert gas generator is shown schematically in Figure 13.25. Units of this type are capable of gas outputs in the range 100–4000 Nm^3/hr at an output pressure of 0.138 bar. Units have also been provided with output pressures of up to 1 bar where required.

In this unit oil is drawn from a storage tank and is pumped by a motor-driven gear pump through a filter and pressure regulator to the pilot and main burners. The necessary air for combustion is delivered by a positive displacement Roots-type air blower.

Oil and air are mixed in the correct proportions in an air atomising burner mounted on the top of a vertical, refractory lined combustion chamber. The burner fires downwards and the products of combustion

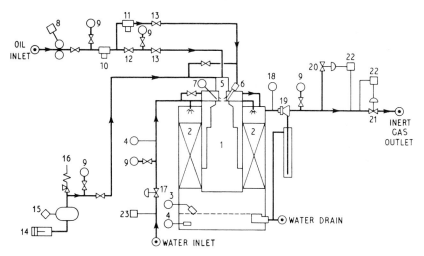

Figure 13.25 Holmes vertical chamber oil-fired inert-gas generator

1. Combustion chamber
2. Cooling annulus
3. Float switch
4. Cooling water thermometer
5. Main burner
6. Pilot burner
7. Flame detector
8. Motor driven oil pump
9. Pressure gauge
10. Oil filter
11. Oil filter-pilot burner
12. Pressure reducing valve

13. Solenoid valve
14. Air filter
15. Air blower
16. Air relief valve
17. Water inlet stop valve
18. Temperature switch
19. Moisture separator
20. Inert gas relief valve
21. Back pressure regulating valve
22. Pressure controllers
23. Pressure switch

leave the combustion chamber at the lower end. They then reverse direction and travel upwards through the cooling annulus surrounding the combustion chamber. The inert gas is cooled by direct contact with seawater in the cooling annulus to a temperature within 2°C of the temperature of the water. This water also keeps the shell of the combustion chamber cool, and in addition removes most of the sulphurous gases.

As the generator is of the fixed output type, a relief valve is fitted to exhaust excess inert gas to atmosphere should there be a reduction in demand. Start up of the generator is fully automatic: all that has to be done is to press a single push button. A programme timer subsequently controls the ignition of the pilot burner, ignition of the main burner and a timed warm-up period. The combustion chamber is then automatically brought up to the correct operating pressure. Operation is continuously monitored for flame or water failure and excessive cooling water level. Should emergency conditions arise, the generator will automatically shut down and an audible alarm sound.

In the Kvaerner Turb inert gas system, inert gas and emergency electrical power generation are both produced from the one packaged unit. This consists of a radial flow gas turbine coupled to a 1200 kW capacity generator. The gas turbine produces an exhaust gas containing approximately 16% of oxygen and, by incorporating an after burner, an exhaust gas with an oxygen content as low as 1% can be produced.

The emergency generator is only capable of producing full output when it is exhausting to atmosphere. When used in conjunction with the after burner the high back-pressure reduces the power output by about 50% but the electrical output is still sufficient to provide power for the cooling water pump used in conjunction with the system and some surplus power will be available which can be fed into the ship's main.

Gas analysis

Fixed oxygen analysers are fitted to ships with inert gas systems. These continuously monitor the oxygen content of the flue gas and the inert gas. Para-magnetic type analysers are frequently used. These feed an electrical signal to the oxygen indicator which is usually calibrated in the range 0 − 10% by volume. The analyser is frequently connected to audible and visual alarms which are activated when the oxygen level rises above the desired value.

Portable oxygen content meters are also frequently carried and if so should be used to check that individual tanks are suitably inerted. Conversely a portable meter can also be used to check whether the atmosphere in a tank will support life and thus permit men to enter the tank. If there is any doubt whatsoever of whether the tank atmosphere is life-supporting then breathing apparatus should be worn when entering.

Combustible gas indicators

A combustible gas indicator can be used to check that an atmosphere will not support combustion. The most common type is the catalytic filament gas indicator. This uses a heated platinum filament to catalyse the oxidation of combustible vapours. Vapour oxidation can take place even outside the limits of flammability and the heat generated will raise the temperature and hence the electrical resistance of the filament. The increase in resistance of the filament can be measured and is nearly proportional to the concentration in vapour in terms of flammability. The instrument can thus be calibrated in terms of percentage flammability − usually 0−100% LFL.

One or two precautions need to be taken with these meters. Firstly the meter will read 0% for over-rich mixtures and the operator needs to be quite sure that the mixture is lean (or below LFL) and not rich (above UFL). A visible sign of this is that, with an over-rich mixture, the meter needle will first swing over maximum before settling on zero. Secondly the filament can be adversely affected by lead compounds or sulphur and after continual exposure to such compounds the meter can give a falsely low reading for hazardous mixtures. To ensure that this does not occur the meters should be checked against the test kits provided every time that they are used. If the test kit indicates that the instrument is inaccurate then the filament should be changed.

BREATHING APPARATUS

In the course of fighting a fire it may be necessary to use breathing apparatus either because of smoke or because a space has been closed down to deprive a fire of oxygen. In some ships the equipment comprises a helmet attached by a length of hose to a bellows or a hand-operated rotary air pump; others are equipped with self-contained units incorporating one or two compressed air cylinders.

The units fitted with air hose and bellows, or handoperated pump, are usually equipped with a helmet which covers the man's head. Operation is quite simple but it is important to memorise a system of signals to indicate to the man operating the pump whether more air is required or if the person wearing the helmet is in trouble. These are displayed on the apparatus and should be carefully noted.

The self-contained units vary in detail but consist essentially of one or two compressed air cylinders fitted to a harness which the man wears on his back. Each cylinder is charged to a pressure of 200 bar and contains enough air to sustain a man for approximately 20 min at a hard-working rate or 40 min if he is at rest. A reducing valve, set at a pressure of about 5.5 bar, is fitted on the cylinder outlet pipe together with a pressure gauge. A by-pass valve is usually incorporated. The bottles are connected to a close-fitting face mask by a length of low pressure air hose and, interposed between the reducing valve and the mask, is a demand valve which the operator adjusts to meet his rate of breathing. This valve may be mounted on the front of the harness or on the back of the harness waist-band. A non-return valve in the face mask permits the expulsion of air. It is essential, when putting the face mask on, to be sure that the mask fits the face properly. This can be checked by stopping the supply of air at the demand valve and breathing in; any leakage around the edge of the mask should then be obvious.

EMERGENCY GENERATING SETS

Diesel-driven emergency generator sets are frequently fitted above the bulkhead deck. These are arranged to provide essential circuits in the event of the main power plant being out of action because of fire, flooding or any other reason. When fitted they must be capable of starting reliably from cold and must have their own independent supply of fuel. In large ships they will be arranged to start automatically if the main supply fails. Machines of up to 60 kW output will be started by a motor drawing its power from an accumulator. This will be normally charged from the mains via a suitable transformer/rectifier. Larger machines may be started by compressed air but this must be from a separate reservoir located above the bulkhead deck.

In some vessels a limited accumulator capacity may be supplemented by a hand-started machine. Many emergency generator prime-movers are air-cooled diesel engines. These have good cold-starting characteristics and there is no liquid cooling system to attend to.

EMERGENCY BILGE PUMPS FOR PASSENGER SHIPS

To meet the contingency of the machinery space being bilged and the power-plant in consequence being put out of action, an emergency bilge pump is required, the power for which is supplied from an emergency generating-set placed high above the deep-water line at which the vessel floats when fully loaded.

The function of the emergency bilge pump is not to endeavour to pump out a compartment which has been badly holed, but to overcome any leakage in adjacent compartments, which may result from straining of the bulkhead structure, small cracks which have developed or fractured pipes. As the compartment in which the pump is placed may be the one which is badly holed, the pump should be capable of working when completely submerged and having to pump out leakage into adjoining compartments.

The pump must be self-priming and capable of overcoming a reasonable amount of air leakage into the bilge-piping system. The emergency bilge pump should be capable of working continuously as an ordinary bilge pump, so as to avoid the necessity of fitting an additional pump.

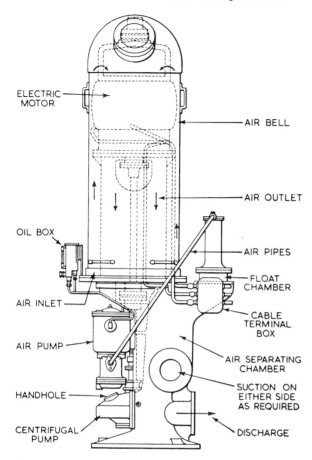

ELECTRIC MOTOR

AIR BELL

AIR OUTLET

OIL BOX

AIR PIPES

FLOAT CHAMBER

AIR INLET

CABLE TERMINAL BOX

AIR PUMP

AIR SEPARATING CHAMBER

HANDHOLE

SUCTION ON EITHER SIDE AS REQUIRED

CENTRIFUGAL PUMP

DISCHARGE

Figure 13.26 Patent emergency bilge pump

An emergency bilge pump is illustrated in Figure 13.26. The motor is placed in an air-bell of proportions such that water cannot reach the motor. A quick-running two-throw air-pump driven by worm gearing runs all the time the pump is being used. This pump takes either air or water or both, and not only primes the centrifugal pump but also deals with air leakage.

POWER-OPERATED WATERTIGHT DOORS

The adequate watertight subdivision of ships is effected by a number of steel watertight bulkheads or transverse divisions extending from the tank top of the double bottom to above the margin line, bulkhead deck or freeboard deck of the ship.

It may be necessary to provide doors in some of the bulkheads. These doors must be properly watertight when closed and capable of being opened and shut locally as well as remotely in the event of an emergency. In some vessels it is considered adequate to provide a simple hand-operated rack-and-pinion system to open these doors. In such cases an extended spindle to a point above the waterline is all that is required to provide remote operation.

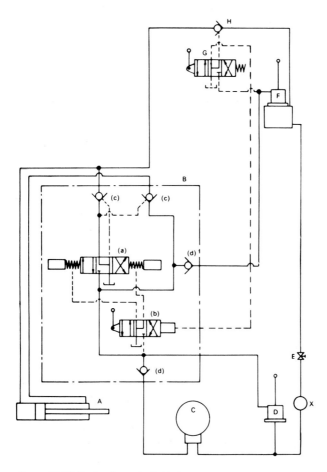

Figure 13.27 Hydraulic watertight door
A *Door cylinder*
B *Door control valve unit, comprising*
    (a) main valve (solenoid); (b) pilot
    valve; (c) pilot operated check valves
    (d) check valves
C *Pump unit*
(Stone Manganese Marine Ltd)

D *Hand pump (local)*
E *Stop valve*
F *Hand pump (remote) and tank*
G *Control valve (remote)*
H *Pilot operated check valve (remote)*
X *Strainer*

In vessels which have a large number of watertight bulkheads, pierced by access doors below the waterline some sort of powered system is necessary. A number of systems are in use, both electrical and hydraulic. A circuit diagram for a hydraulically operated system is shown in Figure 13.27. At each door a motor driven pump supplies oil, at a pressure of 48 bar to a direct acting cylinder through a control valve which is actuated electrically by solenoids when on bridge control. The d.c. supply for the pump motors is obtained, under normal conditions, from an a.c./d.c. transformer rectifier unit common to all doors in the installation. In emergency the supply is taken from the ship's batteries.

Local control

A control is located adjacent to each door for *shut/open* or for *intermediate* positions. The control lever is operated manually from either side of the bulkhead and movement of the lever actuates a pilot valve controlling the pump motor circuit under all conditions of control. Advantage is taken of the area differential of the piston in the door operating cylinder for closing and opening the door. A slightly greater force is available in unsealing (opening) the door.

The control lever is spring loaded to the mid-position to ensure a hydraulic lock within the cylinder, thus preventing any possible movement of the door due to the motion of the ship.

Bridge control

The bridge controller is designed to close (in sequence) a maximum of twenty doors within a specified time of 60 sec. This period includes a 10 sec audible alarm period at each door before the closing movement starts and the alarm continues to sound until each door is fully closed.

Close and *Re-open* controls energise solenoids at the door control valve to start the pump motor. Limit switches, actuated by the movement of each door, control the electric circuit.

Any door which is re-opened locally while the system is under the *Closed* condition from the bridge controller will automatically re-close when the local control lever is released. Indication that power is available to close the doors is given by a *Power-on* amber light, with a test button provided.

Emergency control

Under conditions when no power is available the doors may be

closed and opened by manually operated pump and control valves
at either of two positions:

   (a) Adjacent to each door from either side of the bulkhead, and
   (b) From a position on the bulkhead deck.

Light indication

A coloured light indicator on the bridge shows the position of each
door; *green* for shut and *red* for open. The appropriate lights are
duplicated alongside the manually operated emergency pumps on the
bulkhead deck. To ensure that power is always available all the
indicator lights and alarms are connected direct to the d.c. supply
and to the ship's batteries.

LIFEBOAT DAVITS

Boat davits (Figure 13.28) vary in style very considerably to suit the
load to be handled and the layout of a ship, but the design considerations applying to boat winches is reasonably common to all styles
of outfit.

   When lowering a lifeboat no mechanical assistance apart from
gravity is applied. The only manual function required of the operator
is to release the winch handbrake and hold it at the 'off' position
during the lowering sequence. If the operator loses control of the
brake lever in difficult conditions, the attached weight will provide
a positive means of application and the boat will be held at any
intermediate position until control is recovered. This condition
applies throughout the outboard movement of the boat, from its
stowed position until it is waterborne. Figure 13.30 shows the
arrangement. The main brake (on the left in Figure 13.29) is fitted
with two shoes, pivoted at one end and coupled at the other with the
weighted lever by a link. The lever projects from the casing through a
watertight seal. The shoes are Ferodo-lined and have a normal useful
life of five years or more.

   Figure 13.29 shows in section the main brake described above and
the centrifugal brake (shown on the right-hand side of the figure).
This limits the speed of the boat descending under the increasing
acceleration forces of a load descending by gravity. Shoes of known
weight act upon the inner surface of a stationary drum; these shoes,
held in position by the springs shown (which have no other duty)
describe a circular path whose diameter increases with centrifugal

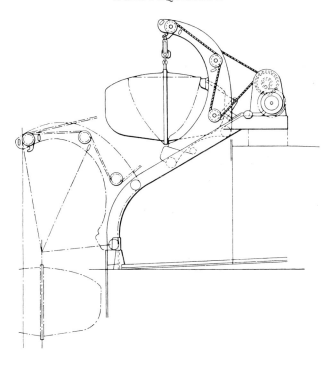

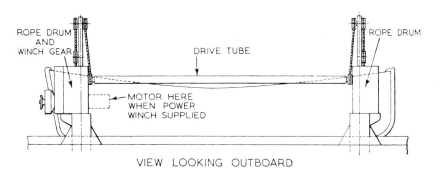

ROPE DRUM
AND
WINCH GEAR

DRIVE TUBE

ROPE DRUM

MOTOR HERE
WHEN POWER
WINCH SUPPLIED

VIEW LOOKING OUTBOARD

Figure 13.28 Overhead gravity davits (Welin Davit & Engineering Co Ltd)

forces (which varies as the square of the speed of rotation) and so limit the speed of the drum shaft; in this way, the lowering speed of the boat can be kept within predesigned limits, say 20–40 m/min. Additional automatic features ensure that the drums will not run in reverse in the event of a power failure when a boat is being hoisted. These brakes require regular inspection for wear and cleaning.

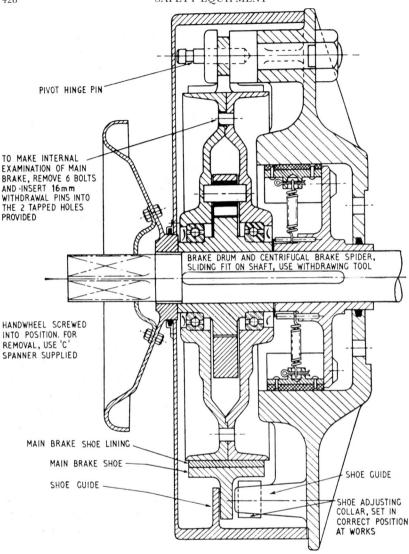

PIVOT HINGE PIN

TO MAKE INTERNAL
EXAMINATION OF MAIN
BRAKE, REMOVE 6 BOLTS
AND ·INSERT 16mm
WITHDRAWAL PINS INTO
THE 2 TAPPED HOLES
PROVIDED

BRAKE DRUM AND CENTRIFUGAL BRAKE SPIDER,
SLIDING FIT ON SHAFT, USE WITHDRAWING TOOL

HANDWHEEL SCREWED
INTO POSITION. FOR
REMOVAL, USE 'C'
SPANNER SUPPLIED

MAIN BRAKE SHOE LINING

MAIN BRAKE SHOE

SHOE GUIDE

SHOE GUIDE

SHOE ADJUSTING
COLLAR, SET IN
CORRECT POSITION
AT WORKS

*Figure 13.29 Welin davit winch. Section of brakes showing centrifugal brake
(Welin Davit & Engineering Co Ltd)*

LIFEBOAT ENGINES

The SOLAS 1974 Convention requires that lifeboat engines, where
fitted, should be compression ignition engines. Both water-cooled
and air-cooled machines will be found installed in lifeboats. Because
of their cold starting characteristics, simplicity and low maintenance

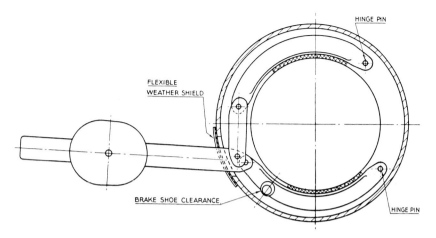

Figure 13.30 Section of main brake (Welin Davit & Engineering Co Ltd)

requirements air-cooled engines are more generally found in open lifeboats. In totally enclosed lifeboats water-cooled engines are used and a small single-pass heat exchanger (usually just a large-bore tube) is arranged on the outside of the lifeboat bottom, via which the freshwater circuit is cooled by the sea. Most air-cooled engines use direct injection, open chamber combustion and, if properly maintained, will start without the use of electric starters, heaters, or ether injection down to −12°C. The units do, however, require an adequate supply of air and when fitted with weather covers care should be taken to ensure that the louvres on these are not blocked while the engine is working.

Running maintenance (see Figure 13.31)

The following is a typical maintenance programme for an air-cooled engine in regular use. Decarbonising may be required more frequently when engines are running on light loads for long periods.

Daily
   (a) Check lubricating oil level and top up if necessary (A).
   (b) Check that the cooling system is in order and free from obstruction.

Every 50 hours
   (a) Clean the air cleaner (oil bath type).
Every 250 hours
   (a) Clean the fuel filter (B).

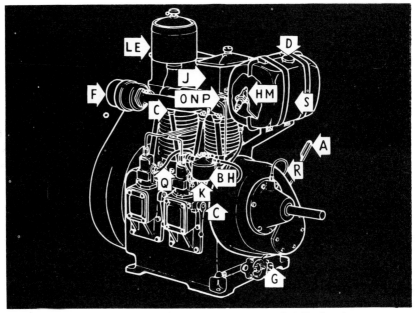

Figure 13.31 Routine maintenance for a two-cylinder air-cooled diesel engine
See 'Running Maintenance', for key to letters

(b) Check all nuts, bolts, etc. for tightness (C).

(c) Make sure the fuel tank filler cap vent hole is clear (D)

(d) Clean the air cleaner (paper element type) (LE).

(e) Clean out deposit from exhaust system (F).

(f) Clean the fuel feed pump strainer.

(g) Drain the sump, flush out with flushing oil and refill with new oil. Clean the strainer. (G) (Paraffin may be used if flushing oil is unobtainable but the engine must not be run with paraffin in the sump.)

(h) Test the fuel system for leaks (H).

(j) Remove the fuel injectors and test spray. If in order replace without further interference.

(k) Check the valve clearance and adjust if necessary (J).

(l) Clean the lubricating oil feed restrictor to rockers.

(m) Fit new lubricating oil filter element and joint ring.

(n) Lubricate the speed control linkage.

Every 500 hours

(a) Fit new fuel filter element (K).

(b) Fit new air cleaner element.

Every 1000 hours

(a) Clean the button filter in the fuel delivery pipe unions at the injector end (M).

Table 13.2 Fault-finding chart for air-cooled diesel engine

Trouble	Reason	Causes	Suggested remedy
Engine will not start	Fuel supply failure Check by operating the fuel pump priming lever and listen for the characteristic squeak in the injector	No fuel in tank	Fill tank and bleed the fuel system
		Air in the pipe line	Bleed the system
		Broken fuel pipe or leaking connection	Repair or renew the pipe and tighten the connection
		Fuel filter choked	Fit new filter element
		Faulty injector nozzle	Fit new nozzle
		Fuel pump plunger sticking	Fit new pump
		Fuel pump tappet sticking	Free and clean the tappet
	Poor compression	Valves sticking	Free the valves
		Cylinder head loose	Tighten all nuts
		Cylinder head gasket blown	Fit new gasket
		Piston rings stuck in grooves	Check rings and clean the piston
		Worn cylinder and piston	Overhaul the engine
		Valves not seating properly	Check valve springs / Grind if necessary / Check valve clearance
	Incorrect lubricating oil		Drain the sump and fill with correct oil
Engine starts but fires intermittently or soon stops	Faulty fuel supply	Air in the fuel lines	Bleed the system
		Water in the fuel	Drain fuel system and fill with clean fuel
		Faulty injector nozzle	Fit new nozzle
		Fuel filter choked	Fit new filter element
	Faulty compression	Broken valve spring	Fit new spring
		Sticking valve	Free the valve
		Pitted valve	Grind or renew
	Dirty engine	Blocked exhaust pipe or similar	Clean out
Engine lacks power and/or shows dirty exhaust	Faulty fuel supply	Faulty fuel pump	Fit new pump
		Faulty injector nozzle	Fit new nozzle
		Unsuitable fuel	Drain the fuel system and fill with correct fuel
	Out of adjustment	Valve clearance incorrect	Adjust
		Fuel timing incorrect	Adjust
	Dirty engine	Blocked exhaust pipe or similar	Clean out
		Dirty air cleaner	Clean out
		Faulty piston ring	Fit new ring
		Excessive carbon on piston and cylinder head	Decarbonise
		Worn cylinder and piston	Overhaul the engine
Faulty running	Knocking	Carbon on piston crown	Decarbonise
		Injector needle sticking	Fit new nozzle
		Fuel timing too far advanced	Adjust timing
		Broken piston ring	Fit new ring
		Slack piston	Fit new piston
		Worn large end bearing	Renew and check lubrication
		Loose flywheel	Refit
		Worn main bearing	Renew and check lubrication
	Overheating	Overload	Reduce the load
		Lubricating oil failure	Fill the sump and check system
		Cylinders giving unequal power	Check and adjust fuel pump setting
		Excessive valve clearance	Adjust
		Cooling system failure	Check that the cooling system is in order and free from obstruction
	Speed surges	Air in fuel pipes	Bleed the system
		Governor sticking	Free the governor
	Sudden stop	Empty fuel tank	Fill tank and bleed the fuel system
		Choked injector	Fit new nozzle
		Fuel pipe broken	Repair or renew
		Seized piston	Fit new piston or, in an emergency, stone down
	Heavy vibration	Loose holding down bolts	Tighten up

(Petter Ltd)

Every 2000 hours
   (a) Decarbonise. Examine valves and grind if necessary (O, N).
   (b) Clean out piston oil return holes. Check cylinder bore wear (P).
   (c) Examine the crankshaft bearings and renew if clearance is excessive (Q).
   (d) Clean out the fuel tank thoroughly (S).
   (e) Wash out lubricating oil system (R).

Hydraulic cranking system

Some lifeboat engines and the larger diesel-driven emergency generator sets may be fitted with a hydraulic motor to crank the engine up. The Startorque system is such a device which uses an automatically charged accumulator to provide power to the hydraulic crankingmotor. The accumulator is precharged with nitrogen gas to 83 bars.

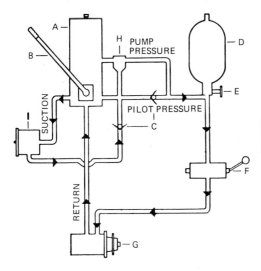

Figure 13.32 Schematic of Startorque system showing principles of operation
A   Oil reservoir and filter
B   Hand pump
C   Non-return valves
D   Hydraulic accumulator
E   Hand shut-off valve
F   Starter operating valve
G   Hydraulic cranking motor
H   Off loading valve
I   Mechanical recharging pump

The system is shown schematically in Figure 13.32 which shows that the accumulator may be re-charged by a hand pump B or an engine driven pump I. The stored energy in the accumulator is released by a hand-operated valve F, assisted by a check valve which allows a small quantity of oil to pass, enabling full engagement of the starter motor pinion with the flywheel. The valve then opens fully allowing full flow to the hydraulic cranking motor which generated enough torque to start the engine. The oil returns to the reservoir where it is pumped back to the accumulator by either the hand pump or the engine driven pump.

An off-loading valve H protects the system from being over-charged and spills at 20.7 bars back to the reservoir on an open circuit maintaining flow through the re-charging pump I at all times. The unit cranks the engine at about 375 rev/min for nine revolutions although these figures may be varied to suit particular engines by modifying the size of the accumulator.

WHISTLES AND SIRENS

Audible signals, to indicate the presence of a ship in poor visibility, or to inform other vessels of the ship's intended movements, have long been used at sea. Steam, air and electric whistles have all been fitted for this duty, some having audible ranges of as much as nine nautical miles.

The air and steam whistles used operate on much the same principle, namely the working fluid causes a diaphragm to vibrate and the sound waves generated are amplified in a horn.

The arrangement of a Super Tyfon air whistle is shown in Figure 13.33. The diaphragm details can be seen in Figure 13.34 and a section of the whistle's control valve is shown in Figure 13.35. Units of this type may be found working from air pressures of 6–42 bar with air consumptions in the range 25–35 litres/sec. Various sized choke plugs 7 are fitted depending on the supply pressure; alternatively an adjustable choke may be provided instead of the plug. It is important that the correct choke setting is selected to

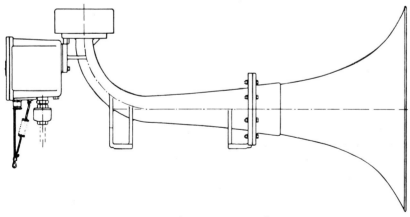

Figure 13.33 Super Tyfon air whistle (Kockums, Sweden)

match the maximum supply pressure. If this is too high for the setting the diaphragm might break; if the pressure is too low the volume will be inadequate.

Primary control of the whistle is afforded by one or more push buttons located at strategic points on the bridge. By pushing the button an electric circuit activates the solenoid pilot valve 1 which then causes piston 4 to move, allowing air to pass to the diaphragm. An automatic device is usually fitted which permits the selection of

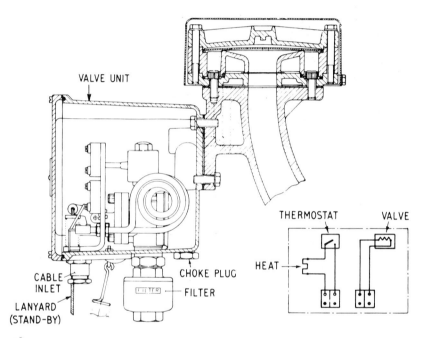

VALVE UNIT

THERMOSTAT   VALVE

HEAT

CABLE
INLET   CHOKE PLUG
                                           FILTER
LANYARD
(STAND-BY)

Figure 13.34 Detailed arrangement of diaphragm (Kockums, Sweden)

one or more automatic period signals. The Tyfon auto-control (Figure 13.36) allows automatic selection of either a 1 or 2 minute cycle. The normal duration of signal is 5 secs. The electric circuit for this equipment is so arranged that, when on automatic signal, depression of the manual button overrides the preselected sequence.

Secondary control of the whistle is by a lanyard which directly operates lever 3 (Figure 13.35) allowing air onto the diaphgram.

Air enters the whistle valve unit via a filter 2 which requires cleaning occasionally. An additional filter is frequently fitted at the lowest point of the air supply line and this will require draining

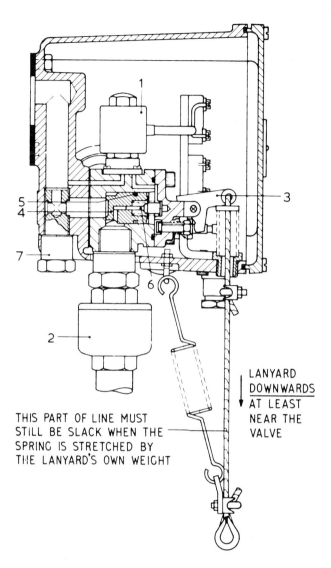

1   3

5
4

7

6

2

THIS PART OF LINE MUST
STILL BE SLACK WHEN THE
SPRING IS STRETCHED BY
THE LANYARD'S OWN WEIGHT

LANYARD
DOWNWARDS
AT LEAST
NEAR THE
VALVE

1. *Pilot valve*
2. *Filter*
3. *Lever*
4. *Piston*
5. *Housing*
6. *Spindle*
7. *Choke plug*

Figure 13.35 Control valve (Kockums, Sweden)

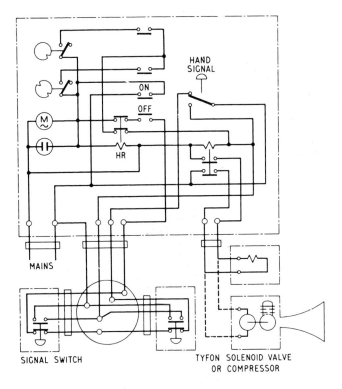

Figure 13.36 Wiring diagram for Tyfon auto-control whistle (Kockums, Sweden)

periodically. A routine inspection of this filter at monthly intervals is recommended.

Should the diaphragm require changing the dirt cover should be removed and the twelve retaining bolts should be unscrewed. It is not necessary to remove the bottom flange. The O ring, on which the diaphragm seats, should always be renewed and it is important to tighten the 12 retaining screws evenly.

Whenever work is to be done on the whistle the air and the electricity supply to the unit should be isolated.

Electric whistles

An Electro-Tyfon whistle is shown in Figure 13.37 from which it can be seen that the electric motor drives a reciprocating piston via a gear train and crank. This generates an air pressure which vibrates the diaphragm.

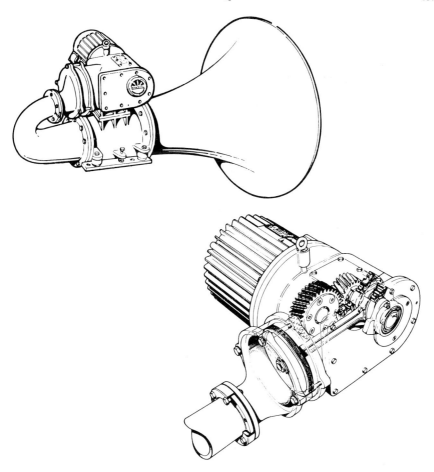

Figure 13.37 Sectional arrangement of electro-Tyfon whistle (Kockums Ltd)

Bibliography

1. *International Conference on Safety of life at Sea*, IMCO (1974)
2. *Fire Prevention and Detection*, Marine Engineers' Review 7 Oct 1980
3. *The Merchant Shipping (Fire Appliances) Rules*.
4. *Survey of Fire Appliances; Instructions for the guidance of Surveyors* (HMSO).

14 Control engineering

The closer control of machinery operating conditions, e.g. cooling water temperatures and pressures, permits machinery to be run at its optimum design conditions, making for fuel economy and reduced maintenance.

Automation can carry out some tasks far more effectively than men. In other areas it is less effective. For example, the monitoring of machinery operating conditions such as temperatures and pressures can be carried out by a solid-state alarm scanning system at the rate of 400 channels/scc., giving a degree of surveillance which would be impossible by human observation. Conversely, the detection of a noisy bearing, a leaky gland or a cracked pipe is scarcely possibly by automatic means. The balance between the

Figure 14.1 Control room of the Ville de Dunketque

438

possible and the necessary would be achieved in this case by combining automatic monitoring of all the likely fault conditions, with routine machinery space inspection say twice daily.

Classification societies

The class notation of a ship granted by a classification society, is a mark of approval of its standard, and it indicates that the vessel has been built to specific rules and thereafter periodically surveyed. It is the practice to include in the notation a special mark for ships designed to operate with periodically unmanned engine rooms, e.g. Lloyds Register add U.M.S. to the 100Al notation to signify approval of operation with Unattended Machinery Spaces.

The granting of special notations is also subject to the ship being built in conformity with rules or recommendations concerning automation. The principal international classification societies are Lloyds Register of Shipping, the American Bureau of Shipping, Det norske Veritas, Bureau Veritas and Germanischer Lloyd.

Basic requirements

If a ship's engine room is to operate unmanned for specific periods, certain requirements must be met. Firstly, control of engine speed and direction (or propeller pitch) must be made available at the bridge. In order to avoid placing any additional work load on the bridge watchkeeper, who is concerned essentially with what is happening outside the ship, the engine remote control system should carry out instructions signalled by the bridge watchkeeper at a simple telegraph type control. No demands for engineering skill should be placed on the bridge personnel. The application of starting air, then fuel and the subsequent rate of acceleration to demanded speed should be functions of the control system rather than of the operator.

Having removed the need for engineers to respond to possible telegraph orders at any time, the next important step is to provide automatic control of main engine services such as cooling water, lubricating oil, fuel and air systems. These functions are carried out by means of automatic controllers and control valves, which maintain system pressures, temperatures etc. at predetermined values despite load changes. The response rate, stability and accuracy of these control loops is dependent not only on the quality of the components but, more important, on the matching of the dynamic characteristics of the control system to the requirements of the particular service.

The third requirement for unattended machinery operation is the provision of an alarm system to monitor all the important operating conditions. These include temperatures and pressures of the fuel, air, lubrication and cooling systems for the main engines and generators, tank level alarms and many others such as bearing temperatures, fuel viscosity, vibration, scavenge fires etc. There is a large range of equipment capable of carrying out these tasks, and some typical systems will be described later.

The most important feature of an unattended machinery complex is the fire alarm system, in which sensors are placed around the engine room and associated spaces in order to detect combustion at the earliest practicable time.

Finally, it is necessary to provide for the continuity of electrical power for essential services in the event of failure of the duty generating equipment. In its simplest form, this may be limited to starting the emergency generator to provide power for essential lighting, but other services such as steering motors, machinery and fire alarm systems may also be provided with emergency power.

These are the five keystones of an automated machinery space, upon which all classification societies agree — remote engine control, automatic control of engine room services, a machinery alarm system, fire alarm system and emergency electrical power.

Planning the system

Planning of the automation system, by which is meant the total complex of remote and automatic controls and plant instrumentation must take account of several basic parameters:

1. The intended service of the ship.
2. The intended manning arrangements.
3. The type of propelling machinery.
4. Ship maintenance policy.
5. Classification society and notation required.
6. Ship resale value.

The above list of 'design inputs' is by no means complete, but represents the major factors which should influence the design of the automation system.

Control system

The simple control loop has three elements, the measuring element, the comparator element and the controlling element. The loop may be effected pneumatically, electronically or hydraulically. In some instances the control loop will be a hybrid system perhaps utilising electronic sensors, a pneumatic relay system and hydraulic or electric valve actuators. Each system has its strengths and weaknesses:

Pneumatics — require a source of clean dry air — can freeze in low temperature, exposed conditions, but equipment is well-proven and widely used.

Electronics — good response speeds with little or no transmission losses over long distances, easily integrated with data logging system, requires to be intrinsically safe in hazardous zones.

Hydraulics — require a power pack, may require accumulator for fail-safe action — compact and powerful and particularly beneficial in exposed conditions.

Measurement of process conditions

The range of parameters to be measured in merchant ships includes temperatures, pressures, level, speed of rotation, flow, electrical quantities and chemical qualities. Instrumentation used for remote information gathering purposes invariably converts the measured parameter to an electrical signal which may be used to indicate the measured value on a suitably calibrated scale, provide input information to a data logger or computer, initiate an alarm or provide a signal for a process controller.

As stated earlier however the more favoured means of providing process control information (as opposed to information display only) is to use a pneumatic system.

Electrical transducers

Any device used to measure one parameter in terms of another (e.g. temperature by change in electrical resistance) is called a transducer. The following are examples of electrical transducers used in shipboard instrumentation systems.

Temperature measurement

Liquid in glass, mercury in steel, liquid and vapour pressure type thermometers have been in use on ships for many years, and need not be described. The three main types employing electrical properties are the resistance thermometer, the thermocouple and the thermistor, and these may be used to provide input to local or panel mounted gauges, data loggers or control systems.

Essentially, a resistance thermometer is a precision resistor with a known temperature coefficient of resistance (i.e. change of resistance with temperature). The majority of resistance thermometers used in marine systems have as their active element a coil of fine platinum wire mounted on a ceramic former. The common standard of calibration is 100Ω at $0°C$, increasing by approx. 0.385Ω per $°C$ up to $100°C$. For more accurate calibration, the manufacturer's temperature/resistance tables should be consulted. Resistance thermometer elements may be housed in many configurations to suit particular applications. The most widely used housing is a stainless steel tube surmounted by a threaded portion and connecting head, for mounting in pipelines or tanks. Another type comprises a length of mineral-insulated, stainless steel or copper covered cable, with the resistance thermometer element built into one end. This design is useful to measure temperatures in difficult locations such as stern-tube outboard bearings.

Thermocouples are formed by the junction of two wires of dissimilar metal. When the free ends of these wires are connected to a measuring circuit, a voltage will appear across the instrument terminals which is a function of the difference in temperature between the junction of the two thermocouple wires (hot junction) and the instrument terminals (cold junction). This voltage is known as a thermo-electric e.m.f., and is different for the various thermocouple materials used. Typical thermocouple combinations using specially developed alloys are copper/constanstan, iron/constanstan and chromel/constanstan, the latter producing the largest signal (approx. 53 mV at $700°C$). In many cases, it is not practicable to run the thermocouple leads back to the measuring instrument without some break point. This may be achieved by:

(a) *Extension leads.* These are simply cables of the same materials as the thermocouple element.

(b) *Compensating cables.* These cables may be made of a cheaper alloy having the same thermo-electric properties as the thermocouple, but not capable of withstanding the same environmental conditions.

(c) *Copper cables.* If copper wires are introduced at an intermediate point in the cable run, then the thermocouple will measure the difference in temperature between the hot junction and the point at which the change to copper wires is introduced.

The main advantages of thermocouples over resistance thermometers are mechanical strength and, when necessary, small dimensions. The disadvantages are a small working signal, the problem of controlling or compensating for the cold junction temperature, and lower accuracy.

The thermistor has many of the advantages of both thermocouples and resistance thermometers. Common types take the form of a small bead of semi-conducting material, from which two measuring leads are led away to a terminal arrangement, with mechanical protection in the various forms suggested for resistance thermometers. The thermistor element exhibits an extremely large negative temperature coefficient in some cases thousands of ohms for a temperature shift of 100°C. These devices can be made very small, very rugged and with extremely accurate resistance/temperature characteristics.

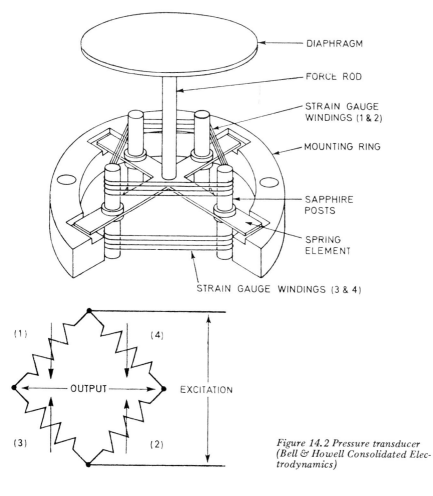

Figure 14.2 Pressure transducer (Bell & Howell Consolidated Electrodynamics)

Pressures

The majority of pressure transducers operate by first producing a mechanical motion proportional to applied pressure, from which is derived an electrical signal by some secondary mechanism.

The types most frequently met are the bourdon tube/potention-meter mechanism, in which the motion of the free end of the tube is used to move the slide of a potentionmeter, and the diaphragm/strain gauge type. There are two generic types of strain gauge, known as bonded and unbonded gauges. Bonded gauges are cemented to the diaphragm, and consist essentially of a grid of conducting material which exhibits a small change in electrical resistance when its shape is changed due to the flexing of the diaphragm. In unbonded types, the active element is generally a wire stretched between rigid supports which are displaced by the motion of the diaphragm (Figure 14.2).

As strain gauges only produce a very small change in resistance, they are normally used in Wheatstone bridge circuits.

Measurement of levels

The majority of engine room service tanks need only be monitored between fixed high and low limits, for which purpose level switches are sufficient. For bunker tanks, the most widely used system is the 'bubbler' type level gauge, in which tubes are led to the tanks. The pressure necessary to pass a small volume of air through the pipes is reflected by manometer tubes calibrated in terms of tank level.

Capacitative type sensors may also be used, in which the level in a tank is measured by the change in capacitance in a circuit comprising two concentric elements, the tank fluid being the dielectric. Pressure transducers may be used for measuring levels in vented tanks.

For more difficult tasks such as boiler drum or condenser hot-well levels, pneumatic level transmitters are generally used.

Flow measurement

Flows of fuel oil, feed water and steam are often required for control purposes and efficiency calculations. Fuel flow is usually measured by a positive displacement meter, in which the process fluid drives a rotor with a known discharge per revolution, the rate of revolution being an analogue of fuel flow.

For feedwater and steam, orifice plates or flow nozzles are used in conjunction with differential pressure transducers, flow being proportional to the square root of the pressure drop across the constrictive element.

TORQUE MEASUREMENT

Many ships have torque-meters and power-meters fitted to their propeller shafts. A number of types have been used, some incorporating optical devices to measure the angular displacement of a section of the shaft; others used electric slip rings for the same purpose. Transducers are now being widely used to measure torque since they are robust and do not require slip rings which are always prone to misreading because of dirt or bad electrical contact.

The Torductor torque transducer

The ring type Torductor torque transducer consists of three identical pole-rings with a number of poles that is always chosen as a multiple of 4 to reduce the influence of the ring joints necessary for easy mounting of the torque transducer around the shaft. The poles are fitted with coils with alternately reversed winding directions. The middle ring is displaced half the pole-pitch relative to the outer rings and the distance between the rings is approximatcly equivalent to

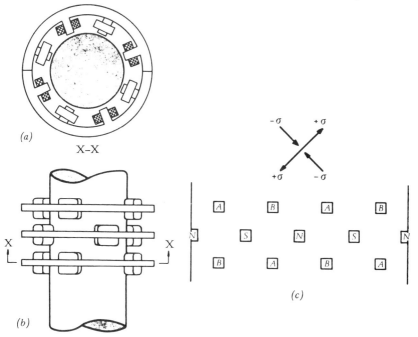

Figure 14.3 Principle of ring type Torductor torque transducer
(a) and (b) Physical arrangement
(c) Development of shaft surface under the Torductor poles

± Principal stresses
N, S. Primary poles
A, B. Secondary poles

half the pole-pitch. The middle ring is normally used as primary and excited with 50 or 60 Hz. The two outer rings are used as secondaries and connected in series, with mutually reversed winding directions, as indicated by the letters A and B in Figure 14.3, which shows the development of the shaft surface under the ring type Torductor torque transducer and the projection of the poles. The primary poles are marked N and S depicting a certain instant in the magnetising cycle.

If the shaft is unloaded and without internal stresses, the magnetic fields between the different N- and S-poles will be symmetrical so that the zero equipotential lines will be situated symmetrically under the secondary poles A and B. The secondary flux and hence the secondary voltage is thus zero at zero stress.

When torque is applied to the shaft, the principal stresses $\pm\sigma$ indicated in Figure 14.3 are obtained. The permeability in the direction of tension, i.e. between the poles B and S and between A and N, is then increased, while the permeability in the direction of compression, i.e. between the poles B and N and between A and S is decreased. Thus all A-poles come magnetically nearer to the N-poles and all the B-poles magnetically nearer to the S-poles. The result is magnetically the same as if the secondary rings had been tangentially displaced in mutually opposite directions and opposite to the torsion of the shaft. The resulting fluxes through the poles co-operate in inducing an output voltage in the series-coupled windings. The output is normally of the order of 10 V and 1 mA, i.e. large enough for an instrument without any amplification.

As the ring type Torductor torque transducer measures almost uniformly around the shaft, the 45° stresses are virtually integrated around the circumference and the modulation of the output voltage is thus reduced to a very small value.

The effective response time of the Torductor is mainly determined by the exciting frequency and the desired degree of filtering of the output signal. For 50 to 60 Hz excitation it can be of the order of 10 to 30 ms., dependent on the circuitry chosen.

The ring type Torductor torque transducer can be statically calibrated with good accuracy. The calibration has to be performed with the torque transducer mounted around the shaft as the sensitivity is dependent on its composition and heat treatment. This is, however, a drawback which the Torductor torque transducer in principle shares with all other torque transducers.

The Torductor tachometer

The tachometer (Figure 14.4) is a three-phase a.c. generator. The

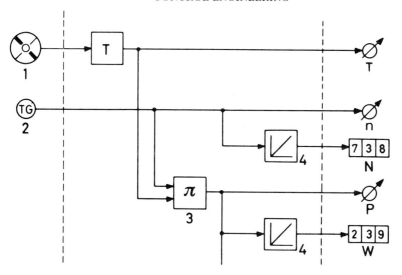

Figure 14.4 Torductor power meter

1. Torque transducer	4. Integrator	N. Total number of revolutions
2. Tacho-generator	T. Torque	P. Shaft horsepower
3. Multiplier	n. Rate of rotation	W. Total power output

frequency, and not the voltage, is used as a measure of the rate of rotation. This avoids the problem of brushes, non-linearity in the amplitude of the output voltage and conductor resistances. The electronic equipment consists of both analogue and digital circuits. It should be mentioned that a very stable oscillator, common to the entire equipment, is used as a reference for generating pulses having a predetermined duration. The operation of the electronic equipment is briefly as follows.

The tachometer signal is converted into a square wave having a constant pulse length and with a frequency proportional to the tachometer frequency. This square wave controls an electronic contact with a constant input voltage. The mean current through this will then be proportional to the tachometer frequency. The current is fed into an a.c. amplifier, whose output voltage is thus directly proportional to the rate of rotation. The direction of rotation, and thus also the polarity of the output signal, is determined through sensing of the phase sequence. The total number of revolutions is indicated on a counter on the engine-room console. A digital circuit imparts to the counter one pulse for every tenth revolution of the shaft.

The shaft (delivered) horsepower is obtained according to the same principle as the rate of rotation. If the constant input voltage of the electronic contact is replaced by a voltage proportional to the torque, the mean current through the contact will be proportional to the shaft horsepower. The current is fed as previously to an a.c. amplifier, which in its turn drives a pointer instrument.

The total power output is obtained through integration of the power signal, which is therefore fed into an amplifier with capacitive feedback serving as an electronic integrator. A level discriminator senses when the output voltage from the amplifier exceeds a certain value. The integrator then receives a resetting pulse via an electronic contact. Each of these pulses represents a certain amount of energy and the total number of pulses the total output of the machinery. Indication is accomplished with an electromechanical counter.

PNEUMATIC CONTROL SYSTEMS

For this discussion, measurement is the value of a process quantity or quality obtained by some suitable measuring device. It may be recorded or indicated by a pen or pointer on a scale. The terms 'pen' and 'measurement' will sometimes be used synonymously. For example a 'change in pen position upscale' means an increase in measurement.

A simple control loop is shown in Figure 14.5. The measuring element determines the pressure downstream of the control valve and transmits this to the comparator element, which is a pneumatic controller. The pneumatic controller compares the measured value with a manually set desired value and generates a correcting signal which causes a control valve to open or close. A Foxboro Model 40

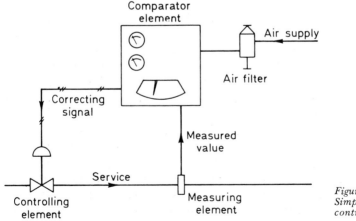

*Figure 14.5
Simple pressure
control loop*

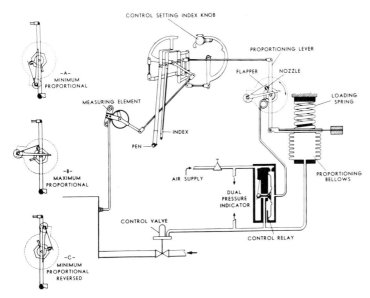

Figure 14.6 Model 40 controller — proportional (Foxboro Yoxall)

Controller is shown schematically in Figure 14.6 which serves to illustrate the way in which pneumatic controllers work. The desired value sought in the controlled system is set by turning the control setting index knob, thus positioning the index. A system of levers and links is arranged so that whatever position the index is placed in, any deviation by the pen from the pre-set index will result in a proportional movement of the horizontal link attached to the proportioning lever.

The particular controller shown in Figure 14.6 is known as a proportional or single term controller. This type of instrument produces a correcting pneumatic signal in the range 0.2—1 bar which is proportional to the deviation of the measured value (pen position) from the desired value (index position). Like most proportional controllers the Model 40 incorporates a device for altering the ratio of actuating signal change to feedback position change; thus the relationship between pen change and output change, called proportional band, is adjustable to suit the process. The effects of proportional band adjustment are illustrated in Figure 14.7.

The basic device used to modify the signal pressure is the flapper-nozzle unit. A regulated air supply is passed through a reducing tube of fine capillary bore and thence to a nozzle of much greater diameter. If the nozzle is unobstructed the pressure in the intervening tubing will be low. If the flapper is re-positioned closer to the nozzle, the pressure will rise in the section between the restriction and nozzle. By connecting the input of an amplifying relay (the control relay of Figure 14.6) between the restriction and the nozzle a flapper movement in the region of 0.015 mm can be made to generate a signal change of 0.83 bar. The three graphs in Figure 14.8 show the behaviour of the process variable after a sudden load change. A wide proportional band brings the variable back to a stable value quickly, but at a new value substantially below the original set point. This effect is known as offset. The moderate and narrow proportional band widths show successively longer stabilisation times and smaller offsets.

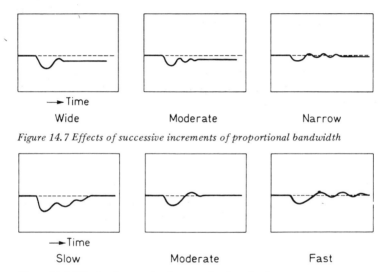

Figure 14.7 *Effects of successive increments of proportional bandwidth*

Figure 14.8 *Effects of successive increments of reset rate*

The proportional band adjustment is therefore very important, as it enables the control system to be matched to the characteristics of the process under control. It is obvious from Figure 14.7 that the essence of band width control is finding the best compromise to prevent excessive cycling after a load change, an undesirable effect.

Two-term controller

In services such as piston and cylinder cooling water, there will be large and frequent thermal load changes during manoeuvring. In such cases the offset produced by single term control may be unacceptable, in which case a two-term controller may be used.

The two-term controller has both proportional and integral actions. Integral action consists of adding to the proportional correcting signal a further increment which will counteract the offset which would otherwise result. The combined effect is to reset the control valve at a rate proportional to the deviation of the measured value from the setpoint value. Thus, for a large load change the speed of reset is rapid initially, slowing down as the deviation reduces.

The ratio between process deviation and speed of reset is adjustable, and the effects of different reset rates are shown in Figure 14.8., the proportional band remaining at a fixed value. The slow reset rate restores the variable to its initial value after a substantial drop following a sudden load change. The moderate rate brings the condition to normal more quickly and with less deviation. With fast reset, the initial deviation is even less, but this further increase in reset rate causes prolonged hunting of the variable about the setpoint. In this illustration the moderate reset rate is therefore the optimum setting.

Reset action creates serious instability during system start up, when the value of the controlled variable is well below the setpoint. Integral action is therefore only brought into effect when the system has reached normal operating conditions.

Three-term controllers

Three term controllers employing derivative action may be encountered occasionally, in which an excess correction is made which is proportional to the rate of change of the controlled variable. This has the effect of reducing the deviation and stabilization time, and may be met in more advanced combustion control systems.

The transmitter

The most common method of generating the pneumatic signal feeding the controller is to use a force balance device. An example of a force balance device used to transmit a pneumatic signal proportional to a process pressure change can be found in Foxboro's Model 13A d/p differential pressure transmitter shown in Figure 14.9.

The basic components — the body, force-balance transmitter system and air connection block — are in separate units. Process piping stresses are thus confined to the body and air piping stresses to the air connection block, so that the force-balance transmitter system is completely isolated from exterior stresses.

The replaceable twin-diaphragm capsule is clamped between separable sections of the body and separates high- and low-pressure chambers in the body. The interior of the capsule is filled with a silicone fluid having a viscosity of 500 centistokes which in combination with restrictions in the capsule backup plate provides a frequency cutoff point at approximately 1 Hz.

452

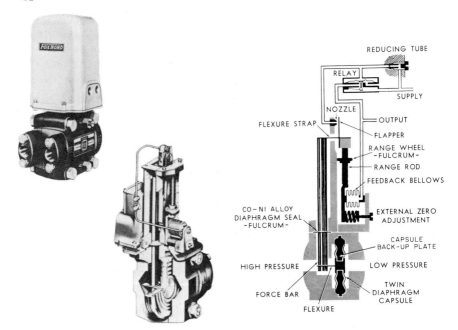

Figure 14.9 Model 13A d/p cell differential pressure transmitter (Foxboro Yoxall)

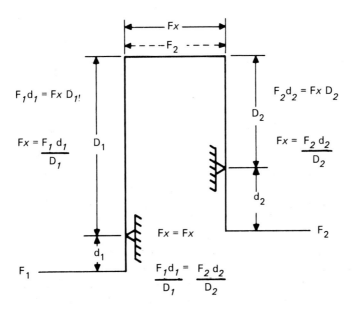

Figure 14.10 d/p cell force-balance moments (Foxboro Yoxall)

A Co-Ni alloy diaphragm, located at the top of the high-pressure chamber, seals off the measuring cavity and also acts as a fulcrum for the force bar. A 'C' flexure connects the twin diaphragms of the capsule and the force bar. The force bar, because of its fixed fulcrum, has a specific mechanical advantage (d_1/D_1) and is connected to the range rod by the flexure strap. A feedback bellows and zero adjustment spring are located at the other end of the range rod. The range wheel, which acts as a fulcrum, will give a different mechanical advantage $(d_2 D_2)$ to the range rod for each different position on the rod.

Pressures are applied to the high- and low-pressure sides of the twin-diaphragm capsule. The difference between these pressures creates tension in the 'C' flexure, and a pulling force (F_1) is exerted on the lower end of the force bar. The effective area of the diaphragm times pressure equals force. The resulting moment $(F_x D_1)$ at the other end of the force bar causes the flapper to tend to move toward the nozzle. The output of the relay is increased, and the force (F_2) of the feedback bellows on the range rod creates a moment that will balance the force (F_1) of the differential pressure across the capsule and its resultant moment on the force bar. This balance of forces results in an output pressure change which is in some proportion to the change in input differential force.

The flapper-nozzle assembly is located at the highest point in the transmitter mechanism. The position of the flapper is controlled by the minute motion of the force bar/range rod combination.

For each value of measurement, the capsule will assume a definite position as will the force bar and flapper and thus produce the same output when the flapper-nozzle is in the throttling band. Total movement at the capsule for full output is less than 0.25 mm.

Figure 14.10 shows the 13A d/p cell transmitter as a double force-bar mechanism — two force bars connected by a flexure strap. The value of force F_x is important in actual calculation but is better represented as an unknown F_1, the maximum force of the differential pressure which may have a value between 0.5 to 21.6 m of water, is variable; d_1/D_1, the mechanical advantage of the force bar, is fixed. F_x, the unknown force, balances F_x. F_2, being the maximum feedback force at 1 bar, is constant. d_2/D_2, the mechanical advantage of the range rod, is variable. In order to increase the range of the transmitter, the range wheel is moved up, increasing d_2, decreasing D_2 — thus increasing the ratio d_2/D_2. Now F_1 must be larger to produce the same balance, requiring an increased range of differential force. The range wheel has a 10 to 1 (maximum to minimum span) adjustment capability.

The externally adjustable zero spring creates a force against the range rod. Its purpose is to preload the range rod to provide a transmitter output of 0.2 bar with no differential across the twin-diaphragm capsule.

In applying this transmitter to some measurements, such as liquid level, density, or reversing flow it becomes desirable to preload force F_x in a direction which will provide span elevation or suppression. The adjustable spring exerts a force through the force bar to the diaphragm capsule, setting minimum output 0.2 bar at the minimum range value desired.

Temperature transmitter

Figure 14.11 shows a pneumatic force-balance device used to transmit a 0.2–1 bar pneumatic signal proportional to measured temperature. A gas filled thermal element, which consists of bulb, capillary and bellows capsule, provides an output signal which is linear with temperature. In the Foxboro Model 12A instrument shown, accuracy is claimed to be ±½ per cent of span for any span within ranges from −73°C to +540°C. Cryogenic systems down to −270°C are available.

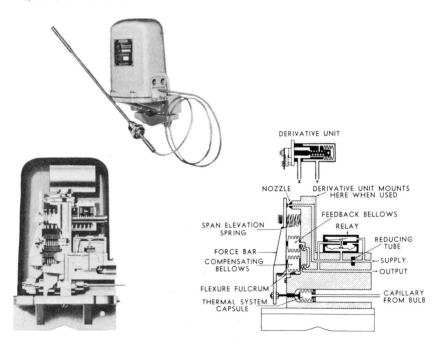

Figure 14.11 Model 12A temperature transmitter (Foxboro Yoxall)

Pneumatic transmitters are also available for various functions such as rate of flow measurement and tank level gauging (see Figure 3.56).

Electronic transmitters

Although pneumatic instruments are the most common signal trans-mitters used in ships there is a growing number of electronic trans-mitters found at sea. These are of varying types but it is interesting to note that Foxboro has adopted the force balance systems used in its pneumatic transmitters for Foxboro electronic systems. Figure 14.12 is a schematic drawing of a Foxboro E13 series electronic differential pressure transmitter suitable for flow, liquid level and low pressure measurement applications.

This instrument measures differential pressure from 0–127 mm H_2O to 0–21.5m H_2O and transmits a proportional 10–50 ma d.c. signal. The similarities in this instrument and that shown in Figure 14.9 are obvious. Instead of the flexure strap and flapper nozzle found in the pneumatic transmitter, however, the force bar pivots about the CoNi alloy seal (4) and transfers a force to the vector mechanism (5).

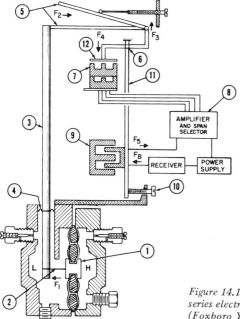

Figure 14.12 Schematic diagram of Foxboro E13 series electronic differential pressure transmitters (Foxboro Yoxall)
For key to numbers, see text

The force transmitted by the vector mechanism to the lever system (11) is dependent on the adjustable angle. Changing this angle adjusts the span of the instrument. At point (6) the lever system pivots and moves a ferrite disk, part of a differential transformer (7) which serves as a detector. Any position change of the ferrite disk changes the output of the differential transformer determining the amplitude output of an oscillator (8). The oscillator output is rectified to a d.c. signal and amplified, resulting in a 10–50 ma d.c. transmitter output signal. A feedback motor (9), in series with the output signal, exerts a force proportional to the error signal generated by the differential transformer. This force rebalances the lever system. Accordingly, the output signal of the transmitter is directly proportional to the applied differential pressure at the capsule.

For any given applied differential pressure, within the calibrated measurement range, the ferrite disk of the detector is continuously throttling, maintaining an output signal from the amplifier proportional to the measurement and retaining the force-balance system in equilibrium.

CONTROL VALVES

The reputation of the diaphragm operated pneumatic control valve as a final element in marine control systems has been established over many years. The control valve consists of four basic parts: the actuator, packing box, valve body and valve trim.

The diaphragm actuator receives pneumatic signals from the controller, which usually has an output in the range 0.20–1.03 bar. Control valves for such services are designed so that a change of signal from 0.20–1.03 bar will cause the valve to travel from the fully open to fully closed position. Actuators may be direct acting, in which an increasing air signal causes the valve stem to move downwards, or reverse acting. They are usually arranged so that failure of the control signals causes the valve to open, and handwheels are provided for manual operation in this event. Sometimes direct-action control valves are used for pressure control applications, in which the process fluid itself is applied to the diaphragm.

Diaphragm motion is transmitted to the valve plug by the stem which passes through the packing box. In most standard valves the packing consists of several V-rings of p.t.f.e., loaded by a spring which provides sealing pressure at low service pressures, and arranged so that as the service pressure rises further sealing pressure is applied see Figure 14.14.

Valve types

The flow characteristic of a valve is determined by the contours of its trim, i.e. its plug and its cage (or its seat). The characteristic may be defined as the relationship between valve lift and the resulting flow. Designs for quick opening, equal percentage, linear and parabolic characteristics are available. The trim design also governs the rangeability of a valve, which is a measure of maximum and minimum controlled flow.

Single-seated valves, guided by the valve plug stem or by vanes on the underside of the valve plug will be found. These offer tight shut-off but have the disadvantage of having to work against the unbalanced force created by the pressure of the process fluid passing between the valve plug and seat. To overcome this unbalanced force some valves will be found in which there are two plugs on the same stem (Figure 14.13). Known as double-ported valves these are arranged so that the two forces act in opposition.

The valve has the disadvantage however that tight shut-off cannot be obtained unless a soft seat is employed (because of difficulties in getting the valve plugs to touch both seats simultaneously).

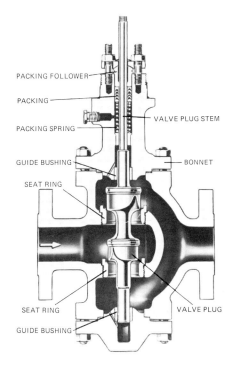

Figure 14.13 Double-ported valve with V-pup valve plug giving equal percentage flow characteristics (Fisher Control Valves Ltd)

It is more general today to find control valve trims of the types shown in Figure 14.14 in which a ported cage is utilised as a valve guide. These can be arranged as shown in Figure 14.14(d) to give tight shut-off and, in the case of the type shown in Figure 14.14(c) without the problems of an unbalanced force acting on the plug. The valve characteristic is governed by the contour of the ports in the valve cage.

Valve selection and sizing

Control valves are selected to suit each particular application, and the process characteristics to be taken into account are pressure, temperature, allowable pressure drop, stability, stream velocity, nature of the process fluid, etc. The effects of these conditions on the individual components must be carefully assessed to ensure that the valve will give many years of trouble free service.

A control valve must be correctly sized in order to control the desired quantity of fluid correctly. If the size is chosen simply to suit pipeline sizes it is most likely that poor control will result due to the valve capacity being too large. The most sophisticated controller will not be able to correct the condition, and in addition the seating surfaces of the valve are likely to wear rapidly. Control valve manufacturers will accept responsibility for the specification and sizing of control valves, providing they are given complete information regarding the service conditions to be met.

ALARM SYSTEMS AND DATA LOGGERS

The first step towards centralised control of marine machinery was simply to extend the conventional control and instrumentation facilities to a central control console, which was sometimes housed in a special control room. Consequently, the resulting consoles were very large and presented a great mass of information on gauges. In later installations and with the widespread use of micro-processors it became the practice to integrate the three basic instrumentation functions — alarm monitoring, display of data and recording — within one electronic system. These systems have been fitted in a large number of ships, including refrigerated cargo vessels where the data logging facility is ideal for recording cargo space temperatures. Alarm scanning is the most important function performed by this type of equipment.

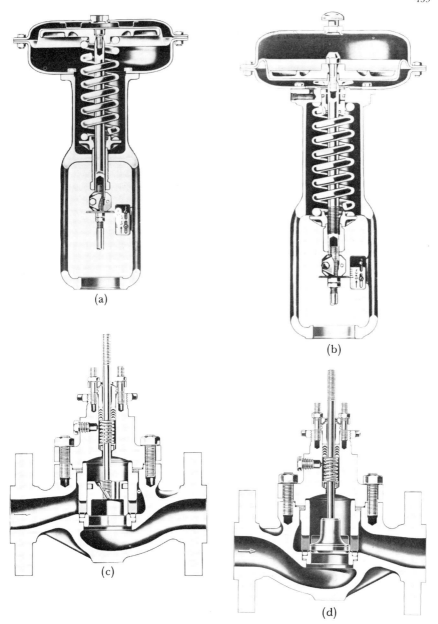

Figure 14.14 Examples of diaphragm operated control valves
(a) direct acting;
(b) reverse acting;
(c) balanced trim. Design ED
(d) Design EC with full-capacity standard
*   valve plug for viscous fluids*
(Fisher Control Valves Ltd.)

Scanning speeds vary between one and 400 channels per second for analogue parameters, and the accuracy of alarm comparison is generally within one per cent of the measurement range. An extremely complex machinery arrangement can be checked for malfunction twice every second, and alarm thresholds can be set very close to normal operating conditions, so giving practicaly instantaneous response to potentially dangerous situations. The development of high-reliability alarm-scanning systems is an important accompaniment to the increasing use of multi-engined propulsion systems, higher b.m.e.p.'s and the growing practice of operating ships with unattended engine rooms.

Several types of equipment are employed in ships, and while the details of operation vary, the basic arrangements are similar. Such equipment may be regarded as comprising four sections; primary measurement, signal selection, signal processing and control of output units. Figure 14.15 shows a typical arrangement.

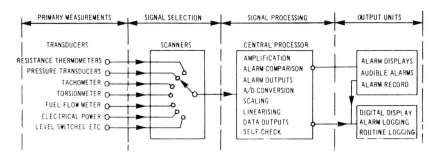

Figure 14.15 Four basic sections of alarm scanning/data logging system.

Primary measurements

The plant conditions to be recorded or checked are measured by transducers of various kinds. The term transducer applies to a range of devices which produce electrical signals proportional to various measurements such as temperature, pressure, etc. Most transducers develop a change in resistance corresponding to the primary measurement, which can be evaluated by the signal processing system. A typical example is the resistance thermometer. Other transducers, such as tachometers and some ship speed logs, generate a voltage.

An analogue transducer gives an output which varies with the magnitude of the measured condition. The term transducer is also more loosely applied to switching devices which only give a change of output at one or more fixed points.

Signal selection

The basic principle of scanning systems is that a number of measurements are evaluated sequentially by one high-quality signal processing system. The transducer signals are selected singly for evaluation by means of relays or solid-state switching networks. The relays or transistor switches are operated by signals from a scan control unit, which is generally regulated by an electronic clock. Scanning speeds vary according to the type of signal selection system used, and on the speed of response of the signal processing equipment.

Relay scanners are usually limited to about 10 channels per second because of limited relay life and the delay required for the signal to stabilise after switching. Solid state scanners do not suffer these limitations to any great extent, and analogue scanning speeds of 400 channels per sec. are thus made possible.

A high scan rate gives the system a very short response time to alarm conditions, which is very important with modern high-rated machinery. It also enables the logging system to tabulate a cascade of faults in proper chronological order, enabling the operator to identify the source of trouble when presented with a complex fault situation.

Signal processing

The first stage in signal processing is amplification of the low-voltage transducer signal, which is typically in the range 0—100 mV, and signal amplifiers will raise this level to, say, 0—5 V. The most important part of signal evaluation is comparison of the signal level with upper and/or lower alarm limits. It therefore follows that this operation should be carried out as early as possible in the signal processing chain, so that failures in other parts of the signal processor do not affect the alarm monitoring function. Alarm comparison is therefore ideally performed while the signal is in analogue form, by comparing the transducer voltage level with alarm limits set on potentiomenters. It should be noted that in some systems this most important function is carried out in the digital part of the system. In this case alarm limits are set by means of pins inserted in matrix boards.

When an alarm limit is exceeded, sub-routines are set into action which control the output units such as alarm display windows, audible alarms and alarm printers.

The digital section of the system converts the analogue data into digital form for presentation on the multi-point indicator or printer.

Each transducer signal is referred to an analogue-to-digital converter (ADC), which reproduces the voltage signal in numerical form. Digital systems use the binary system of numbers by means of which a quantity can be expressed on 1's and 0's, and quantities are transmitted and operated upon while in this form. Binary to decimal conversion occurs at the output units.

The digitised transducer signal is referred to a scaling unit which multiplies the signal by an appropriate constant so that its numerical value corresponds to engineering units such as °C. The scale unit will contain multipliers for each of the transducer measurement ranges in the system's repertoire. For those transducers which produce non-linear signals, the scale unit must apply a multiplier which varies with the magnitude of the transducer signal. Some transducers have offset characteristics, where their reading at, say, 0°C is some value other than 0 V. The scale unit or computer may therefore be called upon to perform addition, multiplication, linearisation and combinations of all three, changing its routine as each transducer signal is processed.

Control of output units

When alarm conditions occur, they are announced by a klaxon and identified by flashing windows bearing the identities of the alarm channels. When an alarm acknowledge button is pressed, the klaxon is silenced and the window remains illuminated until the fault is cleared.

The digital display may be used to indicate the value of any channel, by setting three decade switches to the corresponding channel numbers. Alarm limit settings for each channel may also be shown on the digital display. Where a typewriter is not employed, a handwritten log may be taken by simply switching in the channels required and noting the data presented. Typewriters can be used to print routine logs of all or part of the data processed by the system. Each log entry is time-identified, and means are provided to adjust system time to match ship's time.

The full value of a printer can be realised by incorporating alarm history printing. All alarm conditions, together with value and time, are recorded in red on occurrence and again in black on clearance.

Parallel entry instrumentation systems

One disadvantage of alarm scanning systems is that a failure in the central sections of the equipment can cause loss of all facilities on all channels. For this reason they must be made to a very high standard of reliability which makes them relatively expensive. Also, it is advisable for the ship to carry a fully comprehensive spares kit if it is engaged in deep sea trading.

Generally, the more channels that are monitored by such equipment, the more cost-effective they become, as the cost per channel is reduced. For smaller ships, where the data logging facility is not important, parallel entry instrumentation systems may be employed. The essential difference is that alarm comparators and alarm lamp drive circuits are provided for each channel, instead of being shared by all inputs. The transducer signals may then be manually selected by means of rotary switches, for digitisation, scaling and display at a central facility. Alarm printing may also be incorporated in parallel entry systems.

SPECIAL INSTRUMENTS

In addition to the conventional measurements such as have already been mentioned, small sub-systems are employed for special functions such as fuel viscosity control, turbocharger vibration alarms, being wear-down alarms, flame failure devices for boilers, fire alarm systems, etc. As space does not permit discussion of all these devices, mention will be made only of the viscosity controller.

Viscosity controllers

The basic principles of the Viscotherm viscosity controller are shown in Figure 14.16. A continuous sample of the fuel is pumped at

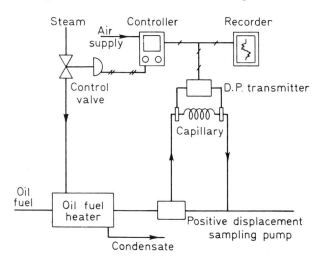

Figure 14.16 Schematic diagram of viscosity controller

constant rate through a fine capillary tube. As the flow through the tube is laminar, the pressure drop across the tube is proportional to viscosity. A differential pressure transmitter thus provides an analogue of viscosity to a pneumatic controller, which regulates the supply of fuel heating steam through a control valve.

BRIDGE CONTROL FOR DIESEL ENGINES

For many years bridge control has been applied to diesel-driven small vessels such as tugs, ferries, coasters, etc., either to the engine or to controllable-pitch propellers, but only comparatively recently has it been applied to larger ships.

Remote control of large diesel engines requires consideration not only of the normal functions of starting and reversing but also the conditions imposed on the engine when reversing while the ship is under way. Account should also be taken of critical running speeds which may be barred because of torsional vibration.

It is essential, particularly with bridge control, that all the operations should take place automatically without intervention by the officer in control, and that he should receive a signal confirming that the order has been obeyed. Movement of the control from stop must first initiate checks that turning gear is disengaged, starting air is available at the correct pressure, cooling water, lubricating-oil and fuel-oil supplies are in order before the starting sequence begins. The following sequence is then initiated:

Camshaft correctly positioned.
Starting air admitted.
Fuel admitted.
Starting air shut off.
Speed adjusted to the value required.

If the initial firing revolutions at starting are not reached within a fixed period, usually about 3 sec, a further period of about 4 sec is allowed to elapse and the cycle repeated, still under automatic control. Usually after three false starts the operation ceases and an alarm functions.

Starting air is usually controlled electrically by solenoid-operated valves which control the air supply to pneumatic actuators or the system may be entirely pneumatic. Fuel control may employ various methods, for example, a pneumatic cylinder with an integral hydraulic dashpot with a closed circuit in which the rate of oil displacement is controlled by flow regulators. This is in turn controlled by the position of the telegraph to stop further movement of the fuel lever when its position matches that of the telegraph.

BOILER CONTROL SYSTEMS

Automatic control of boilers and turbines is a much more complex problem than that of diesels and does not lend itself to any precise directions. Nevertheless it offers the greatest scope for efficiency and economy in both manpower and fuel consumption.

Dealing first with steam raising, the efficiency of modern comp-licated high-efficiency steam and feed-water systems depends on the correct relationship and operation of a large number of indepen-dent controls. They are all inter-related and each variation of main engine load, sea temperature, etc. requires a different combination of values and optimum efficiency is rarely achieved without some form of automatic control. Automated boiler control can therefore be highly cost-effective.

The two basic considerations are:

(a) Maintenance of steam pressure by control of fuel oil.
(b) Optimum combustion control by regulation of combustion air flow and maintenance of correct ratio of fuel and air at all combustion rates.

Fuel consumption is influenced by flue-gas temperatures and excessive oxygen, fan load, low superheat, low boiler pressure, temperature of condensate, feed-water temperature, circulating-water-pump load and the temperature of lubricating oil to the turbines.

Methods and systems adopted for automatic control vary con-siderably, and leading firms in this field have developed and installed them with gratifying results. An analysis of their methods would entail an extensive outline of systems and components and is beyond the scope of this chapter.

The principal items requiring control in an automatic system may be grouped as follows:

Boiler system	*Turbine and reduction gear*
Steam pressure	Speed
Steam temperature (Superheat)	Bleeder valve control
Water level	Lubricating-oil temperature
Feed pump	Overspeed

Feed water temperature   Condensate system
Fuel oil   Temperature of astern turbine
Forced draught fan
Air heater
Smoke density

Note that many subsidiary services not included in this list will need
to be included in a complete system. The subject is dealt with in
greater detail in 'Marine Steam Boilers', another publication in the
Butterworth Marine Series.

Semi-automatic systems

Semi-automatic systems operating over a narrow range have been in
use for many years. Up to the present they have not been capable of
maintaining automatic control during manoeuvring, the main factor
being the turn-down range of burners. However, there are now
available wide-range steam-assisted pressure jet burners which
effectively atomise the fuel even at low pressures, and a steam dump
valve passes surplus steam to an atmospheric condenser at very low
boiler output. The insertion and withdrawal of burners while manoeuvring is thereby rendered unnecessary.

CONTROLS FOR GENERATORS

In unattended machinery installations it is necessary to provide
certain control facilities for the electrical generating plant. These
may vary from simple load sharing and automatic starting of the
emergency generator, to a fully comprehensive system in which
generators are started and stopped in accordance with variations in
load demand.

Medium-speed propulsion plants normally use all-diesel generating
plant. Turbine ships obviously use some of the high quality steam
generated in the main boilers in condensing or back-pressure turbo
generators, with a diesel generator for harbour use. The usual
arrangement on large-bore diesel propulsion systems is a turbo
generator employing steam generated in a waste-heat boiler, plus
diesel generator for manoeuvring, port duty, and periods of high
electrical demand.

Diesel generators

The extent of automation can range from simple fault protection with automatic shut-down for lubricating oil failure, to fully automatic operation. For the latter case the functions to be carried out are:

Preparations for engine starting.
Starting and stopping engines according to load demand.
Synchronisation of incoming sets with supply.
Circuit breaker closure.
Load sharing between alternators.
Maintenance of supply frequency and voltage.
Engine/alternator fault protection.
Preferential tripping of non essential loads.

When diesel generators are arranged for automatic operation, it is good policy to arrange for off-duty sets to be circulated with main engine cooling water so that they are in a state of readiness when required. Pre-starting preparations are then simply limited to lubricating oil priming.

It is necessary to provide fault protection for lubricating oil and cooling services, and in a fully automatic system these fault signals can be employed to start a standby machine, place it on line, and stop the defective set. In some installations, automatic controls carry out the sequence as far as synchronisation, and leave final circuit breaker closure to the engineer.

Turbo-generators

The starting and shut-down sequences for a turbo-generator are more complex than those needed for a diesel-driven set, and fully automatic control is therefore less frequently encountered. However, the control facilities are often centralised in the control room, together with sequence indicator lights to enable the operator to verify each step before proceeding to the next. Interlocks may also be employed to guard against error.

The start up sequence given below is necessarily general, but it illustrates the principal and may be applied to remote manual or automatic control:

Reset governor trip lever.
Reset emergency stop valve.
Start auxiliary L.O. pump.
Start circulating pump.
Apply gland steam.

Start extraction pump.

Start air ejectors.

Open steam valve to run-up turbine.

Where a waste-heat boiler is used to supply steam to a turbo-alternator, control of steam output is normally controlled by a three way valve in the exhaust uptake, the position of which is regulated in accordance with steam demand. Surplus waste-heat is then diverted to a silencer.

TANK CARGOES

Automation is also applicable to the loading and discharging of tankers and to instrumentation of the state of the cargo.

There are three main systems of pipelines in use each requring a different method of handling:

(a) The ring system, in which a ring pipeline has separate suction tail pipes tapped off to each tank and controlled by a gate valve. By opening the appropriate gate valve each tank in turn can be emptied by the main pumps. When loading, the main pumps are by-passed and the ring pipeline becomes a gravity flow line and tanks can be loaded by opening the appropriate gate valve.

(b) The direct system, in which the tanker is divided into three or more 'compartments', each of which in turn is subdivided into individual tanks. One pump is assigned to each 'compartment' by a pipeline, which in turn is connected to tanks by tail pipes and gate valves. The main pipelines are interconnected through cross-over sluice valves.

(c) The free-flow system avoids the use of extensive pipelines, but has the disadvantage that the tanks can only be emptied simultaneously or in a fixed sequence. Oil is drawn through a pump through a suction tail pipe in the after tank and from the remaining tanks through valves in the transverse bulkheads between the tanks. Sometimes the free flow and the direct system are combined so that the direct system is used for port and starboard wing tanks and free-flow for the centre tanks.

It will be seen that a large number of valves must be controlled in a logical sequence, and without centralised remote control this requires a great deal of skilled manpower. In many cases a mixed cargo of different grades of oil is involved and correct handling is important. To avoid dangerous stressing of the hull structure due to unfavourable bending moments as the buoyancy of individual tanks changes it is necessary to operate to a known safe sequence. It is also necessary to take trim and list into account.

It is therefore apparent that this is a fertile field for centralised cargo control and computer operation. Electrical operation of valves would involve danger of explosion, and hydraulic systems are therefore favoured for this purpose. For control and instrumentation purposes electronic circuits are permissible subject to classification approval, provided they are certified as intrinsically safe, i.e. in no circumstances, including fault conditions, can a spark be created which could cause explosion. For protection against incorrect operation some of the valves require sequence interlocking.

When the main pumps have emptied a tank, the sediment and dirt which gather on the bottom of the tank are removed by stripping pumps through a separate pipeline and valve system and, to facilitate the quick turn round now demanded, the stripping system must be brought into use without delay as soon as the cargo pump has finished emptying the tank as far as possible.

POSSIBILITIES FOR COMPUTERS

A number of ships have been fitted with computers which are programmed to carry out a great variety of tasks embracing satellite navigation, ships' housekeeping, crew wages, machinery surveillance, weather routeing etc. General purpose industrial computers have also been employed for the single task of machinery alarm scanning and data logging, but the cost of expert maintenance and repair has been found disproportionate by some operators.

Nevertheless, computers can offer very important and unique benefits if they are applied with due regard to cost-effectiveness. Particular areas in which the computer may excel are the control of advanced steam generating plant, marine gas turbines, and performance monitoring of diesel engines.

Regrettably, the computer has been beset by an aura of mysticism and ill-formed opinion as to its possibilities. The computer does two things extremely well. It can store enormous quantities of data very cheaply, in such a way that retrieval is rapid. It can also carry out arithmetic functions at very high speed. These are its principal attributes, and provided these are recognised the computer may be applied intelligently and cost-effectively.

It is probable that a computer employed on control of a modern sophisticated marine steam generating installation can pay for itself quite easily by fuel economy and the reduced plant maintenance associated with correct control of process conditions. Progress is now being made towards machinery component condition monitoring, which could result in diesel engines only being stripped down for repair of known defects, rather than on the present basis of hours run. If classification societies are prepared, in the fullness of time, to relax periodic survey requirements as a result of this development, then these systems will be widely fitted.

BIBLIOGRAPHY

1. *Automation in Merchant Ships,* Fishing News Books.
2. *Guide for Shipboard Centralised Control and Automation,* American Bureau of Shipping.
3. *Guidance Note NI 134. CN3 Automated Ships,* Bureau Veritas.
4. Gray, D., *Centralised and Automatic Control in Ships,* Pergamon Press.
5. Jones, E.B., *Instrument Technology; Vol. 3 Telemetering and Automatic Control,* Butterworths (1971).
6. *Rules and Regulations for the Construction and Classification of Steel Ships.* Reprint of Chapter 1, Lloyds Register of Shipping.
7. *Rules for the Classification and Construction of Steel Seagoing Ships,* Reprint of Vol. IV, Chapter 9, Germanischer Lloyd.
8. *Rules for the Construction and Classification of Steel Ships,* Reprint of Chapters VII and XV. Det norske Veritas.
9. Wilkinson, P.T.C., *Fundamentals of Marine Control Engineering,* Whitehall Press.
10. BS. 1523. Glossary of terms used in automatic controlling and regulating systems. Section 2 — Process control; Section 3 — Kinetic control, British Standards Institution.

Index